Lecture Notes in Artificial Intelligence   8946

Subseries of Lecture Notes in Computer Science

LNAI Series Editors

Randy Goebel
  University of Alberta, Edmonton, Canada
Yuzuru Tanaka
  Hokkaido University, Sapporo, Japan
Wolfgang Wahlster
  DFKI and Saarland University, Saarbrücken, Germany

LNAI Founding Series Editor

Joerg Siekmann
  DFKI and Saarland University, Saarbrücken, Germany

More information about this series at http://www.springer.com/series/1244

Béatrice Duval · Jaap van den Herik
Stephane Loiseau · Joaquim Filipe (Eds.)

Agents and Artificial Intelligence

6th International Conference, ICAART 2014
Angers, France, March 6–8, 2014
Revised Selected Papers

 Springer

Editors
Béatrice Duval
LERIA - UFR Sciences
Angers
France

Jaap van den Herik
Leiden University
Leiden
The Netherlands

Stephane Loiseau
LERIA - UFR Sciences
Angers
France

Joaquim Filipe
INSTICC
Polytechnic Institute of Setúbal
Setúbal
Portugal

ISSN 0302-9743   ISSN 1611-3349 (electronic)
Lecture Notes in Artificial Intelligence
ISBN 978-3-319-25209-4   ISBN 978-3-319-25210-0 (eBook)
DOI 10.1007/978-3-319-25210-0

Library of Congress Control Number: 2015950896

LNCS Sublibrary: SL7 – Artificial Intelligence

Printed on acid-free paper

Springer International Publishing AG Switzerland is part of Springer Science+Business Media
(www.springer.com)

Preface

The present book includes extended and revised versions of a set of selected papers from the 6th International Conference on Agents and Artificial Intelligence (ICAART 2014), held in Angers, Loire Valley, France, during March 6–8, 2014. The conference was sponsored by the Institute for Systems and Technologies of Information, Control and Communication (INSTICC) in cooperation with the ACM Special Interest Group on Artificial Intelligence (ACM Sigart), the Association for the Advancement of Artificial Intelligence (AAAI), and the Portuguese Association for Artificial Intelligence (APPIA). ICAART was also technically co-sponsored by the IEEE Computer Society.

The purpose of the International Conference on Agents and Artificial Intelligence is to bring together researchers, engineers, and practitioners interested in theory and applications in the areas of agents and artificial intelligence. Two simultaneous related tracks were held, covering both applications and current research work. One track focused on agents, multi-agent systems and software platforms, distributed problem solving and distributed AI in general. The other track focused mainly on artificial intelligence, knowledge representation, planning, learning, scheduling, perception, reactive AI systems, and evolutionary computing.

ICAART 2014 received 225 paper submissions from 48 countries in all continents, of which 15 % were presented at the conference as full papers, and their authors were invited to submit extended versions of their papers for this book. In order to evaluate each submission, a double-blind review was performed by the Program Committee.

We would like to highlight that ICAART 2014 also included six plenary keynote lectures, given by internationally distinguished researchers, namely: Philippe Leray (Université de Nantes, France), Henry Lieberman (MIT Media Laboratory, USA), Matthias Klusch (German Research Center for Artificial Intelligence, DFKI, GmbH, Germany), Claude Frasson (University of Montreal, Canada), Pieter Spronck (Tilburg University, The Netherlands), and Marc-Philippe Huget (University of Savoie, France).

We would like to express our appreciation to all of them and in particular to those who took the time to contribute with a paper to this book.

We must thank the authors, whose research and development efforts are recorded here. We also thank the keynote speakers for their invaluable contribution and for taking the time to synthesize and prepare their talks. The knowledge and diligence of the reviewers and program chairs were essential to ensuring the quality of the papers presented at the conference and published in this book. Finally, a special thanks to all members of the INSTICC team, whose collaboration was fundamental for the success of this conference.

April 2015

Stephane Loiseau
Joaquim Filipe

Organization

Conference Co-chairs

Stephane Loiseau   LERIA, University of Angers, France
Joaquim Filipe   Polytechnic Institute of Setúbal/INSTICC, Portugal

Program Co-chairs

Béatrice Duval   LERIA, University of Angers, France
Jaap van den Herik   Tilburg University, The Netherlands

Organizing Committee

Helder Coelhas   INSTICC, Portugal
Vera Coelho   INSTICC, Portugal
Bruno Encarnação   INSTICC, Portugal
Ana Guerreiro   INSTICC, Portugal
André Lista   INSTICC, Portugal
Andreia Moita   INSTICC, Portugal
Raquel Pedrosa   INSTICC, Portugal
Vitor Pedrosa   INSTICC, Portugal
Susana Ribeiro   INSTICC, Portugal
Sara Santiago   INSTICC, Portugal
Mara Silva   INSTICC, Portugal
José Varela   INSTICC, Portugal
Pedro Varela   INSTICC, Portugal

Program Committee

Jose Aguilar   Universidad de Los Andes, Venezuela
Md. Shamim Akhter   Thompson Rivers University (TRU), Canada and
     American International University Bangladesh
     (AIUB), Bangladesh
Isabel Machado Alexandre   Instituto Universitário de Lisboa (ISCTE-IUL) and
     Instituto de Telecomunicações, Portugal
Vicki Allan   Utah State University, USA
Klaus-Dieter Althoff   German Research Center for Artificial Intelligence /
     University of Hildesheim, Germany
Francisco Martínez Álvarez   Pablo de Olavide University of Seville, Spain
Frédéric Amblard   IRIT - Université Toulouse 1 Capitole, France
Cesar Analide   University of Minho, Portugal

Andreas S. Andreou	Cyprus University of Technology, Cyprus
Diana Arellano	Filmakademie Baden-Württemberg, Germany
Tsz-Chiu Au	Ulsan National Institute of Science and Technology, Republic of Korea
Jean-Michel Auberlet	IFSTTAR (French Institute of Science and Technology for Transport, Development and Networks), France
Snorre Aunet	Norwegian University of Science and Technology, Norway
Florence Bannay	IRIT - Toulouse University, France
Suzanne Barber	The University of Texas, USA
Senén Barro	University of Santiago de Compostela, Spain
Roman Barták	Charles University in Prague, Czech Republic
Teresa M.A. Basile	Università degli Studi di Bari, Italy
Bernhard Bauer	University of Augsburg, Germany
Jacob Beal	BBN Technologies, USA
Punam Bedi	University of Delhi, India
Nabil Belacel	National Research Council Canada, Canada
Orlando Belo	University of Minho, Portugal
Christoph Benzmüller	Freie Universität Berlin, Germany
Carole Bernon	University of Paul Sabatier - Toulouse III, France
Daniel Berrar	Tokyo Institute of Technology, Japan
Djamel Bouchaffra	Oakland University, USA
Bruno Bouchard	LIARA Laboratory, Université du Québec à Chicoutimi, Canada
Sheryl Brahnam	Missouri State University, USA
Ivan Bratko	University of Ljubljana, Slovenia
Paolo Bresciani	European Commission, Italy
Egon L. van den Broek	University of Twente/Radboud UMC Nijmegen, The Netherlands
Stefano Bromuri	University of Applied Sciences Western Switzerland, Switzerland
Aleksander Byrski	AGH University of Science and Technology, Poland
Giacomo Cabri	Università di Modena e Reggio Emilia, Italy
Silvia Calegari	Università Degli Studi Di Milano Bicocca, Italy
Rui Camacho	Universidade do Porto, Portugal
Valérie Camps	IRIT - Université Paul Sabatier, France
John Cartlidge	University of Bristol, UK
José Jesús Castro-Schez	Escuela Superior de Informática, Universidad de Castilla-La Mancha, Spain
Patrick De Causmaecker	Katholieke Universiteit Leuven, Belgium
François Charpillet	Loria - Inria Lorraine, France
Amitava Chatterjee	Jadavpur University, India
Ke Chen	The University of Manchester, UK
Mu-Song Chen	Da-Yeh University, Taiwan
Anders Lyhne Christensen	Instituto Superior das Ciências do Trabalho e da Empresa, Portugal

Paolo Ciancarini	University of Bologna, Italy
Davide Ciucci	Università Degli Studi Di Milano Bicocca, Italy
Helder Coelho	University of Lisbon, Portugal
Diane Cook	Washington State University, USA
Daniel Corkill	University of Massachusetts Amherst, USA
Gabriella Cortellessa	ISTC-CNR, Italy
Paulo Cortez	University of Minho, Portugal
Massimo Cossentino	National Research Council, Italy
Vitor Santos Costa	Universidade do Porto, Portugal
Fabiano Dalpiaz	University of Toronto, Canada
Rajarshi Das	IBM T.J. Watson Research Center, USA
Darryl N. Davis	University of Hull, UK
Scott A. DeLoach	Kansas State University, USA
Andreas Dengel	German Research Center for Artificial Intelligence (DFKI GmbH), Germany
Enrico Denti	Alma Mater Studiorum – Università di Bologna, Italy
Paul Dunne	University of Liverpool, UK
Edmund Durfee	University of Michigan, USA
Béatrice Duval	LERIA, University of Angers, France
Stefan Edelkamp	Universität Bremen, Germany
Thomas Eiter	Technische Universität Wien, Austria
Fabrício Enembreck	Pontifical Catholic University of Paraná, Brazil
Floriana Esposito	Università degli Studi di Bari, Italy
Alessandro Farinelli	University of Verona, Italy
Stefano Ferilli	University of Bari, Italy
Alberto Fernández	University of Rey Juan Carlos, Spain
Edilson Ferneda	Catholic University of Brasília, Brazil
Klaus Fischer	German Research Center for Artificial Intelligence DFKI GmbH, Germany
Roberto Flores	Christopher Newport University, USA
Claude Frasson	University of Montreal, Canada
Kensuke Fukuda	NII, Japan
Naoki Fukuta	Shizuoka University, Japan
Wai-Keung Fung	Robert Gordon University, UK
Leonardo Garrido	Tecnológico de Monterrey, Campus Monterrey, Mexico
Max Gath	Universität Bremen/Center for Computing and Communication Technologies, Germany
Joseph Giampapa	Carnegie Mellon University, USA
João Carlos Gluz	Universidade do Vale do Rio dos Sinos, Brazil
Daniela Godoy	Universidad Nacional del Centro de la Pcia. de Buenos Aires, Argentina
Herman Gomes	Federal University of Campina Grande, Brazil
Madhu Goyal	University of Technology, Sydney, Australia
Dominic Greenwood	Whitestein Technologies AG, Switzerland
Sven Groppe	University of Lübeck, Germany

Adam Slowik	Koszalin University of Technology, Poland
Alexander Smirnov	SPIIRAS, Russian Academy of Sciences, Russian Federation
Marina V. Sokolova	Instituto de Investigación en Informática de Albacete, Spain
Kim Solin	The University of Queensland, Australia
Armando J. Sousa	Universidade do Porto, Portugal
Fernando Da Fonseca De Souza	Centro de Informática - Universidade Federal de Pernambuco, Brazil
Bruno Di Stefano	Nuptek Systems Ltd., Canada
Bernd Steinbach	Freiberg University of Mining and Technology, Germany
Kathleen Steinhofel	King's College London, UK
Daniel Stormont	The PineApple Project, USA
Thomas Stützle	Université Libre de Bruxelles, Belgium
Kaile Su	Peking University, China
Toshiharu Sugawara	Waseda University, Japan
Vijayan Sugumaran	Oakland University, USA
Shiliang Sun	East China Normal University, China
Zhaohao Sun	University of Ballarat, Australia
Boontawee Suntisrivaraporn	Sirindhorn International Institute of Technology, Thailand
Pavel Surynek	Charles University in Prague, Czech Republic
Ryszard Tadeusiewicz	AGH University of Science and Technology, Poland
Nick Taylor	Heriot-Watt University, UK
Patrícia Tedesco	Universidade Federal de Pernambuco/FADE, Brazil
Anastasios Tefas	Aristotle University of Thessaloniki, Greece
Mark Terwilliger	Lake Superior State University, USA
Michael Thielscher	The University of New South Wales, Australia
José Torres	Universidade Fernando Pessoa, Portugal
Paola Turci	University of Parma, Italy
Anni-Yasmin Turhan	Technische Universität Dresden, Germany
Franco Turini	KDD Lab, University of Pisa, Italy
Paulo Urbano	Universidade de Lisboa, Portugal
Visara Urovi	University of Applied Sciences Western Switzerland, Sierre (HES-SO), Switzerland
David Uthus	Google, USA
Eloisa Vargiu	Barcelona Digital Technology Center, Spain
Laurent Vercouter	LITIS, France
Serena Villata	Inria Sophia Antipolis, France
Mirko Viroli	University of Bologna, Italy
Marin Vlada	University of Bucharest, Romania
Yves Wautelet	Hogeschool-Universiteit Brussel, Belgium
Rosina Weber	iSchool at Drexel, USA
Gerhard Weiss	University of Maastricht, The Netherlands
Cees Witteveen	Delft University of Technology, The Netherlands

Stefan Woltran	Technische Universität Wien, Austria
T.N. Wong	The University of Hong Kong, SAR China
Franz Wotawa	Graz University of Technology, Austria
Bozena Wozna-Szczesniak	Jan Dlugosz University, Poland
Feiyu Xu	Deutsches Forschungszentrum für Künstliche Intelligenz (DFKI), Germany
Yiyu Yao	University of Regina, Canada
Li-Yan Yuan	University of Alberta, Canada
Ioannis Zaharakis	Technological Educational Institute of Patras, Greece
Laura Zavala	Megar Evers College of the City University of New York, USA
Hong Zhu	Oxford Brookes University, UK
Alejandro Zunino	ISISTAN-CONICET and UNICEN, Argentina

Additional Reviewers

Sjriek Alers	Maastricht University, The Netherlands
Jhonatan Alves	UFSC, Brazil
Riccardo De Benedictis	CNR - Italian National Research Council, Italy
Alessandro Bianchi	University of Bari, Italy
Claudio Cavalcanti	UFCG, Brazil
Daniel Claes	Maastricht University, The Netherlands
Emir Demirovic	TU Wien, Austria
Ludger van Elst	DFKI, Germany
Nicola Fanizzi	Università degli studi di Bari Aldo Moro, Italy
Markus Goldstein	German Research Center for Artificial Intelligence (DFKI), Germany
Christoph Greulich	Universität Bremen, Germany
Magnus Hjelmblom	University of Gävle, Sweden
Dayana Hristova	University of Vienna, Austria
Ignazio Infantino	Consiglio Nazionale delle Ricerche, Italy
Agnieszka Jastrzebska	Warsaw University of Technology, Poland
Valery Katerinchuk	King's College London, UK
Hung-Ming Lai	King's College London, UK
Florian Lonsing	Vienna University of Technology, Austria
Brian Maher	King's College London, UK
Martin Memmel	DFKI GmbH, Germany
Cecilia Estela Giuffra Palomino	Universidade Federal de Santa Catarina, Brazil
Eanes Pereira	UFCG, Brazil
José Antonio Piedra-Fernandez	University of Almeria, Spain
Josef Pihera	TU Wien, Czech Republic
Hércules Antônio do Prado	Universidade Católica de Brasília, Brazil

David Rajaratnam	The University of New South Wales, Australia
Patrizia Ribino	ICAR- CNR, Italy
Nico Roos	Maastricht University, The Netherlands
Luca Sabatucci	National Research Council - Italy, Italy
Abdallah Saffidine	University of New South Wales, Australia
Julien Saunier	National Institute of Applied Sciences, France
Ognjen Savkovic	Free University of Bozen-Bolzano, Italy
Valeria Seidita	University of Palermo, Italy
Honglei Shi	East China Normal University, China
Christophe Soares	Universidade Fernando Pessoa, Portugal
Albert Brugués de la Torre	University of Applied Sciences Western Switzerland, Switzerland
Mo Yang	Pattern Recognition and Machine Learning Research Group, China

Invited Speakers

Philippe Leray	Université de Nantes, France
Henry Lieberman	MIT Media Laboratory, USA
Matthias Klusch	German Research Center for Artificial Intelligence (DFKI) GmbH, Germany
Claude Frasson	University of Montreal, Canada
Pieter Spronck	Tilburg University, The Netherlands
Marc-Philippe Huget	University of Savoie, France

Contents

Invited Paper

The New Era of High-Functionality Interfaces

Henry Lieberman[1]($\boxtimes$), Christopher Fry[1], and Elizabeth Rosenzweig[2]

[1] Media Laboratory, Massachusetts Institute of Technology,
Cambridge, MA, USA
`lieber@media.mit.edu`
[2] Bentley University, Waltham, MA, USA

Abstract. Traditional user interface design works best for applications that only have a relatively small number of operations for the user to choose from. These applications achieve usability by maintaining a simple correspondence between user goals and interface elements such as menu items or icons. But we are entering an era of high-functionality applications, where there may be hundreds or thousands of possible operations. In contexts like mobile phones, even if each individual application is simple, the combined functionality represented by the entire phone constitutes such a high-functionality command set. How are we going to manage the continued growth of high-functionality computing? Artificial Intelligence promises some strategies for dealing with high-functionality situations. Interfaces can be oriented around the goals of the user rather than the features of the hardware or software. New interaction modalities like natural language, speech, gesture, vision, and multi-modal interfaces can extend the interface vocabulary. End-user programming can rethink the computer as a collection of capabilities to be composed on the fly rather than a set of predefined "applications". We also need new ways of teaching users about what kind of capabilities are available to them, and how to interact in high-functionality situations.

1 High-Functionality (Hi-Fun) Interfaces

As the range of tasks that people want to do with computers expands, and the capability of software grows, we are faced with the development of *high-functionality* (hi-fun) interfaces. We are not going to give a precise definition of hi-functionality interfaces here, but roughly, we mean those that provide large command sets, long menus, large or numerous icon bars, many data types, and complex patterns of use. In many cases, the names of interface operations may not be "obvious" to a beginning user, unless they know the underlying concepts of the application.

Low-functionality (lo-fun) interfaces are much simpler, acting on just a few kinds of data, and providing reasonably small command sets, where the name and effect of each command are expected to be immediately apparent to the user.

Apple's *Preview* is an example of a relatively lo-fun application for images; it can, for example, print, crop, and rotate images, but it has relatively few operations (about 9 top-level operations, 7 menus of 5–15 items, few subsidiary dialogs). Adobe's *Photoshop* is a hi-fun image application (25 top level operations (+ modifier keys on many), 4 palettes of 2–3 tabs each, 8 menus of 10 to > 25 items, many subsidiary

© Springer International Publishing Switzerland 2015
B. Duval et al. (Eds.): ICAART 2014; LNAI 8946, pp. 3–10, 2015.
DOI: 10.1007/978-3-319-25210-0_1

dialogs). It has many different image types, and the total number of operations reaches into the thousands. It has a number of abstract concepts that it is necessary to learn, such as layers and different color models. It is user customizable, can record and play macros, has numerous plug-ins, etc. (Fig. 1).

Fig. 1. Apple's Preview (left) is an example of a "low functionality" application, with a few dozen operations. Adobe's "high-functionality" Photoshop (right) has thousands of operations.

Applications that become popular tend to grow into high-functionality interfaces over time as users desire more features and companies continually try to improve their products. The most successful, like Photoshop or Microsoft Excel, become languages and programming environments in their own right. They become as powerful (and as difficult to learn for new users) as interactive development environments for programming languages.

The UI for high-functionality applications is typically designed for the expert and habitual user. It aims to make all the operations that the expert user would want to use easily accessible. But then the new user doesn't know where to start. And users who try to learn an interface by sequential exploration get confused because they are tempted to try many things for which they won't have use until much later, if at all.

2   User Interface Design for High Functionality Interfaces

Traditional user interface design aims to make interfaces easy to use, especially for beginning users, reduce error rates, and be aesthetically pleasing. The metaphor that most user-interface design tries to promote is that the computer is like a box of tools, like hammers and screwdrivers. Each tool is specialized for a certain job, and it's up to the user to know what each tool does, and which tool is appropriate for which job.

Interfaces strive for an ideal of simplicity and learnability. The menu items and icons that provide access to each tool should be suggestive of the functionality that the tool provides. Don Norman refers to this as "affordances" [11]. Simplicity demands a one-to-one correspondence between controls and tools.

The problem is, as the number of things the person wants to do with the computer grows, you wind up with too many tools in your toolbox. So you wind up with too many controls. Our screens fill up with icons and menu bars. Hierarchical menus, shift keys and other techniques, increase the capacity of the command set, but eventually space runs out (not to mention the user's patience), no matter what you do. The problem is that we're using the UI design principles that are appropriate for low-functionality interfaces to design the interfaces for high-functionality applications.

The alternative is what we call "goal-oriented interfaces". People have goals – things like "plan a trip" or "design a building". But programs don't have goals. What they have is specific functions, which are invoked by menu items, icons and typing. It's up to the user to figure out how to accomplish their goals in terms of the functions that the software provides. But when you have a large number of possible goals, and each goal might require a sequence of steps to be accomplished, planning the interaction becomes a high cognitive load on the user. This is at the root of most problems with usability of high-functionality applications.

The solution is to try to move as much as possible of the burden of translating goals into concrete functions, onto the computer. So Artificial Intelligence has a vital role to play. Natural language and speech recognition interfaces are a good way of interacting with high-functionality applications, since a typical user can use the expressiveness of language to cover a wide range of possible goals. Gestural interfaces, image recognition, and other "natural" interface modalities can expand the vocabulary of interactive elements. Recent technical improvements in this area are making such interfaces increasingly practical. Traditional menu and icon-based interfaces can also be used for high functionality applications, but require a greater degree of context sensitivity and personalization to avoid overwhelming the user (Fig. 2).

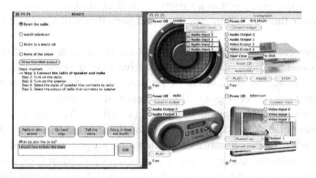

Fig. 2. Roadie is a speech interface for consumer electronics. The user's utterance, "I want to hear the news" is recognized (top left) as one of a number of possible goals (TV, radio). For each goal, a step-by-step plan is constructed (middle left), and a variety of execution options provided (lower left). On the right, interfaces to each of the devices are simulated.

As an example of a high-functionality user interface architecture, Roadie [7] is a framework for goal oriented user interfaces. It provides a goal-oriented speech recognition interface to control a room full of consumer electronics devices like televisions, audio equipment, and other devices. The intent of Roadie is that the user can just speak their goal in relatively unrestricted natural language, and the system performs the operation.

The essential steps in Roadie's operation are:

- Goal recognition;
- Planning;
- Execution (either all at once or step-by-step); and
- (if necessary) Debugging.

Roadie puts a variety of AI technologies to work, detailed in the reference [7]. Roadie uses a commonsense knowledge base to perform the goal recognition. It uses a partial-order planner to compute a plan from each goal. It has introspective knowledge of the capabilities of each device, mapping each function onto possible goals. Diagnostic reasoning modules are invoked in case the user says "Oops" or is otherwise dissatisfied with the results. NLP, planning, and diagnostic reasoning are all well-studied AI technologies, and they result in enabling an easy-to-use, yet high-functionality interface.

3   The Present and Future of Goal-Oriented Interfaces

Significant commercially available examples of goal-oriented interfaces are now beginning to appear. The most well-known of these is perhaps Apple's *Siri*, which provides a speech recognition interface to Apple's iPhone. In the same category is *Google Now*, which leverages the Google search engine to provide personal assistant functions. Microsoft also recently entered this arena with an agent named *Cortana*. IBM has ambitions to transform their successful Jeopardy-playing program, *Watson*, into a broad-spectrum user agent in a variety of vertical markets. These are undoubtably the first shots over the bow of an emerging category that will see rapid growth in the near future.

But at the present time, all these efforts lack some essential capabilities that would be desirable to make a personal assistant effective over a broad range of high functionality applications. The limitations of present efforts provide AI with a research agenda for improving the next generation of personal assistants.

Siri is only designed for one-shot speech interactions. You say something, and if Siri is able to recognize your goal, it invokes a single function in a single application. It can only work with a small number of applications programmed in advance. It can't engage you in a multistep dialogue, execute multistep procedures, or work with more than one application. It is purely a speech agent, and has no visual interface itself. Google Now leverages the high functionality Google search engine (and Siri has a tie to Wolfram Alpha), but again provides its personal assistant functions in a few well-defined areas only, indicated by its "cards", for things like reminders and route planning. There's no way to "debug" Siri or Now if they don't do what you want. Their capacity for true personalization is limited [8].

A key ingredient to making truly high-functionality interfaces is Commonsense knowledge reasoning. In interfaces, Commonsense knowledge can be used to provide intelligent, personalized, context-sensitive defaults; to adapt the interface to the user's expertise, goals, and particulars of a situation, and to provide proactive help. As mentioned above, Roadie consults a Commonsense knowledge base for goal recognition, planning, and debugging. The Open Mind Common Sense effort, which has been running for more than a decade, is collecting millions of statements of Commonsense facts [5], and we have developed a large number of applications in specific areas, from speech recognition to personalizing browsers and Web procedures [1], to enable high-functionality personal assistants.

No less a key ingredient is good user interface design. It is necessary to understand the needs of the user, understand how the user might use the interface to satisfy their goals, and get continual feedback from users on whether they are willing and able to learn, understand, and appreciate the functionality.

4 High-Functionality Interaction is End-User Programming

Part of the key to removing these limitations is the realization that the user of a high-functionality interface is really engaged in what is essentially a programming task. After all, it's a computer; and what other way do we have a telling a computer what to do than to program it? We're just programming it with natural language (and perhaps pointing and typing and other modalities) rather than a conventional programming language. Programming is our high-functionality way of accessing the capabilities of a computer, and it's just that to date, we have artificially imposed on our users the low-functionality interaction paradigm of command and GUI interfaces.

So, we need some way to do everything that you might do in an interactive development environment (IDE). We need sequencing. We need functions with arguments. We need conditionals. We need loops. We need debugging. We even need capabilities that seem advanced or exotic in programming environments, such as introspection. AI also has a tradition of working on various kinds of "Automatic Programming" – generating programs from high-level specifications. Some work has also tried to bring together the commonalities between natural language dialog and programs [10]. Even though there is no conventional programming language, we're really doing end-user programming [6].

Justify [2] is an example. It is a high-functionality decision-support system for online deliberation and argumentation. Arguments are threaded discussions composed of "points", each of which has a type that indicates its role in the argument. It provides automatic summarization at every level, called "assessments", to emphasize their contingent nature. The type system is analogous to that of a programming language, and many programming facilities are provided. It even supports novel AI-style program generation, in providing a facility for Programming by Example [4] (Fig. 3).

Justify has a total of 4808 interface operations, making it comparable to Photoshop (whose documentation index contains 4032 entries). While Photoshop has some programming facilities like macros, it doesn't take its nature is a programming system very seriously. Justify explicitly provides the programming operations and abstractions which give it the true generality that a high functionality system should have.

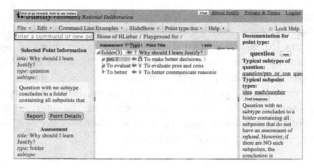

Fig. 3. The *Justify* decision support system. It is a hi-fun system for online discussion, but really, it is a programming environment for decision support. It has a rich type system, and summarization and decision procedures can be programmed.

5   Learning High-Functionality Interfaces

Further, it's not enough just to give a user a high-functionality interface. They've got to be able to learn to use it. Today, many high functionality applications fail because beginning users have a hard time learning them. Some that are successful today are only that way because they started out as low functionality applications, and people gradually learned them over time. Photoshop 1.0 was actually a relatively low functionality application, roughly similar to Preview today. Today's gargantuan Photoshop is only acceptable because the community transitioned slowly to its increasing capabilities. If Photoshop had been first introduced in 2014, most likely it would have been rejected as too difficult to use!

The AI community has had a tradition of research into Intelligent Tutoring Systems (ITS), which hold the promise of being able to resolve the paradox of learning a high functionality system. Contemporary applications seem to have almost given up on the idea of help systems, after experience showing that conventional help is ineffective or that users ignore help. But in the new era of high functionality computing we need to revisit the idea of intelligent tutorials.

We have invented a new kind of interactive tutorial, the *steptorial* ("stepper tutorial") [9] that allows a learner to vary the autonomy of the interaction at every step. A steptorial is a kind of interactive tutorial based on the control structure of a reversible programming language stepper (Fig. 4).

The idea is that the interface steps necessary to complete the introductory example are like a "program" (described by English sentences and/or interaction with the application rather than programming language code). The steptorial allows the user to step through the example, as a programmer steps through code. The steptorial is completely reversible. In extending the stepper metaphor beyond its origins in program debugging, we are enabling learning by end-user debugging of application use-cases.

Not having to choose a fixed level of autonomy in advance means that interactions can be tailored to the level of expertise of the individual user for that particular part of the application, supporting different cognitive styles. Depending on the situation at the

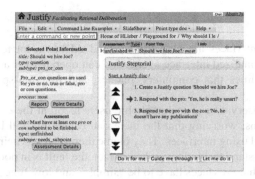

Fig. 4. A *steptorial* ("stepper tutorial") for Justify. The user is stepping through a natural language description of an introductory example. At any point he/she can choose to have the system perform an operation, try it him/herself, be guided through it step-by-step (Kelleher & Pausch 2005) [3], or to back up to a previous state.

moment, the user can choose help either in or out of context. Finally, having a variable level of autonomy reduces the risk, since the user can always go back and choose a different level of autonomy without any penalty.

Again, we see this work in the AI tradition of Intelligent Tutoring Systems, and many opportunities arise for trying to better understand user behavior and provide personalized help.

6 Testing High-Functionality Interfaces

Finally, no discussion of user interface development is complete without attention to testing. User experience (UX) testers should realize that testing a high-functionality interface is not the same as testing a low-functionality interface. In low-functionality interfaces, UX testing may focus on the best way to organize the command set, receive user input, choose and display interface elements, and aesthetics.

In testing high-functionality interfaces, developers need more high-level feedback on the adequacy of the functionality, the flexibility and composability of operations, the learnability for new users, and the effectiveness of personalization to assure efficiency for expert and habitual users. Can users grasp the high functionality that the system can offer? This doesn't need to fully happen in the initial encounter; the new user need only get the sense that there is a world of utility waiting to be discovered. Can the users understand the essential concepts that will enable them to succeed on a simple but interesting example in the time plausible for an introductory session? Will they be enthusiastic about continuing to learn the interface as time goes on? How can the system communicate these essential concepts and offer assistance when users are confused or stuck? Does the system take into account different people's learning styles when providing help in learning the system?

7 Conclusion

High functionality interfaces are here to stay, and AI has a lot to contribute to the development of personalized interface agents that can operate over a broad spectrum of "application" capabilities that computers, phones, and function-specific devices can provide.

Today, some conservative or technophobic users recoil at the ever-growing proliferation of applications, worrying that the increased capability will be plagued by difficulty in learning and using interfaces. They blame themselves as being "non-technical" or too stupid to understand technology, when in fact, the fault is in poor interface design. Let's bust the myth that interfaces have to be "simple" and low-functionality to be easy to use. Human beings are enormously capable, and, at the same time, we have developed ways of interacting with each other that are effective and pleasurable. We should demand no less from our technology.

References

1. Faaborg, A., Lieberman, H.: A goal-oriented web browser. In: Cypher, A., Dontcheva, M., Lau, T., Nichols, J. (eds.) No Code Required: End-User Programming for the Web. Morgan Kaufmann (2010)
2. Fry, C., Lieberman, H.: Decision-making should be more like programming. In: International Symposium on End-User Development, Copenhagen, June 2013
3. Kelleher, C., Pausch, R.: Stencil based tutorials: design and evaluation. In: CHI 2005, pp. 541–550
4. Lieberman, H. (ed.) Your Wish is My Command: Programming by Example. Morgan Kaufmann (2001)
5. Lieberman, H., Liu, H., Singh, P., Barry, B.: Beating common sense into interactive applications. AI Magazine, Association for the Advancement of Artificial Intelligence, Winter 2004–2005
6. Lieberman, H., Paterno, F., Wulf, V. (eds.) End-User Development, Springer (2006)
7. Lieberman, H., Esp, J.: A goal-oriented interface to consumer electronics using planning and commonsense reasoning. Knowl.-Based Syst. J. **20**, 592–606 (2007)
8. Lieberman, H.: Say Hello to Smarter Apps That Fulfill Your Wishes, Wired, UK, p. 70, November 2012
9. Lieberman, H., Rosenzweig, E., Fry, C.: Steptorials: Mixed-initiative learning of high functionality applications. In: ACM Conference on Intelligent User Interfaces (IUI-14), Haifa, Israel, February 2014
10. Liu, H., Lieberman, H.: Metafor: visualizing stories as code. In: ACM Conference on Intelligent User Interfaces (IUI-2005), San Diego, January 2005
11. Norman, D.A.: Affordance, conventions and design. Interactions **6**(3), 38–43 (1999). ACM Press

Agents

Performance of Communicating Cognitive Agents in Cooperative Robot Teams

Changyun Wei$^{(\boxtimes)}$, Koen V. Hindriks, and Catholijn M. Jonker

Interactive Intelligence Group, EEMCS, Delft University of Technology,
Mekelweg 4, 2628 CD Delft, The Netherlands
{c.wei,k.v.hindriks,c.m.jonker}@tudelft.nl

Abstract. In this work, we investigate the effectiveness of communication strategies in the coordination of cooperative robot teams. Robots are required to perform search and retrieval tasks, in which they need to search targets of interest in the environment and deliver them back to a home base. To study communication strategies in robot teams, we first discuss a case without communication, which is considered as the baseline, and also analyse various kinds of coordination strategies for robots to explore and deliver the targets in such a setting. We proceed to analyse three communication cases, where the robots can exchange their *beliefs* and/or *goals* with one another. Using communicated information, the robots can develop more complicated protocols to coordinate their activities. We use the Blocks World for Teams (BW4T) as the simulator to carry out experiments, and robots in the BW4T are controlled by cognitive agents. The team goal of the robots is to search and retrieve a sequence of colored blocks from the environment. In terms of cooperative teamwork, we have studied two main variations: a variant where all blocks to be retrieved have the same color (no ordering constraints on the team goal) and a variant where blocks of various colors need to be retrieved in a particular order (with ordering constraints). The experimental results show that communication will be particularly helpful to enhance the team performance for the second variant, and exchanging more information does not always yield a better team performance.

Keywords: Communication · Multi-robot coordination · Foraging

1 Introduction

In many practical applications, robots are seldom stand-alone systems but need to cooperate and collaborate with one another. In this work, we focus on search and retrieval tasks, which have also been studied in the foraging robot domain [1–3]. Foraging is a canonical task in studying multi-robot teamwork, in which the robots need to search targets of interest in the environment and then deliver them back to a home base. The use of multiple robots may yield significant performance gains compared to the performance of a single robot [1,4]. But multiple robots may also lead to interference between teammates, which can

© Springer International Publishing Switzerland 2015
B. Duval et al. (Eds.): ICAART 2014; LNAI 8946, pp. 13–31, 2015.
DOI: 10.1007/978-3-319-25210-0_2

decrease team performance Therefore, it poses a challenge for a robot teams to develop effective coordination protocols for realising such performance gains.

We are in particular interested in the role of communication in coordination protocols and its impact on team performance. In previous work, e.g., [5,6], it has been reported that more complex communication strategies offer little benefit over more basic strategies. The messages exchanged in [5] among robots, however, are very simple, and they only studied a simple foraging task without ordering constraints on the targets to be collected. As no clear conclusion has been drawn on what kind of communication is most suitable for robot teams [7], it is our aim to gain a better understanding of the impact of more advanced communication strategies where multiple robots can coordinate their behavior by exchanging their *beliefs* and/or *goals*. By communicating beliefs and/or goals, the robots can create a shared mental model to enhance their team awareness.

In this paper, we want to gain in particular a better understanding of the role of communication in the search and retrieval tasks *with and without ordering constraints* on the team goal. The first task without ordering constraints on the team goal is a simple foraging task that requires the robots to retrieve target blocks all of the same color, whereas the second task with ordering constraints on the team goal is a cooperative foraging task that requires the robots to retrieve blocks of various colors in a particular order. To this end, we first analyse a baseline without communication and then proceed to analyse three different communication conditions where the robots exchange only beliefs, only goals, and both beliefs and goals. We use the BW4T simulator as the testbed to carry our experimental study.

The paper is organized as follows. Related work is discussed in Sect. 2. Section 3 introduces the search and retrieval tasks and the BW4T simulator. In Sect. 4, we discuss the coordination protocols for the baseline without communication and for three communication cases. The experimental setup is presented in Sect. 5 and the results are discussed in Sect. 6. Finally, Sect. 7 concludes the paper.

2   Related Work

Robot foraging tasks have been extensively studied and have in particular resulted in various bio-inspired, swarm-based approaches [2,3]. In these approaches, typically, robots minimally interact with one another as in [2], and if they communicate explicitly, only basic information such as the locations of targets or their own locations are exchanged [8]. Most of this work has studied the simple foraging task where the targets to be collected are not distinguished, so the team goal of the robots does not have ordering constraints. Another feature that distinguishes our setup from most related work on foraging tasks is that we use an office-like environment instead of the more usual open area with obstacles. Targets are randomly dispersed in the rooms of the environment, and the robots initially do not know which room has what kinds of targets. Our interest is to evaluate the contribution that explicit communication between robots can make on the time to complete foraging tasks, and to identify the role of communication in coordinating the more complicated foraging task in which the team goal has ordering constraints.

In order to enhance team awareness, we follow the work of [9,10], which claims that shared mental models can improve team performance, but it needs explicit communication among team members. Most of the current research on foraging, however, only has implicit communication for robot teams [7]. The work in [3,5] studies communication in the simple foraging task without ordering constraints on the team goal and its impact on the completion time of the task. The work in [5] compares different communication conditions where robots do not communicate, communicate the main behavior that they are executing, and communicate their target locations. Roughly these conditions map with our no communication, communicating only beliefs, and communicating only goals, whereas we also study the case where both beliefs and goals are exchanged. A key task-related difference is that having multiple robots process the same targets speeds up completion of the task in [5], whereas this is not so in our case. As a result, the use of communicated information is quite different as it makes sense to follow a robot or move directly to the same target location in [5], whereas this is not true in our setting.

The work in [3] studies the conditions where the robots can only exchange messages within certain communication ranges or in nest areas (i.e., the rooms in our case), whereas we do not study the constraints on the communication range; instead, we focus on the communication content. In this work, we assume that a robot can send messages to any of its teammates in the environment, and once a sender robot broadcasts a message, the receiver robots can receive the message successfully.

Several coordination strategies without explicit communication in foraging tasks have been studied in [11], which takes into account the avoidance of interference in scalable robot teams. Apart from the size of robot teams, the authors in [12] consider the size of the environment. In our work, we also use scalable robot teams to perform foraging tasks in scalable environments in our experimental study. We consider a baseline in which the robots do not explicitly communicate with one another, but they can still apply various combinational strategies for exploration and exploitation in performing the foraging tasks. As robots may easily interfere with each other without communication, these combinational coordination strategies in particular take account of the interference in multi-robot teams, and we carry out experiments to study which combinational strategy is the best one for the baseline case.

In this work, we assume that the robots only collide with each other when they want to occupy the same room at the same time in BW4T, and the robots can pass through each other in all the hallways. Once a robot has made a decision to move to a particular room, it can directly calculate the shortest path to that room. Thus, the multi-robot path planning problem is beyond the scope of this paper.

3   Multi-robot Search and Retrieval

General multi-robot teamwork usually consists of multiple subtasks that need to be accomplished concurrently or in sequence. If a robot wants to achieve a specific subtask, it may first need to move to the right place where the subtask can be performed. An example of such teamwork is search and retrieval tasks, which are motivated by many piratical multi-robot applications such as large-scale search and rescue robots [13], deep-sea mining robots [14], etc.

3.1   Search and Retrieval Tasks

Search and retrieval tasks have also been studied in the robot foraging domain, where the team goal of the robots is to search targets of interest in the environment and then deliver them to a home base. At the beginning of the entire task, the environment may be known, unknown or partially-known to the robots. Here the targets of interest correspond to the subtasks of general multi-robot teamwork, and if they can be delivered to the home base concurrently, then the team goal does not have ordering constraints; otherwise, all the needed targets must be collected in the right order.

In this work, the robots do not have prior knowledge about the distribution of the targets. The robots have the map of the rough locations where the targets might be, but they have to explore these locations in order to find the exact dispersed targets. For instance, in the context of searching for and rescuing survivals in a village after an earthquake, even though the robots may have the map information of the village, they are hardly likely know the precise locations of the survivals when starting their work. Moreover, due to the limited carrying capability of robots, we assume that a robot can only carry one target at one time in this work.

3.2   The Blocks World for Teams Simulator

We simulate the search and retrieval tasks using the Blocks World for Teams (BW4T[1]) simulator, which is an extension of the classic single agent Blocks World problem. The BW4T has office-like environments consisting of *rooms* in which colored *blocks* are randomly distritbuted for each simulation (see Fig. 1). One or more robots are supposed to search, locate, and retrieve the required blocks from rooms and return them to a so-called *drop-zone*.

As indicated at the bottom of the simulator in Fig. 1, required blocks need to be returned in a specific order. If all the required blocks have the same color, then the team goal of the task does not have ordering constraints. Access to rooms is limited in the BW4T, and at any time at most one robot can be present in a room or the drop-zone. Robots, moreover, can only carry one block at a time. The robots have the information about the locations of the rooms, but they do

---

[1] BW4T has been integrated into the agent environments in GOAL [15], which can be found from http://ii.tudelft.nl/trac/goal.

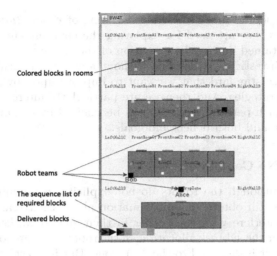

Fig. 1. The Blocks World for Teams Simulator.

not initially know which blocks are present in which rooms. This knowledge is obtained for a particular room by a robot when it visits that room. Each robot, moreover, is informed of the complete required blocks and its teammates at the start of a simulation. Robots in BW4T can be controlled by agents written in GOAL [15], the agent programming language that we have used for implementing and evaluating the team coordination strategies discussed in this paper. GOAL also facilitates communication among the agents.

While interacting with the BW4T environment, each robot gets various percepts that allow it to keep track of the current environment state. Whenever a robot arrives at a place, in a room, or near a block, it will receive a corresponding percept, i.e., `at(PlaceID)`, `in(RoomID)` or `atBlock(BlockID)`. A robot also receives percepts about its state of movement (traveling, arrived, or collided), and, if so, which block the robot is holding. Blocks are identified by a unique ID and a robot in a room can perceive which blocks of what color are in that room by receiving percepts of the form `color(BlockID,ColorID)`.

4   Coordination Protocols

A coordination protocol for search and retrieval tasks consists of three main strategies: deployment, subtask allocation and destination selection strategies. The *deployment* strategy determines the starting positions of the robots. In our settings, all robots start in front of the drop-zone, so we will not further discuss this strategy in our coordination protocols. The *subtask allocation* strategy determines which target blocks the robots should aim for. Once a robot has decided to retrieve a particular target, it needs to choose a room to move towards so that it can get such a target. The *destination selection* strategy determines which rooms the robots should move towards, consisting of *exploration* and *exploitation* sub-strategies that are used for exploring the environment and exploiting the knowledge obtained during the execution of the entire task.

We will first investigate a baseline without any communication and set it as the performance standard that we want to improve upon by adding various communication strategies. And then we are particularly interested in whether, and, if so, how much performance gain can be realised by communicating only beliefs, only goals, and both beliefs and goals.

4.1 Baseline: No Communication

In the baseline, although the robots do not explicitly communicate with one another, they can still obtain some information about their teamwork because the robots may interfere with each other in their shared workspace. Without communication, for the subtask allocation all the robots will aim for the currently needed block until it is delivered to the drop-zone. But for destination selection, the exploration and exploitation sub-strategies can ensure that they will not visit rooms more often than needed (as far as possible), and basically that knowledge is exploited whenever the opportunity arises (i.e., a robot is greedy and will start collecting a known block that is the closest one and has the needed color).

In the baseline, we have identified four dimensions of variation: which room a robot initially will visit, how a robot uses knowledge obtained about another robot through interference, how it selects a (next) room to visit, and what a robot will do when holding a block that is not needed now but needed later.

Initial Room Selection. At the beginning of the task, a robot has to choose a room to explore since it does not have any information about the dispersed blocks in the environment. One possible option labeled (1a) is to choose a *random* room without considering any distance information. Assuming that k robots initially select a room to visit from n available rooms, the probability that each robot chooses a different room to visit is P, and the probability that collisions may occur in the team is $P_c = 1 - P$. Then we can know:

$$P = \begin{cases} \frac{n!}{(n-k)! \cdot n^k} & , k \leq n, \\ 0 & , k > n. \end{cases} \tag{1}$$

This gives, for instance, a probability of 9.38 % that 4 robots select different initial rooms from 4 available rooms, which drops to only 1.81 % for 8 robots performing in the environment with 10 rooms. Working as a team, robots are expected to have as few collisions as possible, and we use P_c to reduce the likelihood that robots may collide with each other. For example, suppose we want the likelihood of collisions for $k = 4$ robots to be less than 5 %, then we need $n > 119$. This tells us that it is virtually impossible to avoid collisions without communication in large robot teams as a very large number of rooms would be required then.

A second option labeled (1b) is to choose the *nearest* room, which means that the robot will take account of the distance from its current location to the room's location. In this case, almost all robots will choose the same initial room given that they all start exploring from more or less the same location according to the deployment strategy in our settings.

Visited by Teammates. Another issue concerns how a robot should use the knowledge about a collision with its teammates. A collision occurs when a robot is trying to enter a room but fails to do so because the room is already occupied by one of its teammates. The first option labeled (2a) is to *ignore* this information. That simply means that the fact that the room is currently being visited by the teammate has no effect on the behavior of the robot.

The second option labeled (2b) is to take this information into account. The idea is to exploit the fact that the robot, even though it still does not know what blocks are in the room, believes that the team knows what is inside the room. Intuitively, there is no urgent need anymore to visit this room therefore. The robot thus will delay visiting this particular room and assign a higher priority to visiting other rooms. Only if there is nothing more useful to do, a robot then would consider visiting this room again.

Next Room Selection. If a robot does not find a block it needs in a room, it has to decide which room to explore next. The available options for this problem are very similar to those for the initial room selection but the situation may actually be quite different as the robots will have moved and most of the time will not locate at more or less the same position anymore. In addition, some rooms have already been visited, which means there are less options available to a robot to select from in this case.

One option labeled (3a) is to *randomly* choose a room from the rooms that have not yet been visited, and a second one labeled (3b) is to visit the room *nearest* to the robot's current position. It is not upfront clear which strategy will perform better. If the robots very systematically visit rooms, because they all start from the same location, this will most likely increase interference. The issue is somewhat similar to the initial room selection problem as it is not clear whether it is best to minimize distance traveled (i.e., choose the nearest room) or to minimize interference (i.e., choose a random room).

Holding a Block Needed Later. When the robots are required to collect blocks of various color with ordering constraints, this issue concerns what to do when a robot is holding a block that is not needed now but is needed later. For instance, robot Alice is delivering a red block to the drop-zone because it believes that the current needed target should be a red one. If robot Bob completes the subtask of retrieving a red block before Alice moves to the drop-zone, and the remaining required targets still need a red block in the future, then Alice comes to this situation.

One option labeled (4a) is to *wait in front of the drop-zone*, and then enter the drop-zone and drop the block when it is needed. The waiting time depends on how long it will take before the block that the robot is holding will be needed. A second option labeled (4b) is to *drop the block in the nearest room*. Since the waiting time in the first option is uncertain, it might be better to store the block in a room where it can be picked up again later if needed and invest time now rather in retrieving blocks that are needed now.

In the baseline case, each dimension discussed above has two options, so we can at most have 16 combinational strategies, some of which can be eliminated for the experimental study (see Sect. 5). We will investigate the best combinational strategy of the baseline, and then we take it as the performance standard to compare with the communication cases.

4.2   Communicating Robot Teams

In decentralized robot teams, there are no central manager or any shared database, for example, in distributed robot teams, so the robots have to explicitly exchange messages to keep track of the progress of their teamwork. In the communication cases, we mainly focus on the communication content in terms of beliefs and goals, and the robots use those shared information enhance team awareness. Since the robots can be better informed about their teammates in comparison with the baseline case, they can have more sophisticated coordination protocols concerning subtask allocation and destination selection.

Constructing Shared Mental Models. By communicating beliefs with one another, robots can be informed about *what other robots have observed in the environment and where they are*. Messages about beliefs are differentiated by the indicative operator ":" from those about goals, whose type is indicated by the imperative operator "!" in GOAL agent programming language. In this work, the robots exchange the following messages in respect of beliefs with associated meaning listed:

- `:block(BlockID, ColorID, RoomID)` means block `BlockID` of color `ColorID` is in room `RoomID`,
- `:holding(BlockID)` means the message sender robot is holding block `BlockID`,
- `:in(RoomID)` means the message sender robot is in room `RoomID`, and
- `:at('DropZone')` means the message sender robot is at the drop-zone.

Each of the messages listed above can also be negated to represent. For example, when a robot leaves a room, it will inform the other robots that it is not in the room using the negated message `:not(in(RoomID))`. Upon receiving a negated message, a robot will remove the corresponding belief from its belief base. A robot sends the first message to its teammates when entering room `RoomID` and getting the percept of `color(BlockID,ColorID)`. Note that only this message does not implicitly refer to the sending robot, and therefore except for the

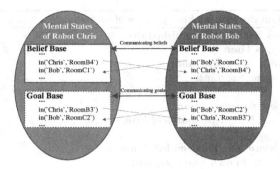

Fig. 2. Constructing a shared mental model via communicating beliefs and/or goals.

first message type, a robot who receives a message will also associate the name of the sender with the message. For instance, if robot Chris receives a message in('RoomC1'in) from robot Bob, Chris will insert in('Bob','RoomC1') into its belief base (see Fig. 2).

By communicating goals with one another, robots can be informed about *what other robots are planning to do.* Robots exchange the following messages in respect of goals with associated meaning listed:

- !holding(BlockID) means the message sender robot wants to hold block BlockID,
- !in(RoomID) means the message sender robot wants to be in room RoomID, and
- !at('DropZone') means the message sender robot wants to be at the drop-zone.

Negated versions of goal messages indicate that previously communicated goals are no longer pursued and have been dropped by the sender. All goal messages received are associated with the sender who sent the message and stored or removed as expected (e.g., see Fig. 2).

As we assume that the robots are cooperative, and they communicate with each other truthfully, the received messages can be used to update their own beliefs and goals. Algorithm 1 shows the main decision making process of an individual robot: how the robot updates its own mental states and at the same time shares them with its teammates for constructing a shared mental model. In its decision making process, the robot first handles new environmental percepts (line 2–5), and then uses received messages to update its own mental states (line 6–8). Based on the updated mental states, the robot can decide whether some dated goals should be dropped (line 9–12) and whether, and, if so, what, new goals can be adopted to execute (line 13–16). When the robot has obtained new percepts about its environment and itself or has adopted or dropped a goal, it will also inform its teammates (line 4, 11 and 15).

As shown in Fig. 2, when decentralized robot teams construct such a shared mental model via communicating belief and goal messages discussed above, even

---

Algorithm 1. Main decision making process of an individual robot.

```
 1: while entire task has not been finished do
 2:     if new percepts then
 3:         update own belief base, and
 4:         send message ":percepts".
 5:     end if
 6:     if receive new messages then
 7:         update own belief base and goal base.
 8:     end if
 9:     if some goals are dated and useless then
10:         drop them from own goal base, and
11:         send messages "!not(goals)".
12:     end if
13:     if new goals are applicable then
14:         adopt them in own goal base, and
15:         send messages "!(goals)".
16:     end if
17:     execute actions.
18: end while
```

---

though they do not have any shared database or centralised manager, they could fully know each other as what they can know in a centralised team. Therefore, such a shared mental model can enable the robots to have more sophisticated coordination protocols.

Subtask Allocation and Destination Selection Strategies. By just communicating belief messages, there is quite a bit of potential to avoid interference since a robot will inform its teammates when entering a room. More interestingly, it is even possible to avoid duplication of effort because a robot can also obtain what block should be picked up next from the information about the blocks that are being delivered by teammates. Robot use the shared beliefs to coordinate their activities for subtask allocation (see item 3 and 4) and destination selection (see item 1 and 2) as follows:

1. A robot will not visit a room for exploration purposes anymore if it has already been explored by a team member;
2. A robot will not adopt a goal to go to a room (or the drop-zone) that is currently occupied by another robot;
3. A robot will not adopt a goal to hold a block if another robot is already holding that block (which may occur when another robot beats the first robot to it);
4. A robot will infer which of the blocks that are required are already being delivered from the information about the blocks that its teammates are holding and will use this to adopt a goal to collect the next block that is not yet picked up and is still to be delivered.

By just communicating goal messages, the robots can also coordinate the activities to avoid, as in the case of sharing only beliefs, interference and duplication of effort. Whereas it is clear that a robot should not want to hold a block that is being held by another robot, it is not clear per se that a robot should not want to hold a block if another already wants to hold it. The focus of our work reported here, however, is not on negotiating options between robots. Instead, we have used a "first-come first-serve" policy here, and the shared goals can be used for subtask allocation (see item 2 and 3) and destination selection (see item 1) as follows:

1. A robot will not adopt a goal to go to a room (or the drop-zone) that another robot already wants to be in,
2. A robot will not adopt a goal to hold a block that another robot already wants to hold, and
3. A robot will infer which of the blocks that are required will be delivered from the information about goals to hold a block from its teammates and will use this to adopt a goal to collect the next block that is not yet planned to be picked up and is still to be delivered.

It should be noted that the difference between item 3 listed above for the use of goal messages and that of item 4 listed above for the use of belief messages is rather subtle. Whereas the information about goals is an indication of what is planned to happen in the near future, the information about beliefs represents what is going on right now. It will turn out that the potential additional gain that can be achieved from the third rule for goals above, because the information is available before the actual fact takes place, is rather limited. Still, we have found that because the information about what is planned to happen in the future precedes the information about what actually is happening, it is possible to almost completely remove interferences between robots.

As the robots are decentralized and have their own decision processes, even though they use first-come first-serve policy to compete for subtasks and destinations, it may happen that they make decisions in a synchronous manner. For instance, both robot Alice and Bob may adopt a goal to explore the same room at the same time, which indeed does not violate the first protocol of shared goals when they make such a decision but may actually lead to an interference situation. In order to prevent such inefficiency, in our coordination protocols the robots can also drop goals that have already been adopted so that they can stop corresponding actions that are being executed. For example, a robot can drop a goal to enter a room if it finds that another robot also wants to enter that room. Apart from this reason, some dated goals should also be removed from the goal base. For example, a robot should drop a goal of retrieving a block if it does not have the currently needed color any more. As can be seen in Algorithm 1, a robot will check dated and useless goals and then drop them (see see line 9–12) in its decision making process.

When robots communicate both beliefs and goals, the coordination protocol combines the rules listed above for belief and goal messages. For example, a robot will not adopt a goal of going to a room if the room is occupied now or another robot already wants to enter it. Similarly, a robot will not adopt a goal to hold a block if the block is already being held by another robot or another robot already wants to hold it. In the case of communicating both beliefs and goals, as the robots should have more complete information about each other, in comparison with the cases of communicating only beliefs and only goals, so they are expected to achieve more additional gains with regard to interference and duplication of effort. But it should be noted that since all the robots have much knowledge to avoid interference, they may need to frequently change their selected destinations or to choose farther rooms, which may result in an increase in the walking time. In our experiments, we will investigate how much performance gain can be realised in these three communication cases.

5   Experimental Design

5.1   Data Collection

All the experiments are performed in the BW4T simulator in GOAL. We have collected data on a number of different items in our experiments. The main performance measure, i.e., *time-to-complete*, has been measured for all runs. In order to gain more detailed insight into the effort needed to finish the task, we have collected data on *duplicated effort* that gives some insight into both the effectiveness of the strategy as well as in the complexity of the tasks. Duplication can be obtained by keeping track of the number of blocks that are dropped by the robots without contributing to the team goal.

Each time when two robots collide, i.e., one robot tries to enter a room occupied by another, is also logged. The *total number of interference* provides an indication of the level of coordination within the team. Finally, to obtain a measure of the cost involved in communication in multi-robot teams, the *number of exchanged messages* is also counted. A distinction is made between messages about the beliefs and messages about goals.

5.2   Experimental Setup

There are many variations in setup that one would like to run in order to gain as much insight as possible into the impact of various factors on the performance

Table 1. Baseline instances based on coordination protocols.

Team size	Single robot		Multiple robots				
Team goal	Either		Blocks of same color		Blocks of random color		
Instances	i	(1a,3a)	i	(1a,2a,3a)	i	(1a,2a,3a,4a)	vi	(1a,2b,3b,4b)
	ii	(1a,3b)	ii	(1a,2b,3a)	ii	(1a,2a,3a,4b)	vii	(1b,2a,3a,4a)
	iii	(1b,3a)	iii	(1a,2b,3b)	iii	(1a,2b,3a,4a)	viii	(1b,2a,3a,4b)
	iv	(1b,3b)	iv	(1b,2a,3a)	iv	(1a,2b,3a,4b)	ix	(1b,2b,3a,4a)
			v	(1b,2b,3a)	v	(1a,2b,3b,4a)	x	(1b,2b,3a,4b)

of a team. Since we want to understand the relative speed up of teams compared to a single robot to measure the effectiveness of various of our coordination protocols, we need to run simulations with a single robot. For multiple robots, we have used team sizes of 5 and 10. We also consider the factor of robots' environmental size, and we use maps of 12 rooms and 24 rooms.

In our experiments, the robots are required to retrieve 10 blocks from their environments where there are total 30 blocks that are randomly distributed for each simulation, but there are two different tasks. Recall that the first task does not have ordering constraints on the team goal and the robots retrieve block of the same color, which is relatively simple and similar to the tasks that many researchers have addressed in the robot foraging domain. Comparatively, the second task is more complicated and is a cooperative foraging task that has ordering constraints on the team goal, requiring the robots to retrieve blocks of various colors. In order to so, in each run, BW4T simulator randomly generates a sequence of blocks of various colors, so the team goal is to collected blocks of random colors in the right order.

We list the baseline instances in Table 1 based on the combinational strategies used by the robots. A single robot's behavior is relatively simple because it does not need to consider interference with other robots, and duplicated effort will never occur. Although there are 4 dimensions in baseline strategies, a single robot only needs to consider the first and the third dimension with regards to initial room selection and next room selection, respectively. Accordantly, the single robot case has 4 setups in each environmental condition. For instance, we use S(iii) in Sect. 6 to indicate the strategy combining (1b, 3a).

When multiple robots participate in the tasks, there are more combinations based on the strategies that the robots may use. As baseline has four dimensions, each of which has two options, we can get at most 16 combinational strategies. Although we cannot directly figure out which combinational strategy is the best one without experiments, we can still eliminate several choices that are apparently inferior to the other ones.

One issue occurs when the strategy combines (2a) and (3b). In case a robot wants to go to room A but one of its teammates arrives earlier, the robot will then reconsider but select the same room again because it will select the nearest room based on the strategy. This behavior will result in very inefficient performance, so we can eliminate those choices combining (2a) and (3b). Another issue arises when a choice combines (1b) and (3b), which will make robots cluster together as they more or less start from the same location, and then they always try to visit the nearest rooms. A cluster of robots will cause inefficiency and interference, so we can further eliminate those choices combining (1b) and (3b). As a result, for the second task, we have 10 setups as shown in Table 1 and, for example, we use M_R(iii) in Sect. 6 to indicate the strategy combining (1a,2b,3a,4a).

When the robots perform the first task, as all the required blocks have the same color, the fourth dimension does not make sense because any holding block can contribute to the team goal until the task is finished. Therefore, we can eliminate this dimension and finally have 5 setups left for this task, and we use M_S(iii) in Sect. 6 to indicate the strategy combining (1a,2b,3b). For the communication cases, we have three setups, communicating only beliefs, only goals, and both beliefs and goals, in each environmental condition. Each setup has been run for 50 times to reduce variance and filter out random effects in our experiments.

6   Results

6.1   Baseline Performance

Figures 3 and 4 show the performance of the various strategies for the baseline on the horizontal axis and the four different conditions related to team size and room numbers on the vertical axis. This results show that the relation between team size and environment size has an important effect on the team performance that also relies on which combinational coordination strategy is used.

Statistically, the performance of the combinational strategies does not significantly differ from any of the others. Even so, from Fig. 3 we can see that the strategy M_S(iii) on average performs better than any of the other ones and has minimal variance which is why we choose this strategy as our baseline to compare with the communication cases. This strategy combines options (1a) which initially selects a room randomly, option (2b) which uses information from collisions with other robots to avoid duplication of effort, and option (3b) which selects the nearest room to go to next. Interestingly, this strategy does not minimize interference. This is because if the robots do not communicate with one another, using the option to go to the nearest rooms for the next room selection, i.e., option (3b), increases the likelihood of selecting the same room at the same time and causes interference.

Similarly, from the data shown in Fig. 4, it follows that strategy M_R(v) on average takes the minimum time-to-complete the second task and again its variance is also less than those of the other strategies. This strategy combines options (1a), (2b), (3b) and (4a). It thus is an extension of the best strategy M_S(iii) for the first task in Fig. 3 with option (4a) which means that robots wait in front

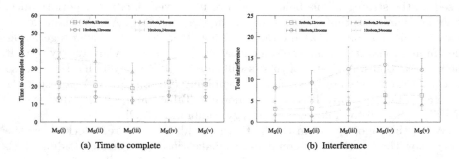

(a) Time to complete   (b) Interference

Fig. 3. Baseline performance for the first task (i.e., team goal without ordering constraints).

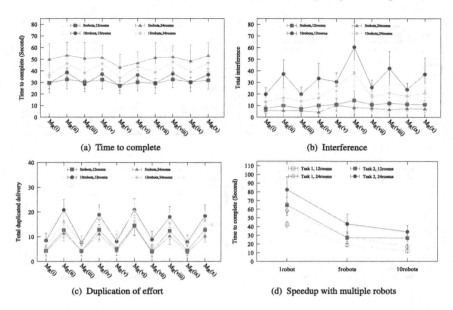

Fig. 4. Baseline performance for the second task (i.e., team goal with ordering constraints) & Speedup.

of the drop-zone if they hold a block that is needed later. We can conclude that even though sometimes option (4a) makes the robots idle away and stop in front of the drop-zone with a block that is needed later, it is more economical than the idea of storing the block in the nearest room so that it can be picked up again later.

6.2 Speedup with Multiple Robots

Balch [5] introduced a speedup measurement, which is used to investigate to what extent multiple robots in comparison with a single robot can speed up a task. We have plotted the performance of the best strategies for single and multiple robot cases to analyse the speedup of various team sizes in Fig. 4(d). The results show that doubling a robot team does not double its performance. For example, 5 robots take 27.26 s on average to complete the second task in 12 rooms, but 10 robots on average need 26.88 s. We therefore conclude that speedup obtained by using more robots is sub-linear, which is consistent with the results reported in [5].

In order to better understand the relation between speedup and the strategies we have proposed, we can inspect Fig. 4(a) again. We can see that speedup depends on the strategy that is used. One particular fact that stands out is that the time-to-complete for the odd numbered strategies in Fig. 4(a) is similar for 5 and 10 robots that are exploring 12 rooms (we find no statistically significant difference). We can conclude that adding more robots does not nec-

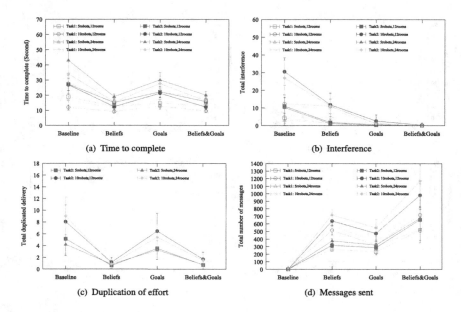

Fig. 5. Performance measures for different communication conditions.

essarily increase team performance because more robots may also bring about more interference among the robots. We can also see in Fig. 4(b) that it is quite clear that the average numbers of interference in 5 and 10 robots exploring 12 rooms are significantly different. Thus, more interference makes the robots take more time to complete the task, but it also depends on the strategies. It follows that using more robots only increases the performance of a team if the right team strategy is used. It is more difficult to explain the fact that both interference and duplication of effort are lower for the odd-numbered strategies than for the even-numbered ones. It turns out that the difference relates to option 4 where robots in all odd-numbered strategies wait in front of the drop-zone if a block is needed later. In a smaller sized environment, adding more robots in the second task means also adding more waiting time which in this particular case cancels out any speedup that one might have expected.

6.3　Communication Performance

Figure 5 shows the results that we obtained for the performance measures for the communication conditions we study here. First, we can see in Fig. 5(a) that communication is much more useful in the second task than the first one. When 5 robots operate in the 12 rooms environment, communicating beliefs yields a 34.15 % gain compared to the no communication case for the first task whereas it yields a 44.5 % gain in the second task. It is also clear from Fig. 5(a) that communication yields more predictable performance as the variance in each of the communication conditions is significantly less than that without communication. Recall that the second task requires robots to retrieve blocks in a particular order, and thus we can conclude that when a multi-robot task consists of multiple subtask that need to be achieved with ordering constraints, communication will be particulary helpful to enhance team performance.

Second, though communicating beliefs is more costly than communicating goals in terms of messages, the resulting performance of time-to-complete is significantly better when communicating only beliefs compared to the performance when only goals are communicated. For example, Fig. 5(a) and (d) show that when 5 robots only communicate beliefs, the team takes 15.13 s and sends 318 messages on average to complete the second task in the 12 rooms environment while only communicating goals takes the same team 22.23 s and 288 messages. This is because communicating beliefs can inform robots about what blocks have been found by teammates, and the environment can become known for all the robots sooner than the case of communicating goals, which can save the exploration time.

Third, the communication of goals yields a significantly higher decrease in interference compared to communicating beliefs. For instance, communicating goals eliminated 91.56 % of the interference present in the no communication condition compared to only 61.6 % when communicating beliefs for 10 robots that perform the second task in 12 rooms. This is because communicating goals can inform a robot about what its teammates want to go, so it can choose a different room as its destination. Comparatively, a robot only inform its teammates when entering a room in the case of communicating beliefs, which cannot effectively prevent other teammates clustering together in front of this room. On the other hand, the communication of goals does not significantly decrease duplication of effort as shown in Fig. 5(c). There is a simple explanation of this fact: even though a robot knows which blocks its teammates want to handle in this case, it does not know what color these blocks have if it did not observe the block itself before. This lack of information about the color of the blocks makes it impossible to avoid duplication of effort in that case.

A somewhat surprising observation is that though communicating only beliefs or only goals would never negatively impact performance, communicating both of them does not always yield a better performance than just communicating beliefs. T-tests show that there is no significant difference between communicating only beliefs and communicating both beliefs and goals with regards to time-to-complete. The reason is that when the robots share both beliefs and goals, they are even better informed about what their teammates are doing, which allows them to reduce interference even more. We can see in Fig. 5(b) that communicating both beliefs and goals can ensure that the robots do not collide with each other anymore. However, this more careful behavior though not often but still sometimes results in a robot choosing rooms that are farther away on average in order to avoid collisions with teammates, which may increases the time to complete the entire task.

7 Conclusions

In this paper, we presented various coordination protocols for cooperative multi-robot teams performing search and retrieval tasks, and we compared the performance of a baseline case without communication with the cases with various communication strategies. We performed extensive experiments using the BW4T simulator to investigate how various factors, but most importantly how the content of communication, impacts the performance of robot teams in such tasks with or without ordering constraints on the team goal. A key insight from our work is that communication is able to improve performance more in the task with ordering constraints on the team goal than the one without ordering constraints. At the same time, however, we also found that communicating more does not always yield better team performance in multi-robot teams because more robots will increase the likelihood of interference that depends on what coordination strategy the robots have used. This suggests that we need to further improve our understanding of factors that influence team performance in order to be able to design appropriate coordination protocols.

References

1. Cao, Y.U., Fukunaga, A.S., Kahng, A.B.: Cooperative mobile robotics: antecedents and directions. Auton. Robot. **4**, 1–23 (1997)
2. Campo, A., Dorigo, M.: Efficient multi-foraging in swarm robotics. In: Almeida e Costa, F., Rocha, L.M., Costa, E., Harvey, I., Coutinho, A. (eds.) ECAL 2007. LNCS (LNAI), vol. 4648, pp. 696–705. Springer, Heidelberg (2007)
3. Krannich, S., Maehle, E.: Analysis of spatially limited local communication for multi-robot foraging. In: Kim, J.-H., Ge, S.S., Vadakkepat, P., Jesse, N., Al Manum, A., Puthusserypady K, S., Rückert, U., Sitte, J., Witkowski, U., Nakatsu, R., Braunl, T., Baltes, J., Anderson, J., Wong, C.-C., Verner, I., Ahlgren, D. (eds.) Progress in Robotics. CCIS, vol. 44, pp. 322–331. Springer, Heidelberg (2009)
4. Farinelli, A., Iocchi, L., Nardi, D.: Multirobot systems: a classification focused on coordination. IEEE Trans. Syst. Man Cybern. **34**, 2015–2028 (2004)
5. Balch, T., Arkin, R.C.: Communication in reactive multiagent robotic systems. Auton. Robot. **1**, 27–52 (1994)
6. Ulam, P., Balch, T.: Using optimal foraging models to evaluate learned robotic foraging behavior. Adapt. Behav. **12**, 213–222 (2004)
7. Mohan, Y., Ponnambalam, S.: An extensive review of research in swarm robotics. In: World Congress on Nature & Biologically Inspired Computing, pp. 140–145. IEEE (2009)
8. Parker, L.E.: Distributed intelligence: overview of the field and its application in multi-robot systems. J. Phys. Agents **2**, 5–14 (2008)
9. Cannon-Bowers, J., Salas, E., Converse, S.: Shared mental models in expert team decision making. In: Castellan, N.J. (ed.) Individual and Group Decision Making, pp. 221–245. Lawrence Erlbaum Associates, Hillsdale (1993)
10. Jonker, C. M., van de Riemsdijk, B., van de Kieft, I. C., Gini, M.: Towards measuring sharedness of team mental models by compositional means. In: Proceedings of 25th International Conference on Industrial, Engineering and Other Applications of Applied Intelligent Systems (IEA/AIE), pp. 242–251 (2012)

11. Rosenfeld, A., Kaminka, G.A., Kraus, S., Shehory, O.: A study of mechanisms for improving robotic group performance. Artif. Intell. **172**, 633–655 (2008)
12. Rybski, P.E., Larson, A., Veeraraghavan, H., Anderson, M., Gini, M.: Performance evaluation of a multi-robot search & retrieval system: experiences with mindart. J. Intell. Robot. Syst. **52**, 363–387 (2008)
13. Davids, A.: Urban search and rescue robots: from tragedy to technology. IEEE Intell. Syst. **17**, 81–83 (2002)
14. Yuh, J.: Design and control of autonomous underwater robots: a survey. Auton. Robot. **8**, 7–24 (2000)
15. Hindriks, K.: The goal agent programming language (2013). http://ii.tudelft.nl/trac/goal

Agent-Based Simulation of Interconnected Wholesale Electricity Markets: An Application to the German and French Market Area

Andreas Bublitz[✉], Philipp Ringler, Massimo Genoese, and Wolf Fichtner

Chair of Energy Economics, Karlsruhe Institute of Technology,
Hertzstraße 16, 76187 Karlsruhe, Germany
andreas.bublitz@kit.edu

Abstract. In the context of wholesale electricity markets, agent-based models have shown to be an appealing approach. Given its ability to adequately model interrelated markets with its main players and considering detailed data, agent-based models have already provided valuable insights in related research questions, e.g. about adequate regulatory frameworks. In this paper, an agent-based model with an application to the German and French market area is presented. The model is able to analyze short-term as well as long-term effects in electricity markets. It simulates the hourly day-ahead market with limited interconnection capacities between the regarded market areas and determines the market outcome as well as the power plant dispatch. Yearly, each agent has the possibility to invest in new conventional capacities which e.g. allows assessing security of supply related questions in future years. Furthermore, the model can be used for participatory simulations where humans take the place of the models agents. In order to adapt to the ongoing changes in electricity markets, e.g. due to the rise of renewable energies and the integration of European electricity markets, the model is constantly developed further. Future extensions include amongst others the implementations of an intraday market as well as the integration of additional market areas.

Keywords: Agent-based simulation · Economic agent models · Electricity markets · Multi-agent systems

1 Introduction

Todays liberalized wholesale electricity markets are generally considered to be highly complex systems. This is due to, among other things, the specific characteristics of the commodity electricity (e.g. instantaneous balancing of supply and demand, limited storability) and the fact that electricity can only be transported by a transmission grid with limited capacities. Other factors that increase the complexity are the various interrelated markets where electricity or related products can be traded (e.g. day-ahead market, future market) and the influence of other volatile markets such as the market for carbon emission allowances.

© Springer International Publishing Switzerland 2015
B. Duval et al. (Eds.): ICAART 2014; LNAI 8946, pp. 32–45, 2015.
DOI: 10.1007/978-3-319-25210-0_3

One important aspect of electricity systems is the reliability which should be ensured at all times. In liberalized European markets electricity generation companies are not obliged to invest in new power plants. Consequently, electricity markets need to be designed in such a way that there are sufficient incentives for adequate investments. The currently often discussed concept to ensure reliability in Europe is called energy only because power plant operators generate their profits mainly from the produced energy but are not compensated for only providing generation capacity that ensures reliability.

In Germany and several other European countries the spot market for electricity, in particular the day-ahead market auctions organized by electricity exchanges, plays an important role as it provides a market place to sell or buy electricity and its price serves as a reference for other markets (e.g. future markets, bilateral contracts). In addition, reserve markets are implemented to ensure the short-term reliability of the electricity system.

Two important developments currently altering the economics of European electricity markets are the increasing electricity generation from renewable energy sources and the European market integration. While for a long time mainly nuclear, coal and oil power plants had been installed in Europe, governments have recognized the decarbonisation potential of the electricity sector and there has been a continuous trend to move towards renewables and gas. Specifically, the introduction of the European Union Emissions Trading System and the creation of various policy programs to support the use of renewable energy sources have contributed to this development.

However, the feed-in of electricity generated from photovoltaic and wind power poses challenges to the electricity markets in their current form because in comparison with thermal power plants the generation from these sources is neither projectable nor exactly predictable and typically enjoys a guaranteed feed-in and compensation, respectively. Consequently, operators of conventional power plants are faced with another source of uncertainty that needs to be considered within the unit commitment problem, where an optimal balance of demand and supply under the various technical constraints of the power plants is to be determined. After determining the day-ahead operation schedule, the intraday market, where electricity can be traded at short notice, offers a possibility to adjust the schedule based on updated information, e.g. forecast of renewable generation. The intraday market is likely to gain importance in the next years, as the generation from renewable energy sources is expected to further increase.

Another important development in the electricity market is that the current borders of the national markets are subject to change; there are ongoing efforts to achieve a single European market. One aspect thereof is the implementation of market-based mechanisms to allocate limited cross-border capacities between European countries. The Central Western Europe (CWE) Market Coupling between Germany, France, Belgium, the Netherlands and Luxemburg serves as one of the most prominent examples. Market coupling maximizes social welfare, leads to price convergence and helps to balance different supply and demand situation in the interconnected market areas. The integration of markets is a matter-of-fact, thus influencing market prices and profitability of power plants in Europe.

Given the electricity systems complexity the relevant actors rely on different types of models for decision support. For instance, models are used by regulatory entities to analyse questions related to market design which is necessary to guarantee system reliability on different levels. Similarly, generation companies rely on electricity market models, for example, in order to examine investment cases. Naturally, market changes need to be reflected appropriately in modelling techniques.

In this paper, the main elements of the detailed bottom-up agent-based simulation model PowerACE are described and current extensions to adjust the model to relevant electricity market developments are presented. The aims of this paper are to present a comprehensive overview of the PowerACE modelling framework for electricity markets and how it can be applied to different research questions.

The paper is organized as follows: Sect. 2 provides a brief overview of the different types of electricity market models and shows the general suitability of agent-based simulation in the context of electricity markets. In Sect. 3, the models main elements with a focus on agents and markets are described. Exemplary results are presented in Sect. 4. Finally, Sect. 5 concludes with a summary and an outlook.

2   Models for Electricity Markets

The models used for electricity markets can be classified into several categories. [14] identify three major categories in electricity market modelling: optimization models, equilibrium models and simulation models. Distinguishing features include the mathematical structure, market representation, computational tractability and main applications.

While in Europe the liberalization of electricity markets started in 1996, electricity market models developed beforehand had been mostly optimizing models incorporating the perspective of a single planner, i.e. the government. Through the liberalization, the integration of a market perspective in models has gained importance, which brought forth the development of alternative models such as agent-based models that are able to adequately represent the current market situation where not one central decision maker is found, but several market players pursue their individual goals. In general, agent-based models which have been developed in quite different disciplines can provide a flexible environment which allows considering inter alia learning effects, imperfect competition including strategic behaviour and asymmetric information among market participants [13].

Nowadays, there exists a large number of agent-based electricity market models. Depending on the research focus, the models in the literature will differ from each other with respect to various criteria.

Each agent-based model features a certain agent definition and architecture which can include several dimensions. In the first place, it is essential to define conceptually what the agent represents in the model. In the field of Agent-based Computational Economics (ACE) agents generally are defined as having a set of data and pre-defined behavioural rules within a computationally constructed world [13]. Agent architecture includes the design of specific agent decision models including adaptive learning algorithms. Market modelling is another large building block of agent-based models. Given the complex nature of electricity wholesale markets and the electricity supply chain, different types of horizontally and vertically integrated markets exist. In order to analyse the existing interrelations between markets, one has to consider these markets with respect to their specific clearing rules. Depending on the spatial coverage of the model, coupling of interconnected areas might be considered as well. Similarly, agent-based models differ with regard to the time resolution as well as time scale of the simulation. The latter aspects includes, for instance, whether short-term behaviour (e.g. spot market bidding strategies) and long-term aspects (e.g. investment decisions) are jointly considered. Another important aspect of electricity market models is the representation of the electricity systems technical constraints (e.g. techno-economic aspects of generation units, grid constraints.

Three comprehensive review papers showing the large body of agent-based models for electricity markets and their distinctive features are provided by [8,12,15]. These literature reviews contain a comparison of the different existing models including the model presented in this paper.

Generally, having an integrated agent and market perspective, as well as a high degree of flexibility, agent-based simulation models can be used for detailed analyses of electricity markets and interactions therein. Potential applications include market power analysis or market design studies while considering the feed-in from renewable energy sources and integrated markets with respect to products, time and region.

3   PowerACE Model

3.1   Model Overview

The development of the PowerACE model started in 2004 and since then the model has been continuously extended and applied to various research questions.

The subject of modelling is the electricity wholesale market which is simulated for each hour of a year. Originally, the model was designed for the German market area. However, Europes electricity markets are all liberalized and set up according to the same fundamental principles. That is why PowerACE can be used to simulate other European market areas as well. Market areas are interpreted as one object in the programming environment featuring different market elements, agents and input data. In order to simulate different market areas, the respective object is instantiated repeatedly.

One of the key features of the model is the integration of both short-term market developments and long-term capacity expansion planning. Thereby, interactions and feedback loops between short-term and long-term output decisions are considered. Decisions regarding the expansion of capacity, i.e. whether to

install a new power plant are influenced by current and future developments in
the daily electricity trading as the main source of income and vice versa.

The key modules are markets, electricity supply, electricity demand and regulatory aspects. The main players participating in the wholesale electricity market are modelled individually; small companies are represented in an aggregated
form. Different types of market participants are modelled as different types of
agents. Each agent takes over certain roles, makes decisions based on specified
functions and either takes part in or sets rules for a respective market. A simplified overview of the model structure with two market areas is given in Fig. 1.

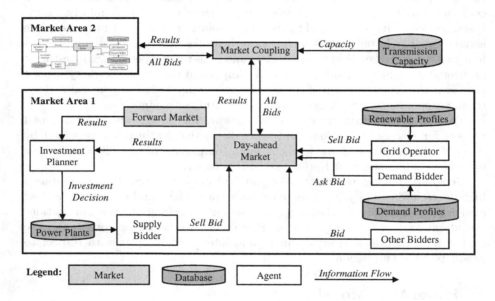

Fig. 1. Simplified structure of the PowerACE model.

In the following sections, the focus is set on the supply side, i.e. on generation
companies which have to decide on the short-term operation of their existing
power plants and on the investment in new ones.

3.2 Short-Term Bidding on Electricity Markets

The short-term operation of power plants is determined by the `SupplyBidder`
agent. The agent evaluates the different markets where energy or capacity of
thermal power plants can be offered and determines the operation schedule and
dispatch of the plants. Within PowerACE the day-ahead market is the main spot
market. In accordance with the current situation in Central Western Europe,
every `SupplyBidder` daily submits for each available power plant electricity
supply bids to the day-ahead market. Besides for thermal power plants, also

supply bids for generation from renewable energy sources, e.g. wind or biomass, are regarded. Since pumped-storage units can produce or consume electricity, they submit either buy or sell bids. The same applies to the electricity exchange with market areas which are not explicitly modelled. After receiving the bids the `DayAheadMarketAuctioneer` determines a uniform price for each hour of the next day considering all submitted supply and demand bids.

`SupplyBidders` are faced with an economic optimization problem, where the offered volume and price of their power plants needs to be determined and which is solved in several steps. Firstly, the available capacity $P_{i,d}$ of a power plant i on a day d needs to be determined. Power plants may not be available at all for a given day due to unexpected issues, e.g. start-up failure, or expected reasons, e.g. maintenance. Since power plants act on other markets (e.g. reserve market) as well, the reserved capacity $P_{r,i,d}$ for these markets is not available anymore for the day-ahead market bidding and needs to be subtracted from the net electrical capacity $P_{net,i}$:

$$P_{i,d} = \begin{cases} P_{net,i} - P_{r,i,d} & \text{if plant } i \text{ is available on day } d \\ 0 & \text{otherwise} \end{cases} \tag{1}$$

Secondly, the bid price is calculated. It consists of three elements: variable costs, start-up costs and a potential mark-up. Variable costs $c_{var,i,d}$ represent the direct costs of producing one unit of electricity and are determined by the fuel price $p_{fuel,i,d}$, the power plants net electrical efficiency η_i, the price of CO_2 emission allowances $p_{CO_2,d}$, the CO_2 emission factor of the fuel EF_{fuel} and the costs for operation and maintenance $c_{O\&M,i}$:

$$c_{var,i,d} = \frac{p_{fuel,i,d}}{\eta_i} + \frac{p_{CO_2,d} \cdot EF_{fuel}}{\eta_i} + c_{O\&M,i} \tag{2}$$

Changing the mode of operation of power plants, i.e. starting up or shutting down, causes additional costs. Firstly, material is stressed mainly by temperature changes reducing life expectancy; secondly, for start-ups fuel is needed in order to reach the operating temperature of a power plant. When determining the bid price the costs from start-up and shutdown processes as an intertemporal restriction can be considered by power plant operators. In the PowerACE model, this means that for base load running power plants also lower market prices are accepted in order to avoid shutting down the power plant. In turn, start-up costs are added to the bid price for peak load power plants in order to earn start-up costs in hours where the plant is expected to be running. To estimate start-up costs a price forecast for the next day is made by an agent. The bid price $p_{i,h}$ including start-up costs in hour h is defined as follows:

$$p_{i,h} = \begin{cases} \max\left(c_{var,i,d} - \frac{c_{s,i}}{t_s}, 0\right) & \text{if } \hat{p}_h < c_{var,i,d} \wedge i \in BL \\ c_{var,i,d} + \frac{c_{s,i}}{t_s} & \text{if } \hat{p}_h > c_{var,i,d} \wedge i \in PL \\ c_{var,i,d} & \text{otherwise} \end{cases} \tag{3}$$

$c_{s,i}$	start-up costs
t_u	number of continuous unscheduled hours per day
t_s	number of continuous scheduled hours per day
$\hat{p}_h$	predicted price for hour h
M	set of all operation-ready power plants
$BL \wedge M$	set of base load power plants
$PL \wedge M$	set of peak load power plants

In addition, SupplyBidders can increase the bid price for their power plants by a mark-up value. According to the standard economic model of perfectly competitive markets, market prices for a respective good are determined by marginal prices at all times. However, in order to cover capital expenditures and fixed costs market prices need to rise above marginal costs of supply at least in some periods. This reasoning is based on the peak-load pricing concept [3]. One potential remedy is to include an additional mark-up factor in the bid price of supply capacity, which is implemented in the PowerACE model.

The value of the mark-up factor depends on the relative scarcity in the market; a higher scarcity induces a higher mark-up, which is added to the bid price:

$$p_{i,d}^{markup} = p_{i,d} + markup_h \tag{4}$$

After determining the offered volume and price for each hour of the following day the bids are submitted to the day-ahead market auctions. A comprehensive and formal description of the original short-term bidding algorithm can be found in [5].

3.3 Coupling of Interconnected Markets

European electricity markets are interconnected via high-voltage transmission lines. Since electricity flows according to physical laws and interconnector capacities are limited, these capacities have to be allocated to market participants otherwise transmission lines might get congested. Management methods are required to avoid congestion and to efficiently use cross-border transmission capacities.

Since 2010, a market coupling approach has been implemented in Central Western Europe which complies with the European Unions general principles of congestion management (e.g. non-discriminatory, market-based). Market coupling describes the implicit auctioning of interconnection capacity through power exchanges for predefined zones (market or bidding areas). The market coupling operator clears the energy markets of the participating market areas simultaneously and determines implicitly the commercial flows between markets areas as well as the prices. The market coupling approach maximizes the social welfare by optimizing the selection of bids while considering limited transmission capacity. The transmission capacity is determined up-front based on defined rules [4].

In accordance with the CWE Market Coupling architecture, market coupling is implemented within PowerACE for the day-ahead market and market participants submit their bid curves to the local power exchanges based on the described method in Sect. 3.2.

In PowerACE the `MarketCouplingOperator` takes over all processes related to the market coupling. For that purpose, the operator receives all day-ahead bids from the local power exchanges. Market coupling itself can be formulated as an optimization problem with the objective to maximize social welfare. Since PowerACE currently only considers hourly bids with a fixed price, the original COSMOS algorithm used for the CWE Market Coupling [1] can be simplified and the mathematical problem is formulated as follows [10]:

$$\underset{q}{\text{maximize}} \sum_{b} \left(\sum_{d} P_{b,d} \cdot Q_{b,d} \cdot q_{b,d} - \sum_{s} P_{b,s} \cdot Q_{b,s} \cdot q_{b,s} \right) \quad (5)$$

subject to

$$q_{b,d}, q_{b,s} \le 1 \quad (6)$$

$$\sum_{d} P_{b,d} \cdot Q_{b,d} \cdot q_{b,d} - \sum_{s} P_{b,s} \cdot Q_{b,s} \cdot q_{b,s}$$
$$+ \sum_{b(to)} Cap_{b,b(to)} - \sum_{b(from)} Cap_{b(from),b} = 0 \quad (7)$$

$$Cap_{b(from),b(to)} \le Cap_{b(from),b(to)}^{\max} \quad (8)$$

where the indices d and s indicate demand and supply variables, respectively. b denotes the market (bidding) area, P_i the bid prices and Q_i the bid volumes. $q_{b,d}$ and $q_{b,s}$ are the acceptance rates of the corresponding demand and supply bids. $Cap_{b(from),b(to)}$ equals the determined capacity between two market areas. $Cap_{b(from),b(to)}^{\max}$ denotes the upper limit for the transmission capacity between to market areas and is given exogenously based on current values from the European Network of Transmission System Operators for Electricity (ENTSO-E).

The constraints ensure that supply and buy bids do not exceed their maximum volume (6), that supply and demand including exports as well as imports in market areas are balanced (7) and that the limitation on the transmission capacity (8) is not violated. In this form, the problem is linear and can be solved with common solvers.

Optimization results are the acceptance rates for each submitted bid and the commercial utilization of transmission capacity. Furthermore, the algorithm determines the market prices of electricity one day- ahead of delivery in the coupled bidding areas and the implicit prices for transmission capacities, which are only different from zero if lines are congested. Prices are sent to the local market areas and processed by the supply agents.

3.4 Long-Term Investment Planning

In the model generation companies can also make decisions regarding their long-term capacity extension through investments in new power plants. The responsible agent is called `InvestmentPlanner`.

The basic methodology is based on a discounted-cash flow valuation of pre-defined technology options. For that purpose the `InvestmentPlanner` makes a forecast of the expected hourly electricity prices during the investment period and calculates the expected yearly gross profit. After accounting for fixed costs and capital expenditures, the net present value is calculated. A formal description is provided in [5].

The quantity of the installed capacity is based on the expected development of market shares within the following five years taking future demand and electricity generation from renewable energy sources into account. As long as the net present value of the investment options is positive and there is need for new capacity, new power plants are built by the `InvestmentPlanner`. After the construction phase, whose length depends on the technology option, the new power plants can generate electricity that can then be sold in the markets.

3.5   Input Data and Technical Implementation

For the considered market areas each thermal power plant with a capacity of at least 10 Megawatt is stored together with its main relevant techno-economic characteristics (e.g. net electrical efficiency, variable and fixed costs, yearly availability) in the database of the model.

The model database includes investment options for different power plant technologies with its relevant characteristics and the electricity feed-in from renewable energy sources. The electricity demand is represented by the aggregated consumption of all consumers connected to the public power supply.

For market coupling, transmission capacities between interconnected market areas are required. Since not all neighbouring countries are always part of a simulation, the electricity exchange with these countries is based on historical values. Prices for fuel and CO_2 emission allowances are required for the calculation of the variable generation costs. Most time series data is stored with hourly values, but sometimes only less detailed values, e.g. for lignite prices, are available.

The models results include the hourly spot market prices in the simulated wholesale markets, the investments in new capacity and the commercial flows between interconnected market areas. Since the model considers the wholesale day-head market as the only trading place for electricity, bilateral day-ahead contracts are not part of models results.

PowerACE is implemented in the object-oriented programming language Java and can simulate each hour of recent historical years as well as future years up to 2050. The simulation runs are comparatively quick in terms of computing time. Yearly runs for one market area last only a few minutes, which is a small fraction of the several hours that optimization models with a similar amount of details may take.

4   Exemplary Applications

The PowerACE model has been used for various research analyses in the past. For instance, [11] find a considerable impact of the subsidised renewable electricity generation in the short run on spot market prices in Germany. The impact of emissions trading on electricity prices is explored by [7]. The authors find for the years under consideration that a large part but not the totality of the CO_2 emission allowance price is added by the generation companies to the variable costs during the bidding process. A thorough analysis of the models capacity to adequately reproduce the main characteristics of the German electricity market can be found, for example, in [5].

In the following sections, additional recent analyses are presented

4.1 Market Coupling Between Germany and France

Based on the algorithm described in Sect. 3.3, effects from a market coupling between the German and French day-ahead electricity markets are analysed. Both markets represent the two largest in Europe in terms of electricity consumption and are part of the CWE Market Coupling. To the authors best knowledge this is the first agent-based approach that includes the coupling of different market areas based on the current market situation.

The simulation of the model coupling is performed for the year 2012. In the *Single Markets* scenario, there is no coupling of the two markets, i.e. no exchange between Germany and France is considered. The *Model Coupling* scenario uses the optimization routine for the coupling of the German and French market areas. The electricity exchange with other countries (e.g. between France and Spain, Germany and Poland) is in both scenarios given exogenously based on historical data.

The *Model Coupling* scenario shows lower average prices than the *Single Market* scenario, while the price decrease is stronger in France than in Germany. The more pronounced effect for France can be explained, to some extent, by the supply curves shapes of the two market areas. The French supply curve has only a gentle slope for a large part of the countrys capacity because of the low variable operating costs of nuclear power stations. However, the small part of the remaining capacity consists of notably more expensive fossil fuel-fired units. These units are often called upon in the *Single Markets* scenario. When coupling the markets, the expensive units in France are less frequently used because cheaper electricity can be imported from Germany.

The change in market prices does not imply that all market participants, buyers and sellers, benefit. The results in this simulation indicate that mainly the consumers benefit from the market coupling which is consistent with expectations given a lower average price. The social welfare (sum of consumer surplus, producer surplus and congestion revenue) increases with market coupling, which could be expected, as the clearing algorithm tries to maximize this value.

In the *Model Coupling* scenario the available transfer capacity is fully used in 65 % of the cases. The high usage of the full capacities and the price effect of the coupling can be seen for a period of 100 h in Fig. 2. Expanding (e.g. doubling) the capacity amplifies the price reduction in both countries; while the additional effect is smaller in France than in Germany, the total price reduction is still stronger in France. In case of sufficient capacity there are identical prices in

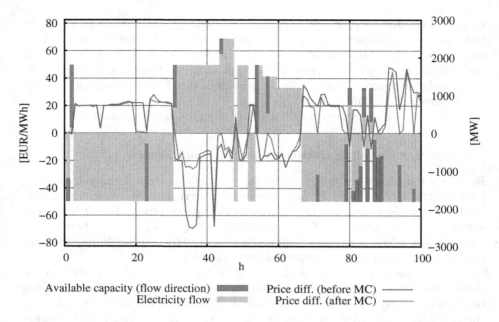

Fig. 2. Simulated electricity flow and price difference (before and after market coupling) between Germany and France for a period of 100 hours in 2012.

all hours, which is equal to the situation of having one completely integrated market.

Regarding only market coupling between two countries, in this case Germany and France, while the exchange with other country is based on historical values, is, of course, a simplification. Germany, for instance, has interconnections with nine countries while France is connected to seven countries. Amongst those countries are some that take part in the market coupling as well, e.g., Austria, Belgium or the Netherlands. Hence, the effects from the market coupling between Germany and France in this paper might be overstated, since either country would exchange electricity with other countries, if this as well is no longer static and less costly than the exchange with Germany or France, respectively.

The presented results also depend on information which is not publicly available and therefore needs to be estimated, such as the operation and maintenance costs of power plants. Deviations between estimated and real world values could, of course, alter the results of the simulation.

4.2   PowerACE LAB

Besides the computational model, there exists a laboratory version, "PowerACE LAB", where real-life participants can assume the tasks of software agents.

Thus, the core agent-based simulation model is supplemented by elements from experimental economics and role-playing games [6].

In literature, two approaches are distinguished in combining agent-based models and role playing games. [2] proposes a parallel existence of agent-based models and role playing games. Hence, the model is rebuilt in a simplified version as a game. The main goal of this approach is to increase the acceptance of the model. [9] develop an agent-based participatory approach, where real participants are integrated into the model by (partly) controlling the agents actions. For this, user interfaces have to be developed. In the PowerACE Lab version, the agent based participatory approach is used.

Currently, in PowerACE Lab two modules exist where human participants can interact. The participants either simulate the power trading or the investments in new generation capacity. In the trading module, the participants receive the same information as the computer agents. Each participant has a list of daily available power plants with all the relevant technical and economic data, e.g. installed capacity, fuel costs and efficiency. In addition, a forecast of the day-ahead prices is presented. Based on this information, the participants submit their bids. When all players have successfully completed their task, the market clearing price is computed analogously to the computational model. The players have the possibility to adopt their strategies in each round in order to maximize profits.

In the investment module, the players can carry out investments according to the power and fuel price forecast and by taking into account the decreasing capacities due to the limited technical lifetime of existing power plants. The players decisions and chosen strategies can be used to improve the behaviour of the computer agents. Computer agents and real participants can coexist as well in the simulations.

5   Conclusions and Outlook

Agent-based simulation in general and the PowerACE model in particular are useful means to analyse different aspects of electricity markets. The market and agent perspective as well as the flexibility of agent-based simulation models allows us to thoroughly analyse electricity markets and interactions therein. The PowerACE model is a detailed bottom-up simulation model which integrates short-term market operations and long-term capacity planning while the most important market participants are represented by different agents. The model has been successfully used for various analyses in the context of electricity markets.

Given the continuously changing economic and regulatory environment in the power sector, several enhancements to the model are currently in progress. In order to reflect the European market integration, the model scope is extended to several market areas which can be simultaneously run and coupled. *Model coupling* clears the energy and capacity markets simultaneously and determines an optimal solution to the plant dispatch in the interconnected market areas considering limited commercial transfer capacities. The model coupling routine presented in this paper offers a socially beneficial opportunity to interconnect electricity markets compared to a situation where no market coupling occurs. The results for Germany and France show that the average market price is lower in both countries, while the price decrease is stronger in France than in Germany.

The methodological approach of PowerACE has nonetheless some limitations. Regarding the supply of electricity, additional technical constraints concerning the operation of power plants (e.g. minimum downtimes or partial efficiency levels) could further improve the model. Furthermore, the perspective is limited to the supply of electricity, which differs from the real world situation where also the heat demand influences the usage of combined heat and power plants.

Given the flexible modelling framework future model extensions could include the development of a generally scalable model version in order to simulate micro-systems as well as larger systems (e.g. Europe) with additional market elements (e.g. intraday market). Concerning the decision making process of agents, the refinement of the investment module and the integration of different aspects of uncertainty is another possibility to extend the model. Regarding the design of electricity markets, the remuneration of power plants by capacity mechanisms in order to ensure system reliability is another topic of research that is currently explored within the model.

References

1. APX-ENDEX, Belpex, EPEX Spot: COSMOS description – CWE Market Coupling algorithm (2010)
2. Barreteau, O., Bousquet, F., Attonaty, J.M.: Role-playing games for opening the black box of multi-agent systems: method and lessons of its application to Senegal river valley irrigated systems. J. Artif. Soc. Soc. Simul. 4(2), 5 (2001)
3. Boiteux, M.: Peak-load pricing. In: Nelson, J.R. (ed.) Marginal Cost Pricing in Practice. Prentice-Hall, Englewood Cliffs (1964)
4. EPEX Spot: Project Document: A report for the regulators of the Central West European (CWE) region on the final design of the market coupling solution in the region, by the CWE MC Project (2010). http://static.epexspot.com/document/7616/01_CWE_ATC_MC_project_documentation.pdf
5. Genoese, M.: Energiewirtschaftliche Analysen des deutschen Strommarkts mit agentenbasierter Simulation. Ph.D. thesis, Universität Karlsruhe (TH), Karlsruhe (2010)
6. Genoese, M., Fichtner, W.: PowerACE LAB: Experimentallabor Energiewirtschaft. Wirtschaftswissenschaftliches Studium 41(6), 338–342 (2012)
7. Genoese, M., Sensfuß, F., Möst, D., Rentz, O.: Agent-based analysis of the impact of CO_2 emission trading on spot market prices for electricity in Germany. Pac. J. Optim. 3(3), 401–424 (2007)
8. Guerci, E., Rastegar, M.A., Cincotti, S.: Agent-based modeling and simulation of competitive wholesale electricity markets. In: Rebennack, S., Pardalos, P.M., Pereira, M.V.F., Iliadis, N.A. (eds.) Handbook of Power Systems II, pp. 241–286. Springer, Heidelberg (2010)
9. Guyot, P., Honiden, S.: Agent-based participatory simulations merging multi-agent systems and role-playing games. J. Artif. Soc. Soc. Simul. 9(4), 8 (2006). http://jasss.soc.surrey.ac.uk/9/4/8.html

10. Meeus, L., Vandezande, L., Cole, S., Belmans, R.: Market coupling and the importance of price coordination between power exchanges. Energy Econ. **34**(3), 228–234 (2009)
11. Sensfuß, F., Ragwitz, M., Genoese, M.: The merit-order effect: a detailed analysis of the price effect of renewable electricity generation on spot market prices in Germany. Energy Policy **36**(8), 3086–3094 (2008)
12. Sensfuß, F., Ragwitz, M., Genoese, M., Möst, D.: Agent-based simulation of electricity markets - a literature review. Working Paper Sustainability and Innovation 5 (2007)
13. Tesfatsion, L.: Agent-based computational economics: a constructive approach to economic theory. In: Tesfatsion, L., Judd, K.L. (eds.) Handbook of Computational Economics: Agent-Based Computational Economics, pp. 831–880. North Holland, Elsevier (2006)
14. Ventosa, M., Ballo, I., Ramos, A., Rivier, M.: Electricity market modeling trends. Energy Policy **33**(7), 897–913 (2005)
15. Weidlich, A., Veit, D.J.: A critical survey of agent-based wholesale electricity market models. Energy Econ. **30**(4), 1728–1759 (2008)

Cooperative Transportation Using Pheromone Agents

Ryo Takahashi[1], Munehiro Takimoto[1]([✉]), and Yasushi Kambayashi[2]

[1] Department of Information Sciences, Tokyo University of Science,
2641 Yamazaki, Noda 278-8510, Japan
{r-takahashi,mune}@cs.is.noda.tus.ac.jp
[2] Department of Computer and Information Engineering,
Nippon Institute of Technology, 4-1 Gakuendai, Miyashiro-machi,
Minamisaitama-gun 345-8501, Japan
yasushi@nit.ac.jp

Abstract. This paper presents an algorithm for cooperatively transporting objects by multiple robots without any initial knowledge. The robots are connected by communication networks, and the controlling algorithm is based on the pheromone communication of social insects such as ants. Unlike traditional pheromone based cooperative transportation, we have implemented the pheromone as mobile software agents that control the mobile robots corresponding to the ants. The pheromone agent has the vector value pointing to its birth location inside, which is used to guide a robot to the birth location. Since the pheromone agent can diffuse with migrations between robots as the same manner as physical pheromone, it can attract other robots scattering in a work field to the birth location. Once the robot finds an object, it briefly pushes the object, measuring the degree of the inclination of the object. The robot generates a pheromone agent with the vector value to pushing point suitable for suppressing the inclination of the object. The process of the pushes and generations of pheromone agents enables the efficient transportation of the object. We have implemented a simulator that follows our algorithm, and conducted experiments to demonstrate the feasibility of our approach.

Keywords: Mobile agent · Multiple robots · Ant colony optimization · Swarm intelligence

1 Introduction

In the last decade, robot systems have made rapid progress not only in their behaviors but also in the way they are controlled. In particular, a control system based on multiple software agents can control robots efficiently [20]. Multi-agent systems introduced modularity, reconfigurability and extensibility to control systems, which had been traditionally monolithic. It has made easier the development of control systems on distributed environments such as multi-robot systems.

B. Duval et al. (Eds.): ICAART 2014; LNAI 8946, pp. 46–62, 2015.
DOI: 10.1007/978-3-319-25210-0_4

Fig. 1. A team of mobile robots are working under control of mobile agents.

On the other hand, excessive interactions among agents in the multi-agent system may cause problems in the multiple robot environments. In order to mitigate the problems of excessive communication, mobile agent methodologies have been developed for distributed environments. In a mobile agent system, each agent can actively migrate from one site to another site. Since a mobile agent can bring the necessary functionalities with it and perform its tasks autonomously, it can reduce the necessity for interaction with other sites. Mobile agent systems are especially useful in an intermittently connected ad hoc network environment. In the minimal case, a mobile agent requires that the connection is established only when it performs migration [2].

The model of our system is a set of cooperative multiple mobile agents executing tasks by controlling a pool of multiple robots as shown in Fig. 1 [8].

The property of inter-robot movements of the mobile agents contributes to the flexible and efficient use of the robot resources. A mobile agent can migrate to the robot that is most conveniently located to a given task, e.g. closest robot to a physical object such as a soccer ball. Since the agent migration is much easier than the robot motion, the agent migration contributes to saving power consumption [20]. Here, notice that any agents on a robot can be killed as soon as they finish their tasks. If the agent has a policy of choosing idle robots rather than busy ones in addition to the power-saving effect, it would result in more efficient use of robot resources.

We have proposed our model in the previous paper [20] and have also shown the effectiveness of saving power consumption and the efficiency of our system for searching targets [1, 14] and transporting them to a designated collection area [16].

In this paper, we focus our attention on transportation of a large object that a single robot alone cannot move. In order to deal with such a large object, several robots have to cooperate to achieve the objective tasks. This task seems to require too artificial behaviors for each robot. For the cooperatively solving of such problem, *swarm-based* approaches have been proposed. A *swarm-based* approach is based on the social insect metaphor. The swarm-based system, which consists of several robots with simple behaviors, can achieve complex tasks, just like ants that behave based on simple rules cooperatively transport a large prey. The swarm of simple robots makes a system more flexible and fault-tolerant, and contributes to suppressing costs of building a large complex system.

We have implemented the ants as actual mobile software agents that control the mobile robots. The ant agent migrates among robots to look for an available one. Once the ant agent finds the available robot, the ant agent physically drives the robot. We also implemented pheromone as mobile software agents, which attract many robots to the large object to convey through diffusing with the migrations. The pheromone agent has a vector value pointing to the birth point. The vector value is modified in order to always point to the birth point, each that the pheromone agent migrates to another robot. Once the pheromone agent migrates to the robot where an ant agent resides, it guides the ant agent to lead the robot to its birth point. Thus, the pheromone agents enable scattering free robots to efficiently attend to the transportation of the object. In our new approach, we take advantage of the pheromone agents not only to collect robots, but also to suppress the rotation of an object during the transportation. We assume that each robot has a simple sensor for checking the degree of the inclination of the object that it is pushing. When each robot slightly pushes the object, it generates pheromones at the suitable location to suppress the increase of the inclination of the object. As a result, the object can be linearly transported to a target area, even if the object has the form such as a stick. Notice that such an object could not be transported without suppressing the rolling.

The structure of the balance of this paper is as follows. In the second section, we describe the background. The third section describes our transport model based on pheromone agents. In the fourth section, we describe the numerical experiments using a simulator that follows our algorithm. In the fifth section, we discuss further improvement of our algorithm and show its effectiveness. Finally, we conclude our discussions in the sixth section and present future research directions.

2   Backgrounds

Making multiple robots cooperatively carry and push common objects has been intensively studied and yet to established the standard way. Many research projects have dealt this topic but few of them have demonstrated on physical multi-robot systems. One of the most demonstrated tasks involving cooperative transport is the pushing objects by multiple robot teams [15,19]. This task is inherently easy to accomplish when comparing to carrying tasks, because a carrying task involves multiple robots' gripping a common object and navigating to a destination in a coordinated fashion [9, 22].

One of the most famous transportation problems is the box-pushing problem [12]. The problem is defined in [7] and consists of cooperatively moving a box, which is relatively large when compared to the size of the multi-robots, from an initial position to a destination using robots that can only perform pushing movements.

On the other hand, algorithms that are inspired by behaviors of social insects such as ants to communicate to each other by an indirect communication called stigmergy are becoming popular. [4,5,19]. Upon observing real ants' behaviors, Dorigo et al. found that ants exchanged information by laying down a trail of a chemical substance (called pheromone) that is followed by other ants. They adopted this ant strategy, known as ant colony optimization (ACO), to solve various optimization problems such as the traveling salesman problem (TSP) [5]. Deneubourg has originally formulated the biology inspired behavioral algorithm that simulates the ant corps gathering and brood sorting behaviors [3]. Wang and Zhang proposed an ant inspired approach along this line of research that sorts objects with multiple robots [21]. Lumer has improved Deneubourg's model and proposed a new simulation model that is called Ant Colony Clustering [11]. His method could cluster similar objects into a few groups. [13] has proposed to implement pheromones as mobile agents in ACO [13]. In their system, ants are implemented as also mobile agents. They repeatedly migrate to robots to searching free robots corresponding to objects. Once the ant agent finds a free robot, it drives the robot to a cluster of robots. The pheromone agent is generated by the ant agent when its driving robot reaches a cluster, and repeatedly migrates to robots to guide ant agents to drive robots to the cluster. The mobile agent based algorithm has been improved to serialize collected robots [17,18].

The applications of ants' behaviors to cooperative transportation have been proposed by Kube and Bonabeau [10]. It transports an object through the interplay of forces. The robots can recognize an object through the light emitted from the object, which works as the pheromone, but its effectiveness is restrictive. Fujisawa et al. have proposed the efficient cooperative transportation using ethanol as physical pheromone [6].

3   Transport Model

We assume that each robot just affects the object they are transporting through pushing. We also assume that each robot has to simultaneously push the object together while other robots to move it. Also, since we assume that our multiple robots system is used for several tasks [14], the robots available for the transportation are just ones not engaging other tasks.

In order to achieve the transportation over the robots shared with several tasks, we introduce two kinds of agents, which are called Ant Agents (AA) and Pheromone Agents (PA). The AA has a role for searching a robot with

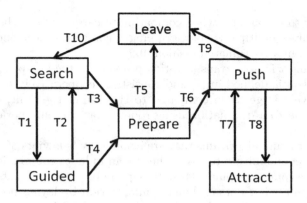

Fig. 2. State transition diagram of ant agent.

no other task, and driving a robot to the object to transport. The PA has a role for efficiently attracting the free robots to the object by guiding AAs on robots instead of directly driving them. In this section, we describe the robot's functionalities required in our system, and the details of two agents.

3.1 Robots

We assume mobile robots such as PIONEER 3-DX, which have two servo-motors with tires, one camera and sixteen sonic sensors [14]. The power is supplied by rechargeable battery. The PIONEER 3-DX has a servo-motor and sensor controller board that sends/receives data to/from a host computer on it through a USB cable. The camera is directly operated by the host. Each robot holds a lap-top computer as a host, which is also used as a server for the migration of our control agents through the wireless LAN.

Also, in our transportation system, we assume that each robot has additional features as follows:

- it can push an object,
- it has the sensor for checking the degree of the inclination of the pushing object,
- it can always know the direction to which the object should be transported, and
- it knows the IP addresses of other robots by identifying them through the camera.

All functionalities for controlling the robot are introduced into each robot through migrations of mobile agents. Once an agent migrates to a host, the agent can communicate with other agents on the same host, so that the user can construct a larger system though migrations of mobile agents to the host.

Fig. 3. State search of AA.   **Fig. 4.** State guided of AA.

3.2   Ant Agent

AA has the list of IP addresses of all the robots in order to traverse them one by one and to find out a free robot. If it has exhausted to check all the robots, it goes back to the home host for administration of the robots, and updates its IP address list in the following cases: (1) some new robots have been added, and (2) some robots have been broken, where notice that the addition or the removal of a robot is recorded on the home host. However, these are so rare that the home host is made to go through in most cases.

Once an AA reaches a free robot, the AA controls the robot along the state transition as shown in Fig. 2. First, the AA drives the robot to the object to transport through two kinds of manners. Initially, the AA makes the robot randomly walk to search the object as shown by state *Search*. Once PA migrates to the same robot, the AA drives the robot to the object along the guidance of the PA as shown by state *Guided*. After that, once the AA finds the object, it prepares for transporting the object as shown by state *Prepare*. Actual transportation is performed by repetition of briefly pushing the object as shown by state *Push* and attracting other robots to the suitable location for balancing the object as shown by state *Attract*. The transportation task of each robot is limited in the specific time, and therefore, after the time-out, the robot is released from the task as shown by state *Leave*. The details of each state is as follows:

State Search: In this state, AA drives the robot at random as shown in Fig. 3. If the robot contacts with other robots, it temporarily stop there, and then take an action for avoidance. Once the robot bumps onto the object to transport, the state of the AA is changed to the state Prepare as shown by *T3* in Fig. 2. On the other hand, if a PA migrates to the robot, the state is changed to the state Guided as shown by *T1*.

State Guided: In this state, AA drives the robot along the guidance of the PA that has migrated to the same robot as shown in Fig. 4. The PA has the vector

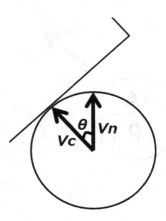

Fig. 5. State prepare of AA.

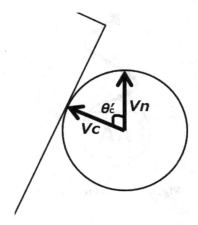

Fig. 6. State attract of AA.

value pointing to the birth location of the PA, which is adjusted to always point to the birth location even if the robot moves or rotates. Since the PA can observe the rotations of wheels, the vector value is modified to neutralize the change caused by the rotation or the approach to the birth point of the robot. AA is guided for driving the robot in the direction of the vector value. Once the robot reaches the destination of the vector, the AA checks whether there are some objects around the robot or not. If some objects are found, the AA changes its state to the state Prepare for transporting the closest object as shown by $T4$; otherwise, the AA abandons the PA guiding it, and then changes its state to the state Search as shown by $T2$. At this time, the PA is killed.

State Prepare: Once an AA finds an object, the AA makes the robot approach to the object till contacting with it. After that, the AA measures the normal vector V_c against the surface of the object to know the degree of the inclination, as shown in Fig. 5. Furthermore, since the AA always holds vector V_n to its nest to which the object is transported, it can calculate θ that is the angle between V_c and V_n. The agents including AA have the vector values in polar coordinate form. Therefore, using the angle θ_c of V_c and θ_n of V_n, θ is straightforwardly calculated as follows:

$$\theta = \theta_c - \theta_n \tag{1}$$

If the θ is more than $-\frac{\pi}{2}$ and less than $\frac{\pi}{2}$, the AA judges that robot has a suitable direction for efficient transportation, and changes its state to Push in order to start the transportation of the object as shown by $T6$. Otherwise, the AA changes its state to Leave in order to abandon the transportation, as shown by $T5$.

State Push: In this state, the AA just briefly pushes the object, and then changes its state to Attract in order to check the change of θ as shown by $T8$, and to attract another robot to the suitable location for balancing the object. Also, the AA releases the robot from the transportation task after specific time. The time-out functionality enables the finite robot resources to be reused for pushing another location, another object or another task, and it is achieved by changing its state to Leave, as shown by $T9$.

State Attract: Figure 6 shows the relation of a robot and an object after the robot pushed the object in the short period in the state Push, where V_c may have change. Therefore AA calculates θ_δ to check the angle θ_c' of new V_c as follows:

$$\theta_\delta = \theta_c' - \theta_c$$

If θ_δ is negative, it means that the object have rotated to the right. In this case, pushing the object at the neighbor location of the robot in the right hand side may suppress the rotation. Conversely, in the case where θ_δ is positive, pushing the object in the left hand side of the robot may suppress the rotation to the left. We call the location to push in order to balance the object *additional push point*. Once the AA knows the additional push point, it generates PA with the vector value pointing to the push point in order to attract another robot. After that, the PA repeatedly migrates some robots to find a free robot to attract. Notice that the AA generates no PA when θ_δ is correctly 0.

The AA generates some PAs in the predefined time, and then changes its state back to the Push, as shown by $T7$. The several transitions between the Push and the Attract achieves the transportation of the object in balance.

State Leave: In this state, the AA makes the robot turn to the back, and then, move it in the specific distance in order to release it from the transportation task.

3.3   Pheromone Agent (PA)

The PA has behaviors imitating the physical pheromone of ants, of which the main role is to attract the ants to preys or their nest. In a PA, the attraction is implemented as a guidance functionality based on the vector value to the birth location. As well as diffusion of physical pheromones, the PA searches the free robot to guide in two steps: *generation* and *propagation* of PA. First, in the generation step, the vector value to the birth location is set to the PA, and then, in the propagation step, it starts migrating to other robots. Finally, it is killed by the abandonment of AA. Also, when PA migrates to the robot on which another PA resides, they are composed in the *composition* step. In the remainder of the section, we describe the details of these steps.

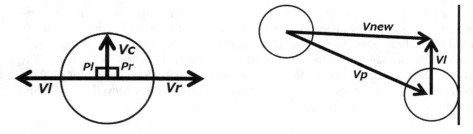

Fig. 7. Generation of PA.   **Fig. 8.** Propagation of PA.

Generation: The PA is generated by AA, of which the vector value is initial-ized to point to the neighbor location of the robot on which the AA resides. For example, when AA founds that the object rotates to the right in the state Attract, it generates a PA with the vector value as shown by V_r in Fig. 7, where V_c is the normal vector against the object. Conversely, if the object rotates to the left, the vector is initialized as shown by V_l. The initial vector has the length corresponding with the size of a robot and 90 degrees against V_c.

Propagation: The purpose of the PA is to find a free robot, and then to lead it to the destination of the vector value. Therefore, as soon as it is generated or arrives at another robot, it checks around the current robot, so that if some robots are found, it generates its clone and makes the clone migrate to the closest one of them. The migration to the closest robot enables the moving distance of the robot guided by the PA to be short. Notice here that in a migration process, the destination of the vector value of new PA after the migration has to be same as one of the old PA before the migration, which is achieved by synthesizing the old vector value V_{old} and the vector value V_p from the destination to source of the migration as follows:

$$V_{new} = V_{old} + Vp$$

Figure 8 shows the first migration of PA after its generation. In this case, V_{new} is set to $V_p + V_l$.

As mentioned above, a PA generates its clone, which migrates to another robot. The number of the generation of a clone is limited to five. A clone is generated one at a time. Also, each PA is finally abandoned by the AA in its state Leave, and therefore, the PA does not continue giving the old information to the system.

Composition: Once the PA migrates to the robot on which another PA resides, they have to be composed. In the case of physical pheromone, the effectiveness of the attraction is strengthened by the composition, whereas in PAs, either of them is alternatively selected. All the vector values of them are checked, so that the PA with shortest vector is selected.

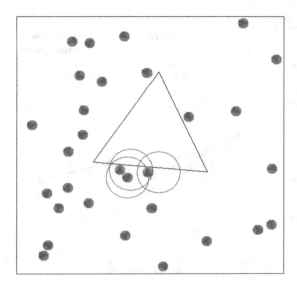

Fig. 9. The snapshot image of the simulator.

This composition strategy enables decreasing the total distance that all robots move, so that it contributes to suppressing energy consumption of the entire robot system. In addition, the strategy also contributes to keeping the degree of the current dispersion, because it is profitable for robots to uniformly scatter considering the case where several objects are simultaneously transported.

4   Experimental Results

In order to demonstrate the effectiveness of our transportation method, we have implemented a simulator. Figure 9 shows the output image of the simulator, through which we can observe the behaviors of the agents and the robots. Numerical statistic data are recorded apart from it.

Each small circle shows a robot, the moving direction of which is shown by a triangle inside it. Each robot is assumed to move from a grid to another grid in 500×500 grids square field. Also, the field is surrounded by a wall, and therefore, robots cannot go out the field. A big red circle with a robot at its center shows the view range of the camera equipped on the robot, and it is also the range where AA or PA can migrate. In the figure, the object to transport is shown by an equilateral triangle with sides 240 grids long.

In order to confirm the accuracy and efficiency of our method, we have conducted eight experiments through applying two kinds of strategies, which are ones with PA and without PA, to four kinds of objects with the shapes of triangle, square, rectangle, and circle. For each experiment, we have measured the

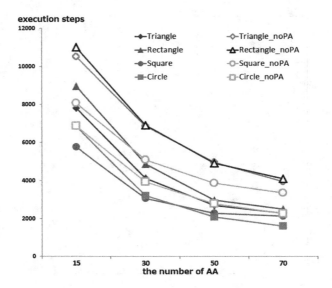

Fig. 10. Transportation time.

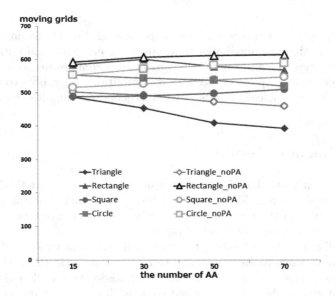

Fig. 11. Transportation distance.

total time and the total distance for transporting an object, and the total angle where the object rotates.

Figure 10 through Fig. 12 show the results of the transportation time, the transportation distance and the total angle of the rotation of the object, respectively. In all figures, the horizontal axes indicate the number of AAs. The vertical

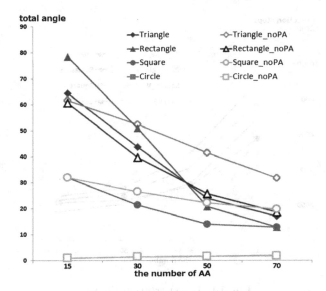

Fig. 12. Rotation angle.

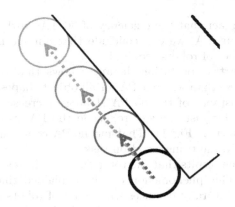

Fig. 13. Generation of PA(Extension).

axis of Fig. 10 indicates the number of the execution steps of the simulator as the transportation time. Each graph of Fig. 11 shows the number of moving grids as the transportation distance. Each graph of Fig. 12 is the total angle in the transportation. In both figures, the thin lines show the results without PA, and the dark lines show the results with PA.

Figure 10 shows the transportation time decreases as AAs increases. This observation means that the more robots attend the transportation task, the more efficiently the task is performed, even though the slope of the graph decreases close to flat around 70 AAs. Comparing the results of without PA and the ones with PA, the transportation with PA is found to be much more efficient. Besides,

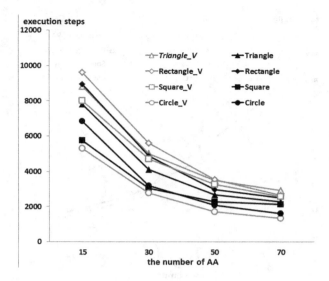

Fig. 14. Transportation time.

as observed from the fact that the efficiency of 30 AAs with PA is the same as one of 70 AAs without PA, we can conclude that our method is effective for suppressing the number of robots required.

The result of the transportation distance shows that the distance of the strategy with PA is less than without PA for all object shapes, and furthermore, the transportation distance of the no PA strategy increases as the number of AAs increase except for just the square, while the PA strategy decreases for all the shapes, as shown in Fig. 11. That means PA contributes to suppressing energy consumption in the transportation task.

Figure 12 shows the results that the rotation angle decreases as AA increases for both strategies. This phenomenon can be explained that the object with the symmetry shape naturally balances when a lot of robots uniformly contacts with one side. Even if such a fact is considered, the effectiveness suppressing the rotation in proportion to the number of AAs is remarkable. However, the degradation around 15 AAs is one of the issues to be solved. The degradation is marked for the rectangle, which likely increases for objects with a shape such as a stick. It is derived from the generation manner of the PA. Since AA generates PA with the vector value to the neighbor location of it, the impact of each attraction is not so strong as in the case of fewer AAs.

5   Discussions

As shown in the experimental results, the transportation system using AAs and PAs works well in most cases. However, in some cases, the results have shown that it may cause rotation or oscillation of an object, which decreases efficiency

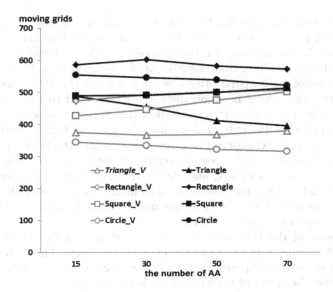

Fig. 15. Transportation distance.

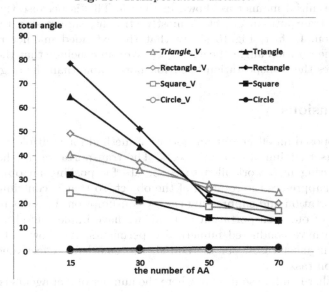

Fig. 16. Rotation angle.

of the transportation task. As mentioned above, the issue is derived from the generation manner of PA, in which AA generates PA with the vector value to the neighbor location of it.

In order to solve the issue, it would be effective to introduce the strategy that determines the destination of PA's vector value in proportion to the degree of the inclination caused in pushing the object as shown in Fig. 13. We have implemented another generation manner of PAs in order to confirm the effectiveness of the strategy. In the manner, a robot generates a PA two robot sizes apart, if $|\theta_\delta|$ is more than 5 and less than 10; also, if $|\theta_\delta|$ is more than 10, it generates the PA three robot sizes apart; otherwise, it generates the PA at its neighbor location. Notice that the generation distance is restricted to three robot sizes because the range where PA can migrate is within three robot size.

In order to observe the extended manner, in addition to the initial experiments, we have conducted experiments that apply it and the initial PA manner to four kinds of objects, and have measured the total time and the total distance for transporting an object, and the total angle where the object rotates. Figures 14, 15 and 16 show the results. In each figure, the results of the extended manner are represented by object shape names with V, such as `Triangle_V`, `Rectangle_V`, `Square_V`, and `Circle_V`.

In the result of the transportation time, as shown in Fig. 14, the time of the initial PA manner was less than the extended manner. This is because the extended manner tends to sparsely attract other robots compared with the initial manner, so that the speed of transportation is decreased.

Inversely, the transportation distance of the extended manner tends to be less than the initial manner as shown in Fig. 15. This is because the extended manner can transport an object in mostly straight, suppressing the rotation and oscillation. In fact, Fig. 16 shows that the extended manner remarkably suppresses the rotation angle of each object. We can conclude that the extended manner makes the transportation behavior more stable than the original one.

6   Conclusions

We have proposed an effective transportation method for multiple robots using mobile agents that imitate social insects. The transportation method enables robots scattering in a work filed to cooperate for pushing an object in balance, which suppresses the rotation of the object, so that it contributes to efficient transportation and suppressing energy consumption. In order to show the effectiveness of our transportation method, we have implemented a simulator, on which we have conducted numerical experiments. As shown in the previous section, in most cases, our method shows remarkable effectiveness for the transportation task.

On the other hand, in some case where the number of Ant agents is small, the effectiveness of our method was restrictive. In order to solve the issue, we gave a simple extension changing the generation locations of PAs in proportion to the degree of the inclination in pushing an object. As the results, we observe that the extension decreases transportation distance, contributing to suppressing energy consumption, but tends to increase transportation time. Most importantly, the extended method achieves the stable transportation behaviors.

References

1. Abe, T., Takimoto, M., Kambayashi, Y.: Searching targets using mobile agents in a large scale multi-robot environment. In: O'Shea, J., Nguyen, N.T., Crockett, K., Howlett, R.J., Jain, L.C. (eds.) KES-AMSTA 2011. LNCS, vol. 6682, pp. 211–220. Springer, Heidelberg (2011)
2. Binder, W., Hulaas, J.G., Villazon, A.: Portable resource control in the J-SEAL2 mobile agent system. In: Proceedings of the Fifth International Conference on Autonomous Agents, AGENTS 2001, pp. 222–223. ACM (2001)
3. Deneubourg, J., Goss, S., Franks, N.R., Sendova-Franks, A.B., Detrain, C., Chreien, L.: The dynamics of collective sorting: robot-like ant and ant-like robot. In: Proceedings of the First Conference on Simulation of Adaptive Behavior: From Animals to Animats, pp. 356–363. MIT Press (1991)
4. Dorigo, M., Birattari, M., Stützle, T.: Ant colony optimization-artificial ants as a computational intelligence technique. IEEE Comput. Intell. Mag. $1(4)$, 28–39 (2006)
5. Dorigo, M., Gambardella, L.M.: Ant colony system: a cooperative learning approach to the traveling salesman. IEEE Trans. Evol. Comput. $1(1)$, 53–66 (1996)
6. Fujisawa, R., Imamura, H., Matsuno, F.: Cooperative transportation by swarm robots using pheromone communication. In: Martinoli, A., Mondada, F., Correll, N., Mermoud, G., Egerstedt, M., Hsieh, M.A., Parker, L.E., Støy, K. (eds.) Distributed Autonomous Robotic Systems. STAR, vol. 83, pp. 559–570. Springer, Heidelberg (2013)
7. Gerkey, B.P., Mataric, M.J.: Pusher-watcher: An approach to fault-tolerant tightly-coupled robot coordination. In: Proceedings of the IEEE International Conference on Robotics and Automation, vol. 1, pp. 464–469 (2002)
8. Kambayashi, Y., Takimoto, M.: Higher-order mobile agents for controlling intelligent robots. Int. J. Intell. Inf. Technol. (IJIIT) $1(2)$, 28–42 (2005)
9. Khatib, O., Yokoi, K., Chang, K., Ruspini, D., Holmberg, R., Casal, A.: Vehicle/arm coordination and mobile manipulator decentralized cooperation. In: Proceedings of the IEEE/RSJ International Conference on Intelligent Robots and Systems, pp. 546–553 (1996)
10. Kube, C.R., Bonabeau, E.: Cooperative transport by ants and robots. Robot. Auton. Syst. $30(1–2)$, 85–101 (2000)
11. Lumer, E.D., Faiesta, B.: Diversity and adaptation in populations of clustering ants, from animals to animats 3. In: Proceedings of the 3rd International Conference on the Simulation of Adaptive Behavior, pp. 501–508. MIT Press (1994)
12. Mataric, M.J., Nilsson, M., Simsarian, K.T.: Cooperative multi-robot box-pushing. In: Proceedings of the IEEE/RSJ International Conference on Intelligent Robots and Systems, vol. 3, pp. 556–561 (1995)
13. Mizutani, M., Takimoto, M., Kambayashi, Y.: Ant colony clustering using mobile agents as ants and pheromone. In: Nguyen, N.T., Le, M.T., Świątek, J. (eds.) ACIIDS 2010. LNCS, vol. 5990, pp. 435–444. Springer, Heidelberg (2010)
14. Nagata, T., Takimoto, M., Kambayashi, Y.: Suppressing the total costs of executing tasks using mobile agents. In: Proceedings of Hawaii International Conference on System Sciences 42 CD-ROM (2009)
15. Rus, D., Donald, B., Jennings, J.: Moving furniture with teams of autonomous robots. In: Proceedings of the IEEE/RSJ International Conference on Intelligent Robots and Systems, pp. 235–242 (1995)

16. Shibuya, R., Takimoto, M., Kambayashi, Y.: Suppressing energy consumption of transportation robots using mobile agents. In: Proceedings of the 5th International Conference on Agents and Artificial Intelligence (ICAART 2013), SciTePress, pp. 219–224 (2013)

17. Shintani, M., Lee, S., Takimoto, M., Kambayashi, Y.: A serialization algorithm for mobile robots using mobile agents with distributed ant colony clustering. In: König, A., Dengel, A., Hinkelmann, K., Kise, K., Howlett, R.J., Jain, L.C. (eds.) KES 2011, Part I. LNCS, vol. 6881, pp. 260–270. Springer, Heidelberg (2011)

18. Shintani, M., Lee, S., Takimoto, M., Kambayashi, Y.: Synthesizing pheromone agents for serialization in the distributed ant colony clustering. In: ECTA and FCTA 2011 - Proceedings of the International Conference on Evolutionary Computation Theory and Applications and the Proceedings of the International Conference on Fuzzy Computation Theory and Applications (parts of the International Joint Conference on Computational Intelligence IJCCI 2011), SciTePress, pp. 220–226

19. Stilwell, D.J., Bay, J.S.: Toward the development of a material transport system using swarms of ant-like robots. In: Proceedings of the IEEE International Conference on Robotics and Automation, pp. 766–771 (1993)

20. Takimoto, M., Mizuno, M., Kurio, M., Kambayashi, Y.: Saving energy consumption of multi-robots using higher-order mobile agents. In: Nguyen, N.T., Grzech, A., Howlett, R.J., Jain, L.C. (eds.) KES-AMSTA 2007. LNCS (LNAI), vol. 4496, pp. 549–558. Springer, Heidelberg (2007)

21. Wand, T., Zhang, H.: Collective sorting with multi-robot. In: Proceedings of the First IEEE International Conference on Robotics and Biomimetics, pp. 716–720 (2004)

22. Wang, Z.D., Kimura, Y., Takahashi, T., Nakano, E.: A control method of a multiple non-holonomic robot system for cooperative object transportation. In: Proceedings of the 5th International Symposium on Distributed Autonomous Robotic Systems on Distributed Autonomous Robotic Systems 4, pp. 447–456 (2000)

Design and Simulation of a Low-Resource Processing Platform for Mobile Multi-agent Systems in Distributed Heterogeneous Networks

Stefan Bosse[✉]

Department of Mathematics & Computer Science,
ISIS Sensorial Materials Scientific Centre,
University of Bremen, Bremen, Germany
sbosse@uni-bremen.de

Abstract. The design and simulation of an agent processing platform suitable for distributed computing in heterogeneous sensor networks consisting of low-resource nodes is presented, providing a unique distributed programming model and enhanced robustness of the entire heterogeneous environment in the presence of node, sensor, link, data processing, and communication failures. In this work multi-agent systems with mobile activity-based agents are used for sensor data processing in unreliable mesh-like networks of nodes, consisting of a single microchip with limited low computational resources, which can be integrated into materials and technical structures. The agent behaviour, based on an activity-transition graph model, the interaction, and mobility can be efficiently integrated on the microchip using a configurable pipelined multi-process architecture based on the Petri-Net model and token-based processing. A new sub-state partitioning of activities simplifies and optimizes the processing platform significantly. Additionally, software implementations and simulation models with equal functional behaviour can be derived from the same program source. Hardware, software, and simulation platforms can be directly connected in heterogeneous networks. Agent interaction and communication is provided by a simple tuple-space database. A reconfiguration mechanism of the agent processing system offers activity graph changes at run-time. The suitability of the agent processing platform in large scale networks is demonstrated by using agent-based simulation of the platform architecture at process level with hundreds of nodes.

Keywords: Multi-agent platform · Sensor network · Mobile agent · Heterogeneous networks · Embedded systems

1 Introduction

Trends are recently emerging in engineering and micro-system applications such as the development of sensorial materials [11] show a growing demand for distributed autonomous sensor networks of miniaturized low-power smart sensors

© Springer International Publishing Switzerland 2015
B. Duval et al. (Eds.): ICAART 2014; LNAI 8946, pp. 63–81, 2015.
DOI: 10.1007/978-3-319-25210-0_5

embedded in technical structures [4]. These sensor networks are used for sensorial perception or structural health monitoring, employed, for example in Cyber-Physical-Systems (*CPS*), and perform the monitoring and control of complex physical processes using applications running on dedicated execution platforms in a resource-constrained manner under real-time processing and technical failure constraints.

To reduce the impact of such embedded sensorial systems on mechanical structure properties, single microchip sensor nodes (in mm^3 range) are preferred. Real-time constraints require parallel data processing inadequately provided by software based systems.

Multi-agent systems can be used for a decentralized and self-organizing approach of data processing in a distributed system like a sensor network [2], enabling information extraction, for example based on pattern recognition [3,12], and by decomposing complex tasks in simpler cooperative agents. Hardware (microchip level) designs have advantages compared with microcontroller approaches concerning power consumption, performance, and chip resources by exploiting parallel data processing (covered by the agent model) and enhanced resource sharing [6], which will be applied in this work.

Usually sensor networks are a part of and connected to a larger heterogeneous computational network [2]. Employing of agents can overcome interface barriers arising between platforms differing considerably in computational and communication capabilities. That's why agent specification models and languages must be independent of the underlying run-time platform. On the other hand, some level of resource and processing control must be available to support the efficient design of hardware platforms.

Hardware implementations of multi-agent systems are still limited to single or a few and non-mobile agents [13,15], and were originally proposed for low level tasks, for example in [8] using agents to negotiate network resources. Coarse grained reconfiguration is enabled by using FPGA technologies [13]. Most current work uses hardware-software co-design methodologies and code generators,like in [14]. This work provides more fine-grained agent reconfiguration and true agent mobility without relying on a specific technology and employs high-level synthesis to create standalone hardware and software platforms delivering the same functional and reactive behaviour.

There is related work concerning agent programming languages and processing architectures, like *APRIL* [10] providing tuple-space like agent communication, and widely used *FIPAACL*, and *KQGML* [7] focusing on high-level knowledge representations and exchange by speech acts, or model-driven engineering (e.g. INGENIAS [9]). But the above required resource and processing control is missing, which is addressed in this work.

There are actually four major issues related to the scaling of traditional software-based multi-agents systems to the hardware level and their design:

- limited static processing, storage, and communication resources, real-time processing, unreliable communication,
- suitable simulation environments for testing distributed and parallel multi-agent processing on functional and operational level,

- suitable simplified agent-oriented programming models and processing architectures qualified for hardware designs with finite state machines (FSM) and
resource sharing for parallel agent execution,
- and appropriate high-level design and simulation tools offering MAS design
on programming level.

Traditionally agent programs are interpreted, leading to a significant decrease
in performance. In the approach presented here, the agent processing is directly
implemented in standalone hardware nodes without intermediate processing levels and without the necessity of an operating system.

This work introduces some novelties compared to other data processing and
agent platform approaches:

- One common agent behaviour model, which is implementable on different
processing platforms (hardware, software, simulator).
- Agent mobility crossing different platforms in a mesh-like network and agent
interaction by using a tuple-space database and global signal propagation aid
solving data distribution and synchronization issues in the design of distributed sensor networks.
- Support for heterogeneous networks and platforms covered by one design and
synthesis flow including functional behavioural and architectural simulation.
- A token-based pipelined multi-process agent processing architecture very well
suited to hardware platforms with the Register-Transfer Level Logic offering
optimized computational resources and speed.
- A Petri-Net representation is used to derive a specification of the hardware
process and communication network, and performing advanced analysis like
deadlock detection. Timed Petri-Nets can be used to calculate computing time
bounds to support real-time processing.
- A fast processing platform simulation on architectural level with large scale
networks (with hundreds of nodes) and simulation of the processing of large
scale MAS (with hundreds of mobile agents).

The next sections introduce the activity based agent processing model, available mobility and interaction, and the proposed agent platform architecture
related to the programming model. Finally, the suitability of the agent processing platform in large scale networks is demonstrated by using a novel agent-based
simulation technique for the simulation of the entire platform and network architecture at process level. The simulator can be integrated in an existing sensor
network offering a simulation-in-the-loop methodology.

2   Programming State-Based Mobile Agents

The implementation of mobile multi-agent systems for resource constrained
embedded systems with a particular focus on microchip level is a complex design
challenge. High-level agent programming and behaviour modelling languages can
aid to solve this design issue. To carry out multi-agent systems on hardware platforms, the activity-based agent-orientated programming language *AAPL* was

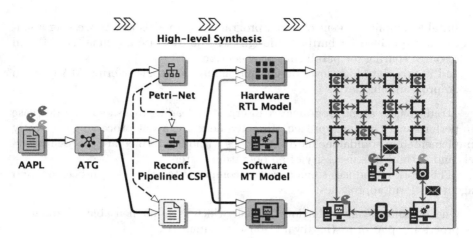

Fig. 1. From one common *AAPL* programming level to heterogeneous distributed networks (*RTL:* Register-Transfer Level, *MT:* Multi-Threading, *CSP:* Communicating Sequential Processes.

designed. Though the imperative programming model is quite simple and closer to a traditional PL it can be used as a common source and intermediate representation for different agent processing platform implementations (hardware, software, simulation) by using a high-level synthesis approach, shown in Fig. 1. Commonly used agent behaviour models based on *PRS/BDI* architectures with a declarative paradigm (2APL, AgentSpeak/Jason), communication models (e.g. FIPA *ACL*, *KQML*), and adaptive agent models can be implemented with *AAPL* providing primitives for the representation of beliefs or plans (discussed later). Agent mobility, interaction, and replication including inheritance are central multi-agent-orientated behaviours provided by *AAPL*.

Definition: *There is a multi-agent system (MAS) consisting of a set of individual agents $\{A_1, A_2,..\}$. There is a set of different agent behaviours, called classes $C=\{AC_1, AC_2,..\}$. An agent belongs to one class. A super class AC_i can be composed of different sub-classes $\{AC_{i,1}, AC_{i,2},..\}$, sharing activities and transitions of the super class. In a specific situation an agent A_i is bound to and processed on a network node $N_{m,n}$ (e.g. microchip, computer, virtual simulation node) at a unique spatial location (m,n). There is a set of different nodes $N=\{N_1, N_2,..\}$ arranged in a mesh-like network with peer-to-peer neighbour connectivity (e.g. two-dimensional grid). Each node is capable to process a number of agents n_i (AC_i) belonging to one agent behaviour class AC_i, and supporting at least a subset of $C' \subseteq C$. An agent (or at least its state) can migrate to a neighbour node where it continues working.*

AAPL Programming Model

The agent behaviour is partitioned and modelled with an activity graph, with activities representing the control state of the agent reasoning engine, and

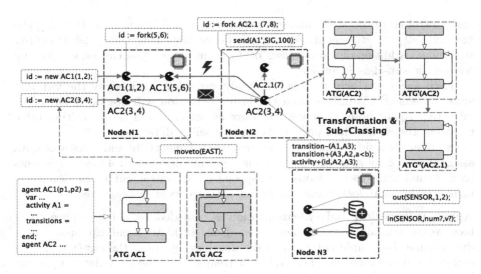

Fig. 2. Effects of AAPL control statements at run-time.

conditional transitions connecting and enabling activities, shown in Fig. 3 (left side). Activities provide a procedural agent processing by sequential execution of imperative data processing and control statements.

The activity-graph based agent model is attractive due to the proximity to the finite-state machine model, which simplifies the hardware implementation. An activity is activated by a transition which can depend on a predicate as a result of the evaluation of (private) agent data related to a part of the agents belief in terms of *BDI* architectures. An agent belongs to a specific parameterizable agent class *AC*, specifying local agent data (only visible for the agent itself), types, signals, activities, signal handlers, and transitions. The class *AC* can be composed of sub-classes, which can be independently selected.

Plans are related to *AAPL* activities and transitions close to conditional triggering of plans. Tables 1 and 2 summarize the available language statements. Their effects on a multi-agent system is shown in Fig. 2.

Agent Identifiers. Some statements like signal propagation require the identification of agents. In a node-local scope this is the token identifier assigned to the agent (Local *LID*). If an agent migrates, the LID can be extended by the relative displacement (Δ) to a unique identifier in the global network scope (GID). But this enforces the blocking of the token on the root node to avoid duplicates. Random generation of agent identifiers can overcome this issue, which is applied in this work.

Instantiation. Parameterized new agents of a specific class *AC* can be created at runtime by agents using the `new AC(v1,v2..)` statement returning a node unique agent identifier. An agent can create multiple living copies of itself with a fork mechanism, creating child agents of the same class with inherited data and control state but with different parameter initialization, done by using the fork(v1,v2,..) statement. Agents can be destroyed by using the kill(id) statement. Additionally, sub-classes of an agent super class can be selected by adding the sub-class identifier.

Each agent has *private data* (the body variables), defined by the var and var* statements. Variables from the latter definition will not be inherited or migrated! Agent body variables, the current activity, and the transition table represent the mobile data part of the agents beliefs database.

Statements inside an activity are processed sequentially and consist of data assignments ($x := \epsilon$) operating on agent's private data, control flow statements (conditional branches and loops), and special agent control and interaction statements, which can block agent processing until an event has occurred.

Agent interaction and synchronization is provided by a tuple-space database server available on each node (based on [10]). An agent can store an n-dimensional data tuple ($v1,v2,..$) in the database by using the out(v1,v2,..) statement (commonly the first value is treated as a key). A data tuple can be removed or read from the database by using the in(v1,p2?,v3,..) or rd(v1,p2?,v3,..) statements with a pattern template based on a set of formal (variable,?) and actual (constant) parameters. These operations block the agent processing until a matching tuple was found/stored in the database. These simple operations solve the mutual exclusion problem in concurrent systems easily. Only agents processed on the same network node can exchange data in this way. Simplified the expression of beliefs of agents is strongly based on *AAPL* tuple database model.Tuple values have their origin in environmental perception and processing bound to a specific node location.

The existence of a tuple can be checked by using the exist? function or with atomic test-and-read behaviour using the try_in/rd functions. A tuple with a limited lifetime (a marking) can be stored in the database by using the mark statement. Tuples with exhausted lifetime are removed automatically (by a garbage collector). Tuples matching a specific pattern can be removed with the rm statement.

Remote interaction between agents is provided by signals carrying optional parameters (they can be used locally, too). A signal can be raised by an agent using the send(ID,S,V) statement specifying the ID of the target agent, the signal name S, and an optional argument value V propagated with the signal. The receiving agent must provide a signal handler (like an activity) to handle signals asynchronously. Alternatively, a signal can be sent to a group of agents belonging to the same class AC within a bounded region using the broadcast(AC,DX,DY,S,V) statement.Signals implement remote procedure calls. Within a signal handler a reply can be sent back to the initial sender by using the reply(S,V) statement.

Timers can be installed for temporal agent control using (private) signal handlers, too. Agent processing can be suspended with the sleep and resumed with the wakeup statements.

Table 1. Summary of the AAPL statements used to define the agent behaviour and control.

Kind	AAPL Statement	Description

Agent Class Definition	`agent AC (arguments) =`
	    definitions
	`end;`
		Defines a new agent class *AC* with optional arguments. The class body consists of variable, activity, and transition definitions.

Creation and Replication	`id := new AC[.C] (args);`
	`id := fork [C] (args);`
	`kill(id);`
		Creates new agents at run-time. They are created from the class template, or forked from the parent agent. A subclass *C* can be selected, too.

Data	`var x,y,z: datatype;` `var* a,b,c: datatype;`	Defines long- and short term agent body variables. The latter ones are not saved on migration or inherited by children.

Activity	`activity [C.]A =`
	    statements
	`end;`
		Defines a new agent activity *A*, which can be bound to a sub-class *C*.

Reconfiguration	`activity+ (id,a1,a2,...);`
	`activity-`
	`(id,a1,a2,...);`
		Adds or removes activities at run-time for a specific agent *id*.

Transition	`transitions [C] =`
	    transitions `end;`
	`ai -> aj: condj; ...`
		Defines transitions at compile time between activities a_i and a_j with predicate *cond$_j$*. Can be used to define a sub-class *C*, too.

Reconfiguration	`transition+`
	`[C] (a1,a2,c);`
	`transition* [C]`
	`(a1,a2,c);`
	`transition- [C] (a1,a2);`
	`(id,..)`
		Changes transitions at run-time (add, replace all,remove all). Can be applied to a sub-class *C* or for a specific agent *id* only.

Mobility	`moveto(Dir);`
	`.. link?(Dir) ..`
	Dir=`{NORTH, SOUTH,`
	`WEST, EAST}`
		Migrates agent to a neighbour node. The connectivity can be tested by using the `link?` operation.

Migration of agents to a neighbour node (preserving the body variables, the processing, and configuration state) is performed by using the `moveto(DIR)` statement, assuming the arrangement of network nodes in a mesh- or cube-like network. To test if a neighbour node is reachable (testing connection liveliness), the `link?(DIR)` statement returning a Boolean result can be used.

Reconfiguration. Agents are capable to *change their transitional network* (initially specified in the transition section) by changing, deleting, or adding (conditional) transitions using the `transition`$\Diamond$(A_i, A_j, `cond`) statements. *This behaviour allows the modification of the activity graph, i.e., based on learning or environmental changes, which can be inherited by child agents.* The modification can be restricted to a sub-class transition set, which is useful for child agent

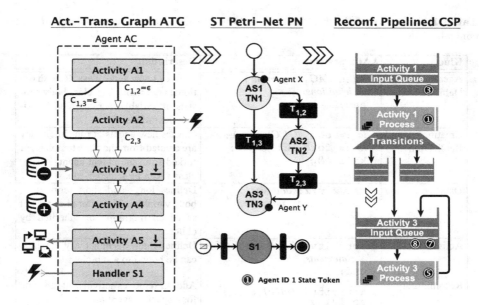

Fig. 3. Pipelined Communicating Sequential Processes Architecture derived from a Petri-Net specification and relationship to the activity-transition graph. Signals are handled asynchronously and independently from the activity processing.

generation. Additionally, the ATG can be transformed by adding or removing activities using the activity$\diamondsuit$(A_i, A_j, ...) statements, which is only applicable for dynamic code-based agents not considered here.

3 Agent Platform Architecture and Synthesis

The *AAPL* model is a common source for the implementation of agent processing in hardware, software, and simulation processing platforms. A database driven high-level synthesis approach [1] is used to map the agent behaviour to these different platforms. The agent processing architecture required at each network node must implement different agent classes and must be efficiently scalable to the microchip level to enable material-integrated embedded system design, which represents a central design issue, further focussing on parallel agent processing and optimized resource sharing.

3.1 The RPCSP Agent Platform

This processing platform - very well matching microchip-level designs - implements the agent behaviour with *reconfigurable pipelined communicating processes* (*RPCSP*) related to the Communicating Sequential Process model (*CSP*) proposed by Hoare (1985). The activities and transitions of the *AAPL* programming model are merged in a first intermediate representation by using

Table 2. Summary of the AAPL statements used for interaction and communication.

Kind	AAPL Statement	Description

Signal	signal S:*datatype*;
	handler $S(x)$ =
	  statements end;
	send(id, S, v);
	reply(S, v);
	broadcast(AC, DX, DY, S, v)
		Defines a signal S which can be processed by a signal handler similar to an activity. Signals are either send to a specific agent id or send to all agents of a specific class within a region.

Tuple Space Database	out($v1, v2, ..$);
	.. exist?($v1, ?, ..$) ..
	in($v1, x1?, v2, x2?, ...$);
	
	rd($v1, x1?, v2, x2?, ...$);
	try in(*timout, v1, ..*);
	try rd(*timeout, v1, ..*);
	mark(*timeout, v1, v2, ..*);
	rm($v1, ?, ..$);
		Synchronized data exchange by agents using the tuple space operations with tuples and patterns. A marking is a tuple with a limited lifetime. Commonly, the first tuple value is treated as a key, e.g. classifying the tuple.

Timer and Blocking	timer+(*timeout, S*); timer-(S); sleep; wakeup;	A timer can be used to raise a signal S. Agents can be suspended and be woken up.

state-transition Petri Nets (*PN*), shown in Fig. 3. This *PN* representation allows the following *CSP* derivation specifying the process and communication network, and advanced analysis like deadlock detection. Timed Petri-Nets can be used to calculate computing time bounds to support real-time processing.

Keeping the *PN* representation in mind, the set of activities $\{A_i\}$ is mapped on a set of sequential processes $\{P_i\}$ executed concurrently. Each subset of transitions $\{T_{i,j}\}$ activating one common activity process P_j is mapped on a synchronous n:1 queue Q_j providing inter-activity-process communication, and the computational part for transitions embedded in all contributing processes $\{P_i\}$, shown in Fig. 3. Changes (reconfiguration) of the transition network at run-time are supported by transition tables, shown in Fig. 4. Body variables of agents are stored in an indexed table set. Activity processes are partitioned into sub-states, at least one computational and one transitional state, discussed below.

Each sequential process is mapped (by synthesis) on a finite-state machine and a data path using a register-transfer architecture (RTL) with mutual exclusive guarded access of shared objects, all implemented in hardware.

This pipeline architecture offers advanced resource sharing and parallel agent processing with only one activity process chain implementation required for each agent class. The hardware resource requirement (digital logic) is divided into a control and a data part. The control part is proportional to the number of supported different agent classes. The data part depends on the maximal number of agents executed by the platform and the storage requirement for each agent.

Token-Based Processing. Agents are represented by tokens (natural numbers equal to the agent identifier, unique at node level), which are transferred by the queues between activity processes depending on the specified transition

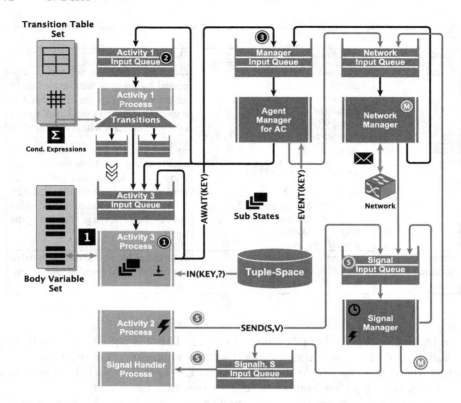

Fig. 4. Interaction of the agent, signal, and network manager with activity processes.

conditions and the enabling of transitions. This multi-process model is directly mappable to Register-Transfer Level (RTL) hardware architectures. Each process P_i is mapped on a finite state machine FSM_i controlling process execution and a register-transfer data path. Local agent data are stored in a region of a memory module assigned to each individual agent. There is only one incoming transition queue for each process consuming tokens, performing processing, and finally passing tokens to outgoing queues, which can depend on conditional expressions and body variables. There are computational and IO/event-based activity statements. The latter ones can block the agent processing until an event occurs (for example, the availability of a data tuple in the database).

Agents in different activity states can be processed concurrently. Thus, activity processes which are shared by several agents may not block. To prevent blocking of IO processes,not-ready processes pass the waiting agent to the agent manager.

Activity Sub-State Partitioning and Event-based Processing. To handle I/O-event and migration related blocking of statements, activity processes executing these statements are partitioned into sub-states $A_i \Rightarrow \{a_{i,1}, a_{i,2}, ..., a_{i,TRANS}\}$ and a sub-state-machine decomposing the process in computational,

I/O statement, and transitional parts, which can be executed sequentially by back passing the agent token to the input queue of the process (sub-state loop iteration). The control state of an agent consists therefore of the actual/next activity A_i/A_{i+n} and the activity sub-state $a_j(A_i)$ to be executed. Agents which wait for the occurrence of an event are passed to the agent manager queue releasing the activity process. After the event occurred, the agent token is passed back to the activity process continuing the execution, shown in Fig. 4. Usually I/O events are related to tuple-space database (*TSDB*) access (*in* -operation is blocked until a matching *out*-operation is performed). For this reason the *TSDB* module is directly connected to the agent manager which is notified about the keys of new tuples stored in the database releasing waiting consumer agents. The following annotated code snippet shows the sub-state partitioning and sub-state transitions ($\rightarrow$:immediate, $\perp$: blocked and passed to the agent manager).

```
activity init =
init₁: dx:= 0; dy := 0; h := 0; → init₂
init₂: if dir <> ORIGIN then
          moveto(dir); ⊥ init₃
          init₃:case dir of
                  | NORTH => backdir:=SOUTH;
                  | SOUTH => backdir:=NORTH;
                  | WEST  => backdir:=EAST;
                  | EAST  => backdir:=WEST;
               end; → init₄
        else  live:=MAXLIVE; backdir:=ORIGIN; → init₄
        end;
init₄: group:= Random(integer[0..1023]);
       out(H,id(SELF),0); → init₅
init₅: rd(SENSORVALUE,s0?); ⊥ init_TRAN
init_TRAN: Transition Computation
```

If there are conditional outgoing transitions which cannot actually be satisfied, the activity process can be suspended (by using the *AAPL* sleep statement) by transferring the agent token to the agent manager. A signal handler of the agent can be used to wake up the agent again (by using the wakeup statement).

Agent and Network Managers. The agent manager is connected with all input queues of the activity processes and with the network managers handling remote agent migration and signal propagation. Agents are associated with control state structures. Agent tokens are injected by the agent manager after agent creation, migration, or resumption.

The agent manager uses agent tables and caches to store information about created, migrated, and passed through agents (req., for ex.,for signal propagation).

Transitions and Reconfiguration. Each activity process has a final transition sub-state $a_{i,\text{TRAN}}$ which tests for enabled transitions in the current context. All possible (enabled and disabled) transitions outgoing from an activity are processed in the transition sub-state of each activity process. If a condition of an enabled transition is true, the agent token is passed to the respective destination queue.

Configuration of the transition network at run-time modifies transition tables, storing the state of each transition {enabled, disabled}. There is one table set for each individual agent which can be divided further in the super class and possible sub-classes.

Though the possible reconfiguration and the conditional expressions must be known at compile time (static resource constraints), a reconfiguration can release the use of some activity processes and enhances the utilization for parallel processing of other agents. The transition network is implemented with selector tables in case of the HW and SW implementations, and with transition lists in case of the SIM implementation.

Reconfiguration can aid to increase and optimize utilization of the activity process network populated by different sub-classed agents using only a sub-set of the activities.

Migration *of agents* requires the transfer of the agent data and the control state of the agent together with a unique global agent identifier (extending the local *id* with the agent class and the relative displacement of its root node) encapsulated in messages.

Messages carrying the state of agents consisting of the body variables (only the long-term part) and the control structure with the current activity, the sub-state which is entered after migration, and agent identifiers (id, Δ). Furthermore messages are used to carry signals. The network managers (input & output) perform message encoding, decoding, and delivery. Migration requires at least one more activity sub-state. After migration, the next sub-state of the last activity is executed.

Tuple-Space Database. Each n-dimensional tuple-space TS^n (storing n-ary tuples) is implemented with fixed size tables in case of the hardware implementation, and with dynamic lists in the case of the software and simulation model implementations. The access of each tuple-space is handled independently. Concurrent access of agents is mutually exclusive. The HW implementation implicates further type constraints, which must be known at design time (e.g. limitation to integer values).

Signal Handling. Signals are handled asynchronously by activating signal handlers, implemented by a process and a signal handler input queue. The signal manager is responsible for the creation and propagation of signals, shown in the bottom of Fig. 4. Signal tokens represent tuple values (*signal, argument, dst-id, src-id, Δ*). Remote signals are processed by the signal and network managers, which encapsulate signals in messages sent to the appropriate target node and agent.

Replication of activity processes sharing the same input queue offers advanced parallel processing of multiple agents for activities with high computing times reducing the mean computational latency.

3.2   Software Platform

The already introduced *RPCSP* architecture can be implemented in software, too. In this case the activity processes are implemented with light weighted processes (threads) communicating through queues, providing token based agent processing, too. The software platform includes the agent and signal managers, tuple space databases, and networking. Software platforms can be directly connected to hardware platforms and vice versa. They are compatible at the interface (message) and agent behaviour level.

Implementing the *RPCSP* architecture in software has the advantage of low-resource requirements and the exploitation of parallelism by multi-processor or multi-core architectures including advanced hyper-threading techniques. The number of threads and resources are known and allocated in advance, which can be mandatory for hard real-time processing systems.

3.3   Simulation Platform

In addition to real hardware and software implemented agent processing platforms there is the capability of the simulation of the agent behaviour, mobility, and interaction on a functional level. The *SeSAm* simulation framework [5] offers a platform for the modeling, simulation, and visualization of mobile multi-agent systems employed in a two-dimensional world. The behaviours of agents are modeled with activity graphs (specifying the agent reasoning machine) close to the *AAPL* model. Activity transitions depend on the evaluation of conditional expressions using agent variables. Agent variables can have a private or global (shared) scope. Basically *SeSAm* agent interaction is performed by modification and access of shared variables and resources (static agents).

Simulation of complex MAS on behavioural level and the methodology using the *SeSAm* simulator was already demonstrated in [16], mapping *AAPL* agents of the MAS one-to-one to *SeSAm* agents. In this work instead the *RPCSP* agent processing platform is simulated with the agent-based *SeSAm* simulation framework, discussed in detail in the following section. This simulation provides the testing and profiling of the proposed processing platform architecture in a distributed network world.

The simulator is also fully compatible with the software and hardware platforms on behavioural and interface level and can be integrated in an existing real-world network, offering simulation-in-the-loop capabilities.

3.4   Synthesis

The database driven synthesis flow (details in [1]) is shown in Fig. 5 and consists of an *AAPL* front end, the core compiler, and several backends targeting different platforms. The *AAPL* program is parsed and mapped on an abstract syntax tree (*AST*). The first compiler stage analyses, checks, and optimizes the agent specification *AST*. The second stage is split in three parts: an activity to process-queue pair mapper with sub-state expansion, a transition network builder, manager generators, and a message generator supporting agent and signal migration. Different outputs can be produced: a hardware description enabling *SoC* synthesis using the *ConPro SoC* high-level synthesis framework (details in [6]), a software description (*C*) which can be embedded in application programs, and the *SeSAm* simulation model (*XML*). The *ConPro* programming model reflects an extended *CSP* with atomic guarded actions on shared global resources. Each process is implemented with a *FSM* and a *RT* data path. The simulation design flow includes an intermediate representation using the *SEM* programming language, providing a textual representation of the entire *SeSAm* simulation model, which can be used independently, too.

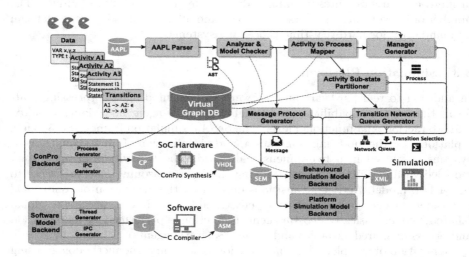

Fig. 5. Simplified overview of the high-level synthesis flow architecture.

All implementation models (HW/SW/SIM) provide equal functional behaviour, and only differ in their timing, resource requirements, and execution environments. Some more implementation and synthesis details follow.

4   Platform Simulation

This section will demonstrate that agent-based simulation is suitable to for the simulation of the *RPCSP* agent processing platform itself and large scale distributed networks, e.g., sensor networks, using the agent-based *SeSAm* simulator [5]. Simulation and analysis of parallel and distributed systems are a challenge. Performance profiling and the detection of race conditions or deadlocks are essential in the design of such systems, where the agent processing platform is a central part. Furthermore, platform simulation allows the estimation and optimization of static resources like agent tables or queues, completed with the ability to study the temporal behaviour of the entire network including communication treated as a distributed virtual machine, e.g., identifying bottlenecks for specific task situations, hard to monitor in technical systems.

Behavioural simulation [1, 16] maps agents of the MAS to be tested directly and isomorphic on agent objects of the simulation model. Platform simulation uses agents to implement architectural blocks like the agent manager or activity processes. Hence, agents of the MAS are virtually represented by the data space of the simulator, and not by the agent objects themselves.

The simulation of the processing platform with large scale networks processing large scale MAS aid to modify and refine the *RCSP* architecture, and to tune the static resource parameters like token pool and queue sizes or activity process replication to optimize timing. The platform simulation allows a fine grained estimation of the required resources.

The networks to be simulated (aka. the simulated world) consist of nodes arranged in a two-dimensional mesh grid, with each node connected to his four neighbour nodes, shown in Fig. 6 for a 10 by 10 sensor network with dedicated computational nodes at the outsides of the network. The entire platform and network system is partitioned into different non-mobile agent and resource classes (a resource is a passive agent with a data state only):

World Agent. The world agent creates all node agents and provides some network wide services. The world agent implements a reduced physical environment, e.g., by creating sensor signals or by disabling (destroying) connections between network nodes. Connections are represented by resources (passive agents providing only a geometric shape and body variables).

Node Agent. Each node is represented by a node agent, basically providing a common interface to data structures and tables required by the node managers and the activity processes. The node agent creates all the platform agents at the beginning of the simulation run.

Manager Agent. There is one "agent manager" agent for each agent class which is supported on the network node platform.

Network Manager Agents. There are two network manager agents. One input network manager agent handling incoming messages from neighbour nodes, decoding messages, creating agent or signal tokens, and finally passing the tokens to the agent or signal manager. The second output network manager agent is responsible for encoding and sending of messages carrying agent states or signals.

Activity Process Agents. For each agent class and each activity process of an agent class there is one activity processing agent performing token-based agent processing. The sub-states of an activity process are implemented by a simple sub-state selector and token loop-backing providing a sub-state *FSM*. Each activity process agent has local storage and an global visible token input queue.

Monitor Agent. There is one monitor agent per world collecting temporal resolved statistical data, finally writing the results to a CSV data file.

Table 3. Comparison of behavioural and platform simulation of the same MAS [16] using the SeSAm simulator.

	Behavioural Simulation	Platform Simulation
Number of Agents and Resources (dynamic = mobile)	Static: 300 agents, 700 res.; dynamic: 130 explorer agents	Static: 1600 agents, 700 res.; dynamic: 130 virtual agent resources
Simulation time including setup of simulation, with a correlated cluster scenario of 8 nodes, until MAS has finished work.	60 simulation steps in 5 s (on 1.2 GHz Intel U9300, 3 GB)	280 simulation steps in 60 s (on 1.2 GHz Intel U9300, 3 GB)

Token and Queues. The agent token queues are implemented with lists in the body variable space of each node agent. The size of the list can be monitored at run-time to detect resource underflow. Mutex guarded operations ($inq, outq$) allow concurrent access to the queues by different agents (manager, activity processes,..). Tokens are record structures with additional descriptive entries like the current queue they are stored in.

Virtual Agent. For visualization and debugging there is a mobile virtual agent resource representing an agent to be processed by a specific agent node platform. The virtual agent references the data and control state of an agent.

The new platform simulation is compared with the behavioural simulation from previous work in Table 3 for the simulation of a self-organizing MAS

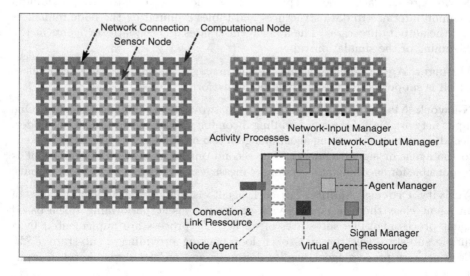

Fig. 6. Simulation world of a sensor network (left) consisting of 10×10 nodes and the network populated with non-mobile platform and virtual mobile agents (right).

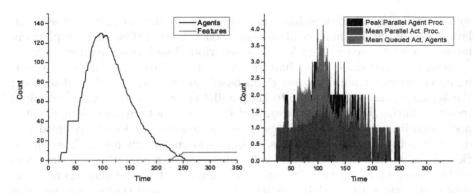

Fig. 7. Analysis of the MAS simulation: Left plot shows the temporal development of the agent population (explorer agent) and the rise of found features in the sensor network, right plot shows the utilization of the platform processes (peak parallel agent processing on one node, mean parallel active processes per node, and mean agent tokens queued per active node).

used for feature extraction in a sensor network. The behaviour model of the MAS is described in detail in [16]. It bases on a distributed divide-and-conquer approach. The number of (non-mobile) agents implementing the processing platform depends mainly on the number of activities decomposing the agent behaviour and the number of agent classes to be supported on the platform. For this example, the platform simulation model requires five times more agents and twenty times more computing time than the behavioural model. But the required resources and computing time for the fine-grained platform simulation is still reasonable and can be handled well with low end computers.

The analysis of a simulation run is shown in Fig. 7. It shows the temporal resolved analysis of the population of explorer agents of the MAS and the utilization of the *PCSP* network for nodes processing actually agents. There are nodes capable to process up to four agents simultaneous (speedup 4, in different activity states and processes). The mean speedup factor is about 1.5 for all nodes actually processing agents. Both platform and behavioural simulation deliver the same computational results of the distributed MAS.

5 Conclusions

A novel **design approach** using mobile agents for reliable distributed and parallel data processing in large scale networks of low-resource nodes was introduced. An agent-orientated programming language *AAPL* provides computational statements and statements for agent creation, inheritance, mobility, interaction, reconfiguration, and information exchange, based on agent behaviour partitioning in an activity graph, which can be directly synthesized to the microchip level by using a high-level synthesis approach. The high-level synthesis tool also enables the synthesis of different processing platforms from a common program source, including standalone hardware and software platforms, as well as simulation models offering functional and behavioural testing. The different platform implementations are compatible at the behavioural and message-interface level.

Agents of the same class share one virtual machine consisting of a reconfigurable pipelined multi-process chain based on the *CSP* model implementing the activities and transitions, offering parallel agent processing with optimized resource sharing. Unique identification of agents does not require unique absolute node identifiers or network addresses, a prerequisite for loosely coupled and dynamic networks (due to failures, reconfiguration, or expansion). The migration of an agent to a neighbour node takes place by migrating the data and control state of an agent using message transfers. Two different agent interaction primitives are available: signals carrying data and tuple-space database access with pattern templates.

Reconfiguration of the activity transition network offers agent behaviour adaptation (which can be inherited by children) at runtime and improved resource sharing for parallel agent processing.

A novel agent-based simulation of the agent processing platform and large-scale networks with a MAS case study demonstrated the suitability of the proposed programming model, processing architecture, and synthesis approach. The platform simulation offers advanced study and visualization of the platform behaviour, performance, and synchronisation issues in a distributed system under real world conditions with respect to the executed MAS. The platform simulation was compared with earlier behavioural agent simulations using the same MAS. Though there is a significant increase of the required data resources and computation time, this simulation approach is well suited for large-scale MAS simulation.

References

1. Bosse, S.: Distributed agent-based computing in material-embedded sensor network systems with the agent-on-chip architecture. IEEE Sens. J., Special Issue MIS **14**, 2159–2170 (2014). doi:10.1109/JSEN.2014.2301938
2. Guijarro, M., Fuentes-Fernandez, R., Pajares, G.: A Multi-Agent System Architecture for Sensor Networks, Multi-Agent Systems - Modeling, Control, Programming, Simulations and Applications (2008)
3. Zhao, X., Yuan, S., Yu, Z., Ye, W., Cao, J.: Designing strategy for multi-agent system based large structural health monitoring. Expert Syst. Appl. **34**(2), 1154–1168 (2008). doi:10.1016/j.eswa.2006.12.022
4. Pantke, F., Bosse, S., Lehmhus, D., Lawo, M.: An artificial intelligence approach towards sensorial materials. In: Future Computing Conference (2011)
5. Klügel, F.: SeSAm: visual programming and participatory simulation for agent-based models. In: Uhrmacher, A.M., Weyns, D. (eds.) Multi-agent Systems - Simulation and Applications. CRC Press, Boca Raton (2009)
6. Bosse, S.: Hardware-software-co-design of parallel and distributed systems using a unique behavioural programming and multi-process model with high-level synthesis In: Proceedings of the SPIE Microtechnologies 2011 Conference, Session EMT 102. doi:10.1117/12.888122

7. Kone, M.T., Shimazu, A., Nakajima, T.: The state of the art in agent communication languages. Knowl. Inf. Syst. **2**(3), 259–284 (2000). doi:10.1007/PL00013712
8. Ebrahimi, M., Daneshtalab, M., Liljeberg, P., Plosila, J., Tenhunen, H.: Agent-based on-chip network using efficient selection method. In: 2011 IEEEIFIP 19th International Conference on VLSI and System-on-Chip, pp. 284–289 (2011)
9. Sansores, C., Pavon, J.: An adaptive agent model for self-organizing MAS. In: Proceedings of 7th International Conference on Autonomous Agents and Multiagent Systems (AAMAS 2008), Estoril, Portugal, 12–16 May 2008, pp. 1639–1642 (2008)
10. McCabe, F.G., Clark, K.L.: APRIL — agent process interaction language. In: Wooldridge, M., Jennings, N.R. (eds.) ECAI 1994. LNCS, vol. 890, pp. 324–340. Springer, Heidelberg (1995)
11. Lang, W., Jakobs, F., Tolstosheeva, E., Sturm, H., Ibragimov, A., Kesel, A., Lehmhus, D., Dicke, U.: From embedded sensors to sensorial materials–the road to function scale integration. Sens. Actuators A Phys. **171**(1), 3–11 (2011)
12. Liu, J.: Autonomous Agents and Multi-agent Systems. World Scientific Publishing, River Edge (2001). ISBN 981-02-4282-4
13. Meng, Y.: An agent-based reconfigurable system-on-chip architecture for real-time systems. In: Proceedings of the Second International Conference on Embedded Software and Systems (ICESS 2005), pp. 166–173 (2005)
14. Jamont, J.-P., Occello, M.: A multiagent method to design hardware/software collaborative systems. In: 12th International Conference on Computer Supported Cooperative Work in Design (2008)
15. Naji, H.: Creating an adaptive embedded system by applying multi-agent techniques to reconfigurable hardware. Future Gener. Comput. Syst. **20**(6), 1055–1081 (2004)
16. Bosse, S.: Design of material-integrated distributed data processing platforms with mobile multi-agent systems in heterogeneous networks. In: Proceedings of the 6th International Conference on Agents and Artificial Intelligence (ICAART 2014) (2014). doi:10.5220/0004817500690080

Behavior Clustering and Explicitation for the Study of Agents' Credibility: Application to a Virtual Driver Simulation

Kévin Darty[1]([⊠]), Julien Saunier[2], and Nicolas Sabouret[3]

[1] Laboratory for Road Operations, Perception, Simulators and Simulations,
French Institute of Science and Technology for Transport,
Development and Networks, 77447 Marne la Vallée, France
kevin.darty@ifsttar.f
[2] Computer Science, Information Processing and Systems Laboratory,
National Institute of Applied Sciences of Rouen,
76800 Saint-étienne-du-rouvray Cedex, France
julien.saunier@insa-rouen.fr
[3] Computer Sciences Laboratory for Mechanics and Engineering Sciences,
National Center for Scientific Research, Paris-Sud University,
91405 Orsay Cedex, France
nicolas.sabouret@limsi.fr

Abstract. The aim of this article is to provide a method for evaluating the credibility of agents' behaviors in immersive multi-agent simulations. It is based on a quantitative data collection from both humans and agents simulation logs during an experiment. These data allow us to semi-automatically extract behavior clusters. In order to obtain explicit information about the behaviors, we analyze questionnaires filled by the users and annotations filled by a second set of participants. It enables to draw user categories related to their behavior in the context of the simulation or of their real life habits. We then study the similarities between behavior clusters, user categories, and participants' annotations. Afterwards, we evaluate the agents' credibility and make their behaviors explicit by comparing human behaviors to agent ones according to user categories and annotations. Our method is applied to the study of virtual driver simulation through an immersive driving simulator.

Keywords: Multi-agent simulation · Behavior clustering · Credibility evaluation

1 Introduction

The validation of the credibility and realism of agents in multi-agent simulations is a complex task that has given rise to a lot of work in the domain of multi-agent simulation (see *e.g.* [4]). When the number of agents increases, Drogoul shows [8] that the validation of such a simulation requires an evaluation of the system at the *macroscopic* level. However, this does not guarantee validity at the

© Springer International Publishing Switzerland 2015
B. Duval et al. (Eds.): ICAART 2014; LNAI 8946, pp. 82–99, 2015.
DOI: 10.1007/978-3-319-25210-0_6

microscopic level, *i.e.* the validity of the behavior of each agent in the system. In some simulations such as virtual reality environments, where humans coexist with simulated agents, the human point of view is purely local and behavior is considered at the microscopic level. Indeed, if the agents' behavior is inconsistent, user immersion in the virtual environment (*i.e.* the human's feeling to belong to the virtual environment) is broken [9,19].

Methods and implementations of behaviors are not directly observable by the user, only the resulting behaviors are. This is why, this notion of credibility at the microscopic level does not depend on the way the behaviors are modeled. The outside observer judges them and this perception depends on many factors including sensory elements (visual rendering, haptic, proprioceptive, *etc.*) [3,29]. The term used in the literature to denote this feeling of realism is called presence effect [30]. The multiple techniques that are used to enhance the presence effect (called immersion techniques) are mainly evaluated on subjective data. Consequently, the evaluation of the presence effect resulting from a virtual reality (*VR*) device is done with methods from human sciences.

In this paper, we propose to evaluate the agents' credibility at the microscopic level. To do so, we combine subjective evaluation methods from human sciences with automated behavior traces analysis based on artificial intelligence algorithms. Section 2 presents the state of the art. Section 3 explains the general method we have developed, which relies on data clustering and comparison, and Sect. 4 gives the details of the underlying algorithms. Section 5 presents its application on an immersive driving simulator and its results.

2 State of the Art

In this section, we first define the notion of behavior. We then present existing subjective and objective approaches.

2.1 Levels of Behavior

Behaviors are a set of observable actions of a person in the environment. There are different levels of human behavior [22]: The lowest level corresponds to simple reflex actions such as going into first gear in a car. These behaviors are similar to the agent's elementary operations. The intermediary level is tactical, it is built on an ordered sequence of elementary behaviors such as a car changing lane on the highway. The highest level of behavior is the strategic level, corresponding to the long term. It is based on a choice of tactics and evolves according to the dynamics of the environment and the mental state of the person [24] as in overtaking a truck platoon or choosing a stable cruise speed. In our study, we evaluate the behavior of the agents at the last two levels (tactical and strategic).

2.2 Subjective Approach

The subjective approach comes from the *VR* field and aims at validating the agents' behavior in simulation. It consists in evaluating the general (or detailed) immersion quality via the presence effect using questionnaires [15]. In our case, the notion of presence is too broad because it includes various elements (visual quality, sound quality, *etc.*) of the device, but does not detail the virtual agents behavior credibility component.

However subparts of the presence effect evaluation are consistent with our goal:

- The behavioral credibility: Users interacting with the agent believe that they observe a human being with his own beliefs, desires and personality [16],
- The psychological fidelity: The simulated task generates for the user an activity and psychological processes which are similar to those generated by the real task [21]. The simulator produces a similar behavior to the one required in the real situation [14].

In this article, we focus on the behavioral credibility and especially on its qualitative and quantitative evaluation. A solution is to set up a mixed system where humans control avatars in the virtual environment. The evaluation of presence or of behavioral credibility is subjective. This is why it is sensitive to psychological phenomena such as the inability to explain one's judgments [12]. Moreover this evaluation does not necessarily explain missing behaviors nor the faults of the behaviors judged as not credible.

That is why we propose to complete these subjective studies with an objective analysis of simulation data.

2.3   Objective Approach

The objective approach is generally used in the field of multi-agents systems: It consists in comparing quantitative data produced by humans with data produced by different categories of virtual agents [4]. It aims at verifying that the behavior of the agents is identical to the one observed in reality and therefore at evaluating the realism of the simulation. When the number of agents increases, objective evaluation is generally done at the macroscopic level because real data are both more readily available and easier to compare with simulation outputs [7,18].

This macroscopic validation is necessary for the *VR* but not sufficient to validate the agents' behavior. A valid collective behavior does not imply that the individual behaviors that compose it are valid. Thus, an analysis at the microscopic level is required, although microscopic data analysis and comparison is complex. Some tools are available to summarize interactions of a multi-agent system for manual debugging [28]. Nevertheless, as simulation data involving participants consist of more than just message exchange variables, these tools are not directly applicable to complex and noisy data. A solution for data analysis, adopted by [10] for driving tasks consists of classifying participants according to variables. However, our method deals with a larger amount of both variables and participants, increasing the clustering task difficulty. It also provides explicit high-level behaviors via external annotation.

To the best of our knowledge, there is no tool to analyze strategic behavior in simulation combining both a validation of behavioral credibility and similarities between humans and virtual agents. Subjective and objective approaches complement each other in two different ways: human expertise and raw data.

3   Objective and Approach

Our goal is the evaluation of the agents of a multi-agents simulation at a microscopic level, in the context of virtual environments. The method we propose is based on the aggregation of individual data (for both agents and human participants) into behavior clusters that will support the actual behavior analysis. In this view, behavior clusters act as abstractions of individual traces. This paper details the computation of such clusters (Sect. 4) and their use for behavior analysis (Sect. 5). The originality of our model is that we use the two available types of data: objective data with logs and subjective data via questionnaires.

3.1   General Approach

The general architecture of the method is described in the Fig. 1 and the data processing is detailed in Fig. 4. It consists of 4 main steps: collection of data in simulation, annotation of this data, automatic clustering of data, and clusters comparison.

The first step of our method is to collect data about human participants. We consider both subjective data, using questionnaires about their general *habits* and their adopted behaviors in the given task, and objective data, using immersive (or participatory) simulation in the virtual environment. The raw data from

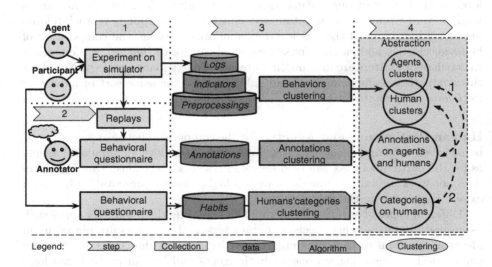

Fig. 1. An architecture for behavior analysis and evaluation.

participants' experiments in the simulator is called *logs* and the answers to questionnaires is called *habits* in the figure.

The second step is to refine this data by 1) producing new simulations (or "replays") in which the human participant has been replaced by a virtual agent; and 2) having all replays (with human participants and with virtual agents) being annotated by a different set of participants, using the behavior questionnaire. This step produces a set of *annotations*.

Our objective is twofold. First, we want to study the correlation between participants' categories and their behaviors observed in the simulation so as to verify that the automated clustering of observation data is related to task-related high-level behavior. Second, we need to compare participants' behavior and agents' behavior so as to report on the capability of agents to simulate human behavior. In both cases, this cannot be done on raw data (should it be questionnaires or data logs). Logs, especially in the case of participants, are noisy: two different logs can represent the same type of tactic or strategic behavior. This is the reason why, in order to generalize the analysis of our logs to a higher behavior level, we propose to use behavior categories (called *abstractions* in the figure). These categories serve as abstraction to the logs by gathering together, within the same cluster, different logs representative of the same high level behavior. This is done using automatic unsupervised clustering methods (because supervised algorithms require labeling by an expert of a large amount of logs). In the same way, we use clustering methods on the two questionnaires *habits* and *annotations*.

The comparison of these abstractions is our final step. We both evaluate the similarity between agents and humans logs and the annotated behaviors (dashed arrow number 1); and between the logs and self-reported *habits* for humans (dashed arrow number 2). While the first comparison allows us to evaluate the level of credibility of our virtual agents in the simulation, the second one is used to verify that the logs automated clustering corresponds to task-related high-level behaviors. If there is a strong similarity between the composition of behavior clusters and participant self-reported categories (*habits*), it then means that behavior clusters are meaningful in terms of participant typology. Note that this comparison is meaningful if and only if we use the same sort of indicators for *habits* and *annotations*.

Human Participants and Agents. For the comparison between participants' behavior clusters and agents' ones, we collect the same logs for simulated agents as for the participants. As will be discussed in Sect. 4.2, the clustering algorithm does not work directly on raw data: we use higher-level representation based on expert knowledge on the field.

Different types of agents are generated by exploring the parameter space such as normativity, experience, decision parameters ... The agents are placed in an identical situation to that presented to participants, so that the same logs are collected. The clustering is done on both agents and human participants logs, gathered together in the general term of *main actors* (see Sect. 4.2).

Fig. 2. Driving simulator device with 3 screens, steering wheel, gearbox and a set of pedals.

For the evaluation step, it is possible to distinguish three cluster types:

1. The clusters containing both human and agent main actors; they corresponds to high-level behaviors that are correctly reproduced by the agents.
2. Those consisting of simulated agents only; they correspond to behaviors that were produced only by the agents. In most cases, it reflects simulation errors, but it can also be due to a too small participants sample.
3. Those consisting of participants only; they correspond to behaviors that have not been replicated by the agents, and are thus either lacks in the agent's model, or due to a too small agents sample in the parameter space.

In the end, we combine this agent-human comparison with the *annotation-habits* analysis: The participants' behavior clusters are correlated to their *habits* categories. Furthermore, the composition of the behavior clusters in term of simulated agents and participants allows us to give explicit information about those agents.

3.2 Case Study

Our method was tested in the context of driving simulators. We want to evaluate the realism and credibility of the behavior of the *IFSTTAR*'s road traffic simulator's agents (see Fig. 2) by using the *ARCHISIM* driving simulator [6]. To do this, the participants drive a car on a road containing simulated vehicles. The circuit (shown in Fig. 3) provides a driving situation on a single carriage way with two lanes in the same direction. It corresponds to about 1 min of driving. The main actor encounters a vehicle at low speed on the right lane.

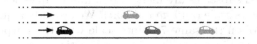

Fig. 3. Scenario: The main actor (in black) is driving on a single carriage way with two lanes in the same direction with a smooth traffic flow. Then, a vehicle at low speed on the right lane (in dark gray) disturbs the traffic.

Our method is illustrated in the following sections with this application to the study of driving behavior. However, the presented method may be used in any kind of participatory simulation, by choosing relevant task-related questionnaires.

4   Data Processing Method

In this section, we detail the different elements of our behavioral validation method and the algorithms that we use.

4.1   Clustering of Main Actors Categories

We first describe the *habits* questionnaire and the *annotations* questionnaire applied to the driving simulation, and then detail the clustering algorithm.

Participants *Habits*. In the first place it is necessary to submit a behavior questionnaire specific to the field before the experiment to characterize the general behavior of the participant in the studied activity. In the context of our application to driving simulators, we chose the Driving Behavior Questionnaire (*DBQ*) [26]. It provides a general score, but also scores on 5 subscales: *(1) slips*; *(2) lapses*; *(3) mistakes*; *(4)unintended violations*; and *(5) deliberate violations*. In addition, it supplies 3 subscales related to the accident risk: *(1) no risk*; *(2) possible risk*; and *(3) definite risk*.

Annotation of Main Actors Behaviors. An adopted behavior in a precise situation may not correspond to the participant's general behavior. For example, in driving simulators, the general driving behavior captured by the *DBQ* may not correspond to the participant's behavior in the precise studied situation. Furthermore, the general behavior questionnaire is completed by the driver himself about his own behaviors. This adds a bias due to introspection.

This is why we need to use a second questionnaire called *annotations*. This questionnaire is completed by a different set of participants. It avoids the introspection bias. Furthermore, having a population which is observing the situation allows us to collect situation specific information. The questions are rated on a *Likert*-type scale [17]. In our application to driving simulators, the questionnaire contains a question rated on a 7 points scale (and *no opinion*) from *no* to *yes* for each of the 5 *DBQ* subscales.

The 3 risk-related subscales are merged in a unique question named *accident risk* rated on a 3 points scale (and *no opinion*). We also add a question related to the perceived control on the same 7 points scale with the purpose of evaluating the main actors control in general. At last, a question asking if the main actor is a human or a simulated agent is added in order to compare how the behavior clustering and the annotators separated the participants from the agents.

From Data to Categories. In the general case (independently from the application domain), using behavior questionnaires, we obtain qualitative data on *Likert*-type scales. The answers are transformed into quantitative data via a linear numeric scale. Scale scores of questionnaires are then calculated by adding the scale-related questions, and normalized between 0 and 1. Once data are processed, we classify the participants' scores using a clustering algorithm to obtain drivers categories. This allows us to obtain clusters corresponding to participants' *habits* and how they are annotated. As seen in Sect. 3.1, the algorithm must be unsupervised with a free number of clusters. Several algorithms exist in the literature to this purpose, such as *Cascade K-means* with the *Variance Ratio Criterion* (*Calinski-Harabasz*) [5], *X-means* based on the *Bayesian Information Criterion* [23], or *Self-Organizing Maps* [13].

We chose to use the *Cascade K-means* algorithm which executes several *K-means* for $K \in \{1, \ldots, N\}$. The classic *K-means* algorithm uses K random initial centroids. It then proceeds those two steps alternatively until convergence: *(1)* The assignment step which assigns each main actor ma to the cluster C_i whose mean yields the least within-cluster sum of squares m_i at time t (see Eq. 1); *(2)* The update step which calculates the new means m to be the centroids of the main actors in the new clusters at time $t+1$ (see Eq. 2).

$$\forall\, j \in \{1, \ldots, k\}$$

$$C_i^{(t)} = \left\{ ma_p : \left\| ma_p - m_i^{(t)} \right\|^2 \leq \left\| ma_p - m_j^{(t)} \right\|^2 \right\} \tag{1}$$

$$m_i^{(t+1)} = \frac{1}{|C_i^{(t)}|} \sum_{ma_j \in C_i^{(t)}} ma_j \tag{2}$$

The initialization of the clusters is done with *K-means++* [1] which allows a better distribution of clusters' centers in accordance with the data. To do so, the centroid of the first cluster is initialized randomly among the main actors. Until having K clusters, the algorithm computes the distance of each main actor to the last selected centroid. Then, it selects the centroid of a new cluster among the main actors. The selection is done randomly according to a weighted probability distribution proportional to their squared distance.

Finally, we must select the "best" number of clusters with respect to our clustering goal. This is done using the *Variance Ratio Criterion* which takes into account the inter-distance (*i.e.* the within-cluster error sum of squares W) and intra-distance (*i.e.* the between-cluster error sum of squares B) of the clusters [5] (see Eq. 4). Let $|C_k|$ be the number of elements in the cluster C_k, $\overline{C_k}$ be the barycenter of this cluster and $\overline{\mathbb{C}}$ be the barycenter of all main actors (*i.e.* the clustering). Then, the *Variance Ratio Criterion* CH for K clusters is as described bellow (in Eq. 3):

$$CH(K) = \frac{B/(K-1)}{W/(N-K)} \tag{3}$$

$$\text{where } B = \sum_{k=1}^{K} |C_k| \|\overline{C_k} - \overline{\mathbb{C}}\|^2 \text{ and } W = \sum_{k=1}^{K} \sum_{n=1}^{N} \|ma_{k,n} - \overline{C_k}\|^2 \tag{4}$$

4.2   Clustering of Behaviors

This section describes how raw data logs are turned into clusters, within a series of pre-processing and clustering methods. The Fig. 4 shows the pre-processing applied to the logs in order to obtain clusters. Squares indicate the data name and its shape with the number of variables (X), the number of indicators (K), the number of main actors (N), and the time (T). The used algorithms are in squircles above arrows. The Sect. 4.2 (on the top of the figure) describes the logs of the main actors; the Sect. 4.2 (on the middle of the figure) explains the pre-processing; and the Sect. 4.2 (on the right of the figure) explains the clustering algorithm.

Main Actors Logs. During the simulation we collect the logs of the main actor (participant or agent), of neighboring agents and of the environment. These logs are then used for the clustering of tactical and strategic behaviors. The data to be recorded must be defined by experts in the domain of application.

In our traffic simulation example, we collect each $300\,ms$ from 8 to 13 variables. The variables shared by all the main actors are the time, the milepost, the road, the gap and the cap to the lane axis, the speed, the acceleration, and the topology. Specific variables to the driven vehicles are added: the wheel angle, the pressure on pedals (acceleration, brake and clutch), and the gear.

The road traffic experts chose the following indicators: some high-level variables like the inter-vehicles distance and time, the jerk (the derivative of acceleration with respect to time), the time to collision (under the assumption of constant speeds for both vehicles), and the number of lane changings (which is not a temporal indicator); as well as some low-level variables such as speed, acceleration, and lateral distance to the road axis.

Pre-processing. Some significant indicators dependent on the application field cannot be directly obtained. For this reason, field experts are consulted to identify important indicators. Then we calculate the indicators from the logs for those that could not be collected.

In the context of a dynamic simulation, most of the indicators are temporal. The data to classify are thus ordered sequences of values for each main actor. In

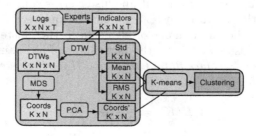

Fig. 4. Logs pre-processing and clustering.

order to classify those data, two ways exist: to use an algorithm taking temporal data as input or to use flat data by concatenating temporal indicators related to a participant on a single line. The first solution significantly increases the algorithms' complexity because they must take into account the possible temporal offsets of similar behaviors. The second one ignores temporal offsets but permits the application of classic algorithms.

We chose a hybrid solution of data pre-processing which allows us both to have a single set of attributes for each participant and to take into account temporal offsets. To do this, we generate as many vectors as main actors (participants and virtual agents). Each vector contains the following information extracted from the indicators identified by the field experts: *(a)*mean values; *(b)* standard deviations; *(c)* root mean squares; and *(d)* temporal aggregations. Temporal indicators are compared with an algorithm taking into account temporal offsets.

The adopted solution for the pre-processing of temporal offsets is to use a pattern matching algorithm such as Dynamic Time Warping (DTW) or Longest Common Subsequence (LCS). We chose the DTW algorithm which calculates the matching cost between two ordered sequences (*i.e.* indicators ind^a and ind^b) in order to measure their similarity [2]. Let T be the number of simulation time steps. The algorithm computes a $T \times T$ matrix. It initializes the first row and the first column to *infinity*, and the first element to 0. It then computes each elements of the matrix $M_{i,j} \forall (i,j) \in \{2, \ldots, T+1\}^2$ according to the distance between the two sequences at this time t and to the matrix element neighborhood (see Eq. 5). As DTW complexity is $O(N^2)$, we use an approximation of this algorithm: the *FastDTW* algorithm which has order of $O(N)$ time and memory complexity [27].

$$DTW[i,j] \leftarrow distance(ind^a_i, ind^b_j) + \qquad (5)$$
$$min(DTW[i-1,j], DTW[i,j-1], DTW[i-1,j-1])$$

As DTW calculates the similarity between two instances of a temporal variable, the less the instances are similar, the more the cost increases. Let $inds$ be the set of indicators and $K = |inds|$ be the number of indicators. For each indicator $ind \in inds$, we calculate the $N \times N$ mutual distances matrix D^{ind}_{DTW}, where N is the number of main actors (participants and agents).

In order to include DTW similarities as new variables describing the main actors, we use a Multi-Dimensional Scaling algorithm (MDS) to place each main actor in a dimensional space. The algorithm assigns a point to each instance in a multidimensional space and tries to minimize the number of space dimensions. The goal is to find N vectors ($coord_1, \ldots, coord_N$) $\in \mathbb{R}^N$ such that $\forall (i,j) \in N^2, \|coord_i - coord_j\| \approx D^{ind}_{DTW}(i,j)$.

As DTW is a mathematical distance, the MDS algorithm applied to each D_{DTW} is able to minimize the number of space dimensions to 1 (*i.e.* a vector of coordinates). Then we have as many vectors of coordinates as indicators.

Indicators' coordinates may be correlated among each others but the K-*means* algorithm uses a dimensional space of which the axes are orthogonal to each other. In order to apply this algorithm, we need to project the data on an orthogonal hyperplane of which the axes are two by two non-correlated.

The Principal Component Analysis (PCA) calculates the non-correlated axes which give a maximal dispersion of the data. It is then possible to reduce the number of dimensions avoiding redundant information by compressing them. Data are represented in a matrix of coordinates C with K random variables $\{ind_1, \ldots, ind_K\}$ containing N independent realizations. This matrix is standardized according to the center of gravity $(\overline{ind_1}, \ldots, \overline{ind_K})$ (with $\overline{ind}$ the arithmetic mean) and to the standard deviation σ of the random variables. It is then possible to calculate the correlation matrix: $\frac{1}{N} \cdot \widetilde{C}^T \cdot \widetilde{C}$. The PCA looks for the axis u which maximizes the variance of the data. To do so, it calculates a linear combination of the random variables in order to project the data on this axis: $\pi_u(C) = C \cdot u$. We keep the same number of axes K' for the projected indicators as for the indicators (K).

$$\widetilde{C} = \begin{bmatrix} \frac{ind_{1,1} - \overline{ind_1}}{\sigma(ind_1)} & \cdots & \frac{ind_{1,K} - \overline{ind_K}}{\sigma(ind_K)} \\ \vdots & \ddots & \vdots \\ \frac{ind_{N,1} - \overline{ind_1}}{\sigma(ind_1)} & \cdots & \frac{ind_{N,K} - \overline{ind_K}}{\sigma(ind_K)} \end{bmatrix} \tag{6}$$

Behavior Clusters. Finally, we apply on the PCA projected indicators the same K-*means* algorithm as the one applied on the questionnaire's scores in order to classify these data (normalized between 0 and 1). We thus obtain behavior clusters of main actors, as shown in Fig. 4.

4.3   Clusterings Comparison

Now that we have *annotations* clustering, *behaviors* clustering on main actors and *habits* clustering on participants, we want to compare the clusters composition between the *annotations* and the *behaviors*, and between the *habits* and the *behaviors*.

As we want to compare clusterings, we need a similarity measure between two clusterings $\mathbb{C}_1$ and $\mathbb{C}_2$. We use the *Adjusted Rand Index* (ARI) [11] - a well known index recommended in [20] - which is based on pair counting: a) N_{00}: the number of pairs that are in the same set in both clusterings (agreement); b) N_{11}: the number of pairs that are in different sets in both clusterings (agreement); and c) N_{01} and N_{10}: the number of pairs that are in the same set in one clustering and in different sets in the other (disagreement) and vice-versa. The *Rand Index* $RI \in [0, 1]$ is described in Eq. 7 [25]. The *Adjusted Rand Index* $ARI \in [-1, 1]$ is calculated using a contingency table $[n_{ij}]$ where n_{ij} is the number of agreements between instances i and j: $n_{ij} = |\mathbb{C}_1^i \cap \mathbb{C}_2^j|$ (see Eq. 8). It is a corrected-for-randomness version of the RI: Where the expected value of RI for two random clusterings is not constant, the expected value of ARI is 0.

$$RI(\mathbb{C}_1,\mathbb{C}_2) = \frac{N_{00} + N_{11}}{N_{00} + N_{11} + N_{01} + N_{10}} \tag{7}$$

$$ARI(\mathbb{C}_1,\mathbb{C}_2) = \frac{RI(\mathbb{C}_1,\mathbb{C}_2) - E\left[RI(\mathbb{C}_1,\mathbb{C}_2)\right]}{1 - E\left[RI(\mathbb{C}_1,\mathbb{C}_2)\right]} \tag{8}$$

where

$$E\left[RI(\mathbb{C}_1,\mathbb{C}_2)\right] = \left[\sum_i \binom{\sum_k n_{ik}}{2} \sum_j \binom{\sum_l n_{lj}}{2}\right] / \binom{n}{2}$$

5   Experimentation

The participants to our driving simulation experiment are regular drivers aged from 24 to 59 (44 % female). Our experiment is carried out on a device comprising a steering wheel, a set of pedals, a gearbox and 3 screens allowing sufficient lateral field of view (see Fig. 2). These screens are also used to integrate the rear-view mirror and the left-hand mirror. 23 participants used this device.

Firstly, the Driver Behavior Questionnaire is submitted before the simulation. Secondly, a first test without simulated traffic is performed for the participant to get accustomed to the functioning of the simulator and to the circuit. Then, the participant performs the scenario, this time in interaction with simulated vehicles. It should be noted that as the behavior of simulated vehicles is not scripted, situations differ more or less depending on the main actor behavior. The data are then recorded for the processing phase. A video is also made for the replay. Finally, another population of 6 participants fills the *annotations* questionnaire after viewing the video replay of the simulation in order to evaluate the adopted behaviors of the main actors (*i.e.* 23 participants and 14 agents).

One participant had simulator sickness but was able to finish the experiment, and one annotator had dizziness and ceased watching.

5.1   Results

We have compared the *habits* clustering, the annotations clustering, and the behaviors clustering. The composition of the clusterings is illustrated with three graphs. Agents are represented with rectangles and are named $a\#$. Participants are represented with ellipses and are named $s\#$. The main actors of one clustering are grouped together within rectangles containing the cluster number on the top. The others clustering's main actors are grouped together by color and the cluster number is written just bellow the main actors names.

***Habits and Annotations* Clusters.** We want to compare the *DBQ* scales and the summarized *DBQ* scales of our *annotations* questionnaire. The Fig. 5 shows the *habits* clustering (within rectangles) and the *annotations* clustering (grouped

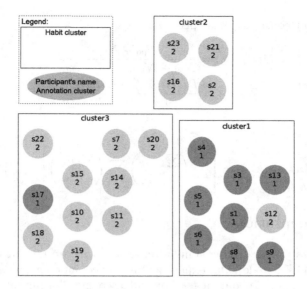

Fig. 5. Comparison of participants between *habits* clustering (within rectangles) and *annotations* clustering (grouped together by color) (Color figure online).

together by color), and their similarity. As the *habits* clustering from the *DBQ* questionnaire is only filled by simulation participants, we do not display the agents from the *annotations* clustering. The *habits* clustering contains 3 clusters which are close to the 2 clusters of the *annotations* clustering. *cluster1* contains nearly all participants of the cluster (1) (excepted *s*12). *cluster2* is composed of cluster (2) participants only. Also, *cluster3* is mainly composed of cluster (2) participants (excepted *s*17). The rand index is 0.71 and the adjusted rand index is 0.42. This means that our summarized *DBQ* scales in the *annotations* questionnaire is meaningful with regard to driver behavior *habits*.

Behavior Clusters and *Annotations* Clusters. With the behavior clustering on main actors, we are able to analyze how many human behaviors are reproduced by the agents, how many human behaviors are not adopted by the agents, and how many agent behaviors are not adopted by participants. We are also capable of making explicit those behaviors via the similarity with *annotations* clusters if relevant.

The Fig. 6 presents the behaviors clustering (within rectangles) and the *annotations*[1] clustering grouped together by color. The number of clusters is similar in both clusterings (3 behaviors versus 2 types of *annotations*). The rand index is 0.59 and the adjusted rand index is 0.17.

– *cluster1* contains one participant and nearly all the agents (excepted *a*5).
  Most of its main actors are annotated in the same way (*i.e.* in cluster (1)). So,

---

[1] except for the *human or simulated agent* question which is not related to the adopted behavior.

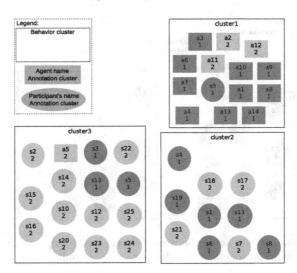

Fig. 6. Comparison of main actors between behavior clustering within rectangles and *annotations* clustering grouped together by color (Color figure online).

the main actors of the *cluster 1* adopted a similar driving behavior and were annotated in the same way, *i.e.*: the highest score on perceived control question and the lowest scores on the questions related to driving issues. Therefore, they are judged as careful drivers.

- *cluster2* is only composed of participants which are mixed between the two *annotations* clusters. These participants do not overtake the vehicle at low speed on the right lane. Some of them do not reach the disturbed traffic zone due to low speed at the start, others decide not to overtake the vehicle at low speed but rather to follow it. This behavior is characterized by a lower speed and no lane change.
- *cluster3* is mainly composed of participants (and the agent *a5*). Those participants are largely annotated in the same cluster (2), which has the lowest score on the perceived control question and the highest scores on the questions related to driving issues, meaning that they are judged as unsafe drivers.

Behavior Clusters and *Habits*. We have compared the drivers *habits* using the *DBQ* questionnaire with the adopted behavior (see Fig. 7). As the *habits* clustering from the *DBQ* questionnaire is only filled by simulation participants, we do not displayed the agents from the *annotations* clustering. The *Variance Ratio Criterion* selects 3 clusters. The rand index is low (0.48) as is the adjusted rand index (0.07). The clustering contains a singleton cluster and two other clusters, each containing a mixture of all the *DBQ* clusters, meaning that the behavior clustering does not correspond to the *habits* clustering. It validates the use of *annotations* by observers, which are closer to data clustering results.

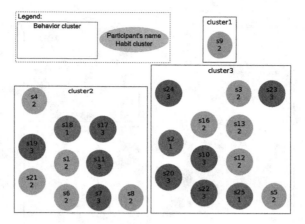

Fig. 7. Comparison of main actors between behavior clustering within rectangles and *habits* clustering grouped together by color (Color figure online).

5.2 Discussion

Firstly, there is no significant similarity between *habits* clustering and behaviors clustering. This might be due to the general approach of the *DBQ* questionnaire. The driver's *habits* may differ from the adopted behavior. The *DBQ* is filled by the driver itself. The introspection bias may be the reason of the differences. This is also an issue for us because we cannot apply it to the agent. The 3 scales dealing with the *risk* could be another problem: 8 % of the participants had some high scores on the *No risk* scale and the *Definite risk* scale but a low score one the *Possible risk* scale. Another problem is that the same type of participant in term of *DBQ* cluster can adopt different behaviors for the same situation, leading to different logs. Similarly, the same behavior can be adopted by different types of participants. This is an issue to analyze the similarity between the *habits* clustering and the behaviors clustering with the *ARI* measure. A solution could be to merge the subsets for which all main actors are also in the same cluster in the other clustering.

Secondly, we have a significant similarity between *annotations* and behavior clusterings, meaning that we are able to classify our logs data into high-level behavior clusters which are meaningful in term of driving *annotations*. Nevertheless the two clusterings are not identical with regard to the clusters composition nor with regard to the clusters number. This could be due to the few number of annotators, we are currently increasing this population. Furthermore, the behavior clustering is done on noisy indicators for human participants and on smooth indicators for agents. A solution might be to add a noise on the agents data or to smooth the participants data. This problem may come from the clustering algorithm which is a classic but basic one. In order to better take account of the data temporal aspect than by aggregating them, we have to test with temporal algorithms.

In the comparison of *annotations* and *behaviors*, one agent was in a mainly human composed cluster. Does this mean that it is able to simulate the majority of driver's behaviors of this cluster which is *cautious* ? If it is, we can then consider that this *cautious* behavior is an agents ability. To verify this assumption, we would need a specific test in which the parameter set that was used for the agent $a5$ is confronted to different situations, and compared with logs and *annotations* of cautious human drivers. Conversely, one participant was in the mainly agent composed cluster judged as dangerous for a majority of them. This requires further study to understand what was specific in this subject's driving behavior that was similar to the agents' behaviors. *cluster2* does not contain any agent, meaning that the agent's model is not able to reproduce this human driving behavior (*i.e.* this behavior is lacking in the agent's model). Another type of cluster - which does not appear in this experiment - is composed of agents only. In that case, we can consider - as no participant adopted this behavior - that the agents behavior is inaccurate (*i.e.* is an error) and should be investigated further.

6   Conclusion and Perspectives

This paper presents a method to study the agents' behavioral credibility through an experiment in a virtual environment. This validation is original in coupling a subjective analysis of the agents' behavioral credibility (via human sciences questionnaires and annotations) with an objective analysis of the agents' abilities. This analysis is based on behaviors clustering which allows us to obtain behaviors categories at a higher level than raw data. The method is generic for mixed simulation where agents and humans interact. When applied to a new domain, some of the tools have to be adapted, such as the choice of the behavior questionnaire which is domain-specific. The method is fully implemented, built on the *Weka* toolkit. The software shall be made available in the future.

Our validation method was applied to the road traffic simulation. This experiment showed that the methodology is usable for mixed and complex *VEs* and that it is possible to obtain high-level behaviors from the logs via our abstraction. A larger annotators population should provide more evidence of the method's robustness.

Several tracks for further work remain to explore. On the clustering part, the evaluation of multiple algorithms should enable to better assess their relevance. To do so, the use of the results of the comparison between *annotations* clusters and observed behavior clusters allows us to choose the most pertinent algorithm depending on the application. Another research open issue - as *annotation* are similar to behaviors whereas *habits* differ - is how the behaviors clustering evolve through multiple situations of a longer scenario, whether the participants clusters remain stable or change in number or composition.

References

1. Arthur, D., Vassilvitskii, S.: k-means++: The advantages of careful seeding. In: Proceedings of the Eighteenth Annual ACM-SIAM Symposium on Discrete Algorithms, pp. 1027–1035. Society for Industrial and Applied Mathematics (2007)
2. Berndt, D., Clifford, J.: Using dynamic time warping to find patterns in time series. KDD workshop. 10, 359–370 (1994)
3. Burkhardt, J.M., Bardy, B., Lourdeaux, D.: Immersion, réalisme et présence dans la conception et l'évaluation des environnements virtuels. Psychol. française 48(2), 35–42 (2003)
4. Caillou, P., Gil-Quijano, J.: Simanalyzer: Automated description of groups dynamics in agent-based simulations. In: Proceedings of the 11th International Conference on Autonomous Agents and Multiagent Systems, Vol. 3, pp. 1353–1354. International Foundation for Autonomous Agents and Multiagent Systems (2012)
5. Caliński, T., Harabasz, J.: A dendrite method for cluster analysis. Commun. Stat.-Theory Methods 3(1), 1–27 (1974)
6. Champion, A., Éspié, S., Auberlet, J.M.: Behavioral road traffic simulation with archisim. In: Summer Computer Simulation Conference, pp. 359–364. Society for Computer Simulation International (1998, 2001)
7. Champion, A., Zhang, M.Y., Auberlet, J.M., Espié, S.: Behavioral simulation: Towards high-density network traffic studies. In: ASCE (2002)
8. Drogoul, A., Corbara, B., Fresneau, D.: Manta New experimental results on the emergence of (artificial) ant societies. In: Gilbert, N., Conte, R. (eds.) Artificial Societies the Computer Simulation of Social Life, pp. 190–211. UCL Press, London (1995)
9. Fontaine, G.: The experience of a sense of presence in intercultural and international encounters. Presence: Teleoperators Virtual Environ. 1(4), 482–490 (1992)
10. Gonçalves, J., Rossetti, R.J.F.: Extending sumo to support tailored driving styles. In: 1st SUMO User Conference, DLR, Berlin Adlershof, Germany, vol. 21, pp. 205–211 (2013)
11. Hubert, L., Arabie, P.: Comparing partitions. J. classif. 2(1), 193–218 (1985)
12. Javeau, C.: L'enquête par questionnaire: manuel à l'usage du praticien. Editions de l'Université de Bruxelles (1978)
13. Kohonen, T.: The self-organizing map. Proc. IEEE 78(9), 1464–1480 (1990)
14. Leplat, J.: Simulation et simulateur: principes et usages. Regards sur l'activité en situation de travail: contribution à la psychologie ergonomique, pp. 157–181 (1997)
15. Lessiter, J., Freeman, J., Keogh, E., Davidoff, J.: A cross-media presence questionnaire: the itc-sense of presence inventory. Presence: Teleoperators Virtual Environ. 10(3), 282–297 (2001)
16. Lester, J.C., Converse, S.A., et al.: The persona effect: affective impact of animated pedagogical agents. In: Proceedings of the SIGCHI Conference on Human Factors in Computing Systems, pp. 359–366. ACM (1997)
17. Likert, R.: A technique for the measurement of attitudes. Arch. psychol. 22, 1–55 (1932)
18. Maes, P., Kozierok, R.: Learning interface agents. In: Proceedings of the National Conference on Artificial Intelligence, pp. 459–459. Wiley (1993)
19. McGreevy, M.W.: The presence of field geologists in mars-like terrain. Presence: Teleoperators and Virtual Environ. 1(4), 375–403 (1992)
20. Milligan, G.W., Cooper, M.C.: A study of the comparability of external criteria for hierarchical cluster analysis. Multivar. Behav. Res. 21(4), 441–458 (1986)

21. Patrick, J.: Training: Research and Practice. Academic Press, London (1992)
22. Pavlov, I.P., Anrep, G.V.: Conditioned reflexes. Dover Publications, New York (2003)
23. Pelleg, D., Moore, A., et al.: X-means: Extending k-means with efficient estimation of the number of clusters. In: Proceedings of the Seventeenth International Conference on Machine Learning, vol. 1, pp. 727–734. San Francisco (2000)
24. Premack, D., Woodruff, G., et al.: Does the chimpanzee have a theory of mind. Behav. Brain Sci. **1**(4), 515–526 (1978)
25. Rand, W.M.: Objective criteria for the evaluation of clustering methods. J. Am. Stat. Assoc. **66**(336), 846–850 (1971)
26. Reason, J., Manstead, A., Stradling, S., Baxter, J., Campbell, K.: Errors and violations on the roads: a real distinction? Ergonomics **33**(10–11), 1315–1332 (1990)
27. Salvador, S., Chan, P.: Toward accurate dynamic time warping in linear time and space. Intell. Data Anal. **11**(5), 561–580 (2007)
28. Serrano, E., Muñoz, A., Botia, J.: An approach to debug interactions in multi-agent system software tests. Inf. Sci. **205**, 38–57 (2012)
29. Stoffregen, T.A., Bardy, B.G., Smart, L.J., Pagulayan, R.J.: On the nature and evaluation of fidelity in virtual environments. In: Hettinger, L.J., Hass, M.W. (eds.) Virtual and Adaptive Environments Applications, Implications, and Human Performance Issues, pp. 111–128. Erlbaum, Mahwah (2003)
30. Witmer, B.G., Singer, M.J.: Measuring presence in virtual environments: a presence questionnaire. Presence **7**(3), 225–240 (1998)

Artificial Intelligence

Knowledge Gradient for Online Reinforcement Learning

Saba Yahyaa[✉] and Bernard Manderick

Computer Science Department, Vrije Universiteit Brussel,
Pleinlaan 2, 1050 Brussels, Belgium
{syahyaa,bmanderi}@vub.ac.be
https://ai.vub.ac.be

Abstract. The most interesting challenge for a reinforcement learning agent is to learn online in unknown large discrete, or continuous stochastic model. The agent has not only to trade-off between exploration and exploitation, but also has to find a good set of basis functions to approximate the value function. We extend offline kernel-based LSPI (or least squares policy iteration) to online learning. Online kernel-based LSPI combines feature of offline kernel-based LSPI and online LSPI. Online kernel-based LSPI uses knowledge gradient policy as an exploration policy to trade-off between exploration and exploitation, and the approximate linear dependency based kernel sparsification method to select basis functions automatically. We compare between online kernel-based LSPI and online LSPI on 5 discrete Markov decision problems, where online kernel-based LSPI outperforms online LSPI according to the optimal policy performance.

Keywords: Online reinforcement learning · Trade-off between exploration and exploitation · Knowledge gradient exploration policy · Value function approximation · (Kernel-based) least squares policy iteration · Approximate linear dependency kernel sparsification

1 Introduction

A Reinforcement Learning (RL) agent has to learn to make optimal sequential decisions while interacting with its environment. At each time step, the agent takes an action and as a result the environment transits from the current state to the next one while the agent receives feedback signal from the environment in the form of a scalar reward.

The mapping from states to actions that specifies which actions to take in states is called a policy π and the goal of the agent is to find the optimal policy π^*, i.e. the one that maximises the total expected discounted reward, as soon as possible. The state-action value function $Q^\pi(s, a)$ is defined as the total expected discounted reward obtained when the agent starts in state s, takes action a, and follows policy π thereafter. The optimal policy maximises these $Q^\pi(s, a)$ values.

When the agent's environment can be modelled as a Markov Decision Process (MDP) then the Bellman equations for the state-action value functions, one per

© Springer International Publishing Switzerland 2015
B. Duval et al. (Eds.): ICAART 2014; LNAI 8946, pp. 103–118, 2015.
DOI: 10.1007/978-3-319-25210-0_7

state-action pair, can be written down and can be solved by algorithms like policy iteration or value iteration [1]. We refer to Sect. 2.1 for more details.

When no such model is available, the Bellman equations cannot be written down. Instead, the agent has to rely only on information collected while interacting with its environment. At each time step, the information collected consists of the current state, the action taken in that state, the reward obtained and the next state of the environment. The agent can either learn *offline* when firstly a batch of past experience is collected and subsequently used and reused to learn the optimal policy π^*, or *online* when it tries to improve its behaviour at each time step based on the current information.

Fortunately, the optimal Q-values can still be determined using Q-learning [1] which represents the actions-value $Q^\pi(s, a)$ as a lookup table and uses the agent's experience to build the $Q^\pi(s, a)$. Unfortunately, when the state and/or the action spaces are large finite or continuous space, the agent faces a challenge called the curse of dimensionality, since the memory space needed to store all the Q-values grows exponentially in the number of states and actions. Computing all Q-values becomes infeasible. To handle this challenge, function approximation methods have been introduced to approximate the Q-values, e.g. [2] have proposed Least Squares Policy Iteration (LSPI) to find the optimal policy when no model of the environment is available. LSPI is an example of both approximate policy iteration and offline learning. LSPI approximates the Q-values using a linear combination of predefined basis functions. The used predefined basis functions have a large impact on the performance of LSPI in terms of the number of iterations that LSPI needs to converge to a policy, the probability that the converged policy is optimal, and the accuracy of the approximated Q-values.

To improve the accuracy of the approximated Q-values and to find a (near) optimal policy, [3] have proposed Kernel-Based LSPI (KBLSPI), an example of offline approximated policy iteration that uses Mercer kernels to approximate Q-values [4]. Moreover, kernel-based LSPI provides automatic feature selection by the kernel basis functions since it uses the approximate linear dependency sparsification method described in [5].

[6, 7] have adapted LSPI, which does offline learning, for online reinforcement learning and the result is called *online LSPI*. A good online learning algorithm must quickly produce acceptable performance rather than at the end of the learning process as is the case in offline learning. In order to obtain good performance, an online algorithm has to find a proper balance between exploitation, i.e. using the collected information in the best possible way, and exploration, i.e. testing out the available alternatives [1]. Several exploration policies are available for that purpose and one of the most popular ones is ϵ-greedy exploration that selects with probability $1 - \epsilon$ the action with the highest estimated Q-value and selects uniformly, randomly with probability ϵ one of the actions available in the current state. To get good performance, the parameter ϵ has to be tuned for each problem. To get rid of parameter tuning and to increase the performance of online LSPI, [8] have proposed using Knowledge Gradient (KG) policy [9] in the online-LSPI.

To improve the performance of online-LSPI and to get automatic feature selection, we propose online kernel-based LSPI and we use the knowledge gradient (KG) as an exploration policy. The rest of the paper is organised as follows: In Sect. 2 we present Markov decision processes, LSPI, the knowledge gradient policy for online learning, kernel-based LSPI and the approximate linear dependency test. While in Sect. 3, we present the knowledge gradient policy in online kernel-based LSPI. In Sect. 4 we give the domains used in our experiments and our results. We conclude in Sect. 5.

2   Preliminaries

In this section, we discuss Markov decision processes, online LSPI, the knowledge gradient exploration policy (KG), offline kernel-based LSPI (KBLSPI) and approximate linear dependency (ALD).

2.1   Markov Decision Process

A finite Markov decision process (MDP) is a 5-tuple (S, A, P, R, γ), where the state space S contains a finite number of states s and the action space A contains a finite number of actions a, the transition probabilities $P(s, a, s')$ give the conditional probabilities $p(s'|s, a)$ that the environment transits to state s' when the agent takes action a in state s, the reward distributions $R(s, a, s')$ give the expected immediate reward when the environment transits to state s' after taking action a in state s, and $\gamma \in [0, 1)$ is the discount factor that determines the present value of future rewards [1,10].

A deterministic policy $\pi : S \rightarrow A$ determines which action a the agent takes in each state s. For the MDPs considered, there is always a deterministic optimal policy and so we can restrict the search process to such policies [1,10]. By definition, the state-action value function $Q^\pi(s, a)$ for a policy π gives the expected total discounted reward $\mathbb{E}_\pi(\sum_{i=t}^{\infty} \gamma^t r_t)$ when the agent starts in state s, takes action a and follows policy π thereafter. The goal of the agent is to find the optimal policy π^*, i.e. the one that maximizes Q^π for every state s and action a: $\pi^*(s) = argmax_{a \in A} Q^*(s, a)$ where $Q^*(s, a) = max_\pi Q^\pi(s, a)$ is the optimal state-action value function. For the MDPs considered, the Bellman equations for the state-action value function Q^π are given by

$$Q^\pi(s, a) = R(s, a, s') + \gamma \sum_{s'} P(s, a, s') Q^\pi(s', a') \tag{1}$$

In Eq. 1, the sum is taken over all states s' that can be reached from state s when action a is taken, and the action a' taken in next state s' is determined by the policy π, i.e. $a' = \pi(s')$. If the MDP is completely known then algorithms such as value or policy iteration find the optimal policy π^*. Policy iteration starts with an initial policy π_0, e.g. randomly selected, and repeats the next two steps until no further improvement is found: (1) *policy evaluation* where the current policy π_i is evaluated using Bellman equation 1 to calculate the corresponding value function Q^{π_i}, and (2) *policy improvement* where this value function is used to find an improved new policy π_{i+1} that is greedy in the previous one, i.e. $\pi_{i+1} = argmax_{a \in A} Q^{\pi_i}(s, a)$ [1].

For finite MDPs, the action-value functions Q^{π} for a policy π can be represented by a lookup table of size $|S| \times |A|$, one entry per state-action pair. However, when the state and/or action spaces are large, this approach becomes computationally infeasible due to the curse of dimensionality and one has to rely on function approximation instead. Moreover, the agent does not know the transition probabilities $P(s, a, s')$ and the reward distributions $R(s, a, s')$. Therefore, it must rely on information collected while interacting with the environment to learn the optimal policy. The information collected is a trajectory of samples of the form (s_t, a_t, r_t, s_{t+1}) or $(s_t, a_t, r_t, s_{t+1}, a_{t+1})$, where s_t, a_t, r_t, s_{t+1}, and a_{t+1}, are the state, the action in the state, the reward, the next state, and the next action in the next state, respectively. To overcome these problems, least squares policy iteration (LSPI) uses such samples to approximate the Q^{π}-values [2].

More recently, [6] have adapted LSPI so that it can work online and [8] have used the knowledge gradient (KG) policy in this online LSPI. Since we are interested in the most challenging RL problem: online learning in a stochastic environment of which no model is available. Therefore, we are going to compare the performance of online-LSPI with the proposed algorithm using KG policy.

2.2   Least Squares Policy Iteration

LSPI approximates the action-value Q^{π} for a policy π in a linear way [2]:

$$\hat{Q}^{\pi}(s, a; w^{\pi}) = \sum_{i=1}^{n} \boldsymbol{\phi}_i(s, a) w_i^{\pi} \tag{2}$$

where n, $n << |S \times A|$, is the number of basis functions, the weights $(w_i^{\pi})_{i=1}^{n}$ are parameters to be learned for each policy π, and $\{\boldsymbol{\phi}_i(s, a)\}_{i=1}^{n}$ is the set of predefined basis function vectors. Let $\boldsymbol{\Phi}$ be the basis matrix of size $|S \times A| \times n$, where each row contains the values of all basis functions in one of the state-action pairs (s, a) and each column contains the values of one of the basis function vector ϕ_i in all state-action pairs and let $\boldsymbol{w}^{\pi}$ be a column weight vector of length n.

Given a trajectory of length L of samples $(s_t, a_t, r_t, s_{t+1})_{t=1}^{L}$. Offline-LSPI is an example of approximated policy iteration and repeats the following two steps until no further improvement in the policy is obtained: (1) *Approximate policy evaluation* that approximates the state-action value function Q^{π} of the current policy π, and (2) *Approximate policy improvement* that derives from the current estimated state-action value functions $\hat{Q}^{\pi}$ a better policy π', i.e. $\pi' = argmax_{a \in A} \hat{Q}^{\pi}(s, a)$

Using the least square error of the projected Bellman's equation, Eq. 1, the weight vector $\boldsymbol{w}^{\pi}$ can be approximated as follows [2]:

$$\hat{\boldsymbol{A}} \hat{\boldsymbol{w}}^{\pi} = \hat{\boldsymbol{b}} \tag{3}$$

where $\hat{A}$ is a matrix and $\hat{b}$ is a vector. Offline-LSPI updates the matrix $\hat{A}$ and the vector $\hat{b}$ from all available samples as follows:

$$\hat{A}_t = \hat{A}_{t-1} + \phi(s_t, a_t)[\phi(s_t, a_t) - \gamma\phi(s_{t+1}, \pi(s_{t+1}))]^T$$

$$\hat{b}_t = \hat{b}_{t-1} + \phi(s_t, a_t)r_t \tag{4}$$

where T is the transpose and r_t is the immediate reward that is obtained at time step t. After iterating over all collected samples, the weight vector $\hat{w}^\pi$ can be found. [6] have adapted offline-LSPI for online learning. The changes with respect to the offline algorithm are twofold: (1) online-LSPI updates the matrix $\hat{A}$ and the vector $\hat{b}$ after each time step t. Then, after every few samples K_θ obtained from the environment, online-LSPI estimates the weight vector $\hat{w}^\pi$ for the current policy π, computes the corresponding approximated $\hat{Q}$-function, and derives an improved new learned policy π', $\pi' = argmax_{a \in A}\hat{Q}^\pi(s, a)$. When $K_\theta = 1$, online-LSPI is called fully optimistic and when $K_\theta > 1$ is a small value, online-LSPI is called partially optimistic. (2) online-LSPI needs an exploration policy and [8] proposed using KG policy as an exploration policy instead of ϵ-greedy policy. [8] have shown that the performance of the online-LSPI is increased, e.g. the average frequency that the learned policy is converged to the optimal policy. Therefore, we are going to use KG policy in our algorithm and experiments.

2.3 KG Exploration Policy

Knowledge gradient KG [9] assumes that the rewards of each action a are drawn according to a probability distribution and it takes normal distributions $N(\mu_a, \sigma_a^2)$ with mean μ_a and standard deviation σ_a. At time step t, the current estimates, based on the rewards obtained so far, are denoted by $\hat{\mu}_a(t)$ and $\hat{\sigma}_a(t)$. The root-mean-square error (RMSE) of the estimated mean reward $\hat{\mu}_a(t)$, given $n_a(t)$ rewards resulting from action a, at time step t is given by $\bar{\hat{\sigma}}_a(t) = \hat{\sigma}_a/\sqrt{n_a}$. And, the change in the RMSE of action a at time step t is $\tilde{\sigma}_a(t)$, $\tilde{\sigma}_a(t) = \sqrt{(\hat{\sigma}_a(t)/\sqrt{n_a})^2 - (\hat{\sigma}_a(t+1)/\sqrt{n_a(t+1)})^2}$, where $\hat{\sigma}_a(t+1)$ is the estimated standard deviation of the mean of action a at time step $t+1$ and $n_a(t+1)$ is the number of times action a is selected at time step $t+1$. The KG is an index strategy that determines for each action a the index $V^{KG}(a)$ and selects the action with the 'highest' index. The index $V^{KG}(a)$[1] is calculated as follows:

$$V^{KG}(a) \hat{=} \tilde{\sigma}_a f\left(-|\frac{\hat{\mu}_a - \max_{a' \neq a} \hat{\mu}_{a'}}{\tilde{\sigma}_a}|\right). \tag{5}$$

In this equation, $f(x) \hat{=} \phi_{KG}(x) + x\Phi_{KG}(x)$ where $\phi_{KG}(x) = 1/\sqrt{2\pi} \exp(-x^2/2)$ is the density of the standard normal distribution and $\Phi_{KG}(x) = \int_{-\infty}^{x} \phi(x')dx'$ is its cumulative distribution. Then KG selects the next action according to:

$$a_{KG} \hat{=} \underset{a \in A}{argmax}\left(\hat{\mu}_a + \frac{\gamma}{1-\gamma}V^{KG}(a)\right) \tag{6}$$

---

[1] In order not to overload the notation we omit the time step t when it does not cause confusion.

1. Input: current state s_t;discount factor γ;the current state-action value estimate $\hat{Q}(s_t,a_i)$;the current RMSEs $\hat{\sigma}_q^2(s_t,a_i)$ for all actions a_i in state s_t

2. For each action $a_i \in A_{s_t}$

3. $\acute{Q}(s_t,a_i) \leftarrow \underset{a_j \in A_{s_t}, a_j \neq a_i}{\text{argmax}} \hat{Q}(s_t,a_j)$

4. End for

5. For each action $a_i \in A_{s_t}$

6. $\zeta_{a_i} \leftarrow -\text{abs}((\hat{Q}(s_t,a_i)-\acute{Q}(s_t,a_i))/\hat{\sigma}_q(s_t,a_i); f(\zeta_{a_i}) \leftarrow \zeta_{a_i}\Phi_{KG}(\zeta_{a_i}) + \phi_{KG}(\zeta_{a_i})$

7. $V^{KG}(a_i) \leftarrow \hat{Q}(s_t,a_i) + \frac{\gamma}{1-\gamma} \hat{\sigma}_q(s_t,a_i)f(\zeta_{a_i})$

8. End for

9. Output: $a_t \leftarrow \underset{a_i \in A}{\text{argmax}} V^{KG}(a_i)$

Fig. 1. Algorithm: (Knowledge gradient).

where the second term in the right hand side is the total discounted index of action a. For more details about KG policy, we refer to [11]. KG prefers those actions about which comparatively little is known. These actions are the ones whose RMSE (or spread) $\hat{\sigma}_a$ around the estimated mean reward $\hat{\mu}_a$ is large. Thus, KG prefers an action a over its alternatives if its confidence in the estimated mean reward $\hat{\mu}_a$ is low.

For discrete MDPs, [8] estimated the Q-values, $\hat{Q}(s_t,a_i)$ and the RMSE, $\hat{\sigma}_q^2$ of the estimated Q-value to calculate the index $V^{KG}(a_i)$ for each available action $a_i, a_i \in A_{s_t}$ in the current state s_t, where A_{s_t} is the set of actions in state s_t[2]. The pseudocode algorithm of the KG exploration policy is shown in Fig. 1. KG is easy to implement and does not have parameters to be tuned like ϵ-greedy or *softmax* action selection policies [1]. KG balances between exploration and exploitation by adding an exploration bonus to the estimated Q-values for each available action a_i in the current state s_t and this bonus depends on all estimated Q-values $\hat{Q}(s_t,a_i)$ and the RMSE of the estimated Q-value $\hat{\sigma}_q^2$ (steps: 2–8 in Fig. 1). The RMSE $\hat{\sigma}_q^2$ are updated according to [11].

2.4 Kernel-Based LSPI

Kernel-based LSPI [3] is a kernelized version of offline-LSPI. Kernel-based LSPI uses Mercer's kernels in the approximated policy evaluation and improvement [4]. Given a finite set of points, i.e. $\{z_1, z_2, \cdots, z_t\}$, where z_i is the state-action pair, with the corresponding set of basis functions, i.e. $\phi(z) : z \rightarrow \mathcal{R}$. Mercer theorem states the kernel function $\boldsymbol{K}$ is a positive definite matrix, i.e. $K(z_i, z_j) = <\phi(z_i), \phi(z_j)>$.

Given a trajectory of length L of samples and an initial policy π_0. Offline kernel-based LSPI (KBLSPI) uses the approximate linear dependency based

---

[2] Note that, [8] used a variant of the KG policy. [8] used the RMSE $\hat{\sigma}$ instead of the change in the RMSE $\tilde{\sigma}$ to calculate the KG index V^{KG} to get better trade-off between exploration and exploitation.

sparsification method to select a part of the data samples and consists a dictionary Dic elements set, i.e. $Dic = \{(s_i, a_i)\}_{i=1}^{|Dic|}$ with the corresponding kernel matrix $\boldsymbol{K}_{Dic}$ of size $|Dic \times Dic|$ [5]. Kernel-based LSPI repeats the following two steps: (1) *Approximate policy evaluation*, kernel-based LSPI approximates the weight vector $\hat{\boldsymbol{w}}^\pi$ for policy π, Eq. 3 from all available samples as follows:

$$\hat{\boldsymbol{A}}_t = \hat{\boldsymbol{A}}_{t-1} + \boldsymbol{k}((s_t, a_t), j)[\boldsymbol{k}((s_t, a_t), j) - \gamma\boldsymbol{k}((s_{t+1}, \pi(s_{t+1})), j)]^T$$

$$\hat{\boldsymbol{b}}_t = \hat{\boldsymbol{b}}_{t-1} + \boldsymbol{k}((s_t, a_t), j)r_t, \quad j \in Dic, \ j = 1, \cdots, |Dic| \tag{7}$$

where $\boldsymbol{k}(., .)$ is a kernel vector with the kernel function $k(., .)$ between two points (a state-action pair (s, a) and j, where j is the state-action pair z_j that is element in the dictionary Dic, i.e. $j \in \{z_1, z_2, \cdots, z_{|Dic|}\}$). The matrix $\hat{\boldsymbol{A}}$ should be initialized to a small multiple of the identity matrix to calculate the inverse of $\hat{\boldsymbol{A}}$ or using the pseudo inverse. After iterating for all the collected samples, the weight vector $\hat{\boldsymbol{w}}^\pi$ under policy π can be found and the approximated Q^π-values for policy π is the following linear combination:

$$\hat{Q}^\pi(s, a) = \hat{\boldsymbol{w}}^\pi \boldsymbol{k}((s, a), j), \ j \in Dic, \ j = 1, 2, \cdots, |Dic| \tag{8}$$

(2) *Approximate policy improvement*, KBLSPI derives a new learned policy which is the greedy one, i.e. $\pi'(s) = \text{argmax}_{a \in A} \hat{Q}^\pi(s, a)$. The above two steps are repeated until no change in the improved policy or a maximum number of iterations is reached.

2.5   Approximate Linear Dependency

Given a set of data samples D from a MDP, i.e. $D = \{z_1, \ldots, z_L\}$, where z_i is a state-action pair and the corresponding linear independent basis functions set $\boldsymbol{\Phi}$, i.e. $\boldsymbol{\Phi} = \{\boldsymbol{\phi}(z_1), \cdots, \boldsymbol{\phi}(z_L)\}$. Approximate linear dependency ALD method [5] over the data samples set D is to find a subset Dic, i.e. $Dic \subset D$ whose elements $\{z_i\}_{i=1}^{|Dic|}$ and the corresponding basis functions are stored in $\boldsymbol{\Phi}_{Dic}$, i.e. $\boldsymbol{\Phi}_{Dic} \subset \boldsymbol{\Phi}$.

The data dictionary Dic is initially empty, i.e. $Dic = \{\}$ and ALD is implemented by testing every basis function vector $\boldsymbol{\phi}$ in the basis function matrix $\boldsymbol{\Phi}$, one at time. If the basis function $\boldsymbol{\phi}(z_t)$ can not be approximated, within a predefined accuracy v, by the linear combination of the basis functions of the elements that stored in Dic_t, then the basis function $\boldsymbol{\phi}(z_t)$ will be added to $\boldsymbol{\Phi}_{Dic_t}$ and z_t will be added to Dic_t, otherwise z_t will not be added to Dic_t and $\boldsymbol{\phi}(z_t)$ will not be added to the dictionary basis functions $\boldsymbol{\Phi}_{Dic}$. As a result, after the ALD test, the dictionary basis functions $\boldsymbol{\Phi}_{Dic}$ can approximate all the basis functions of $\boldsymbol{\Phi}$.

At time step t, let $Dic_t = \{z_j\}_{j=1}^{|Dic_t|}$ and the corresponding basis functions are stored in the dictionary basis functions $\boldsymbol{\Phi}_{Dic_t}$, i.e. $\boldsymbol{\Phi}_{Dic_t} = \{\boldsymbol{\phi}(z_j)\}_{j=1}^{|Dic_t|}$ and z_t is a given state-action pair at time t. The ALD test on the basis function vector $\boldsymbol{\phi}(z_t)$ supposes that the basis functions are linearly dependent and uses least squares error to approximate $\boldsymbol{\phi}(z_t)$ by all the basis functions of the elements in the dictionary set Dic_t, for more detail we refer to [12]. The least squares error is:

$$error = \min_c || \sum_{j=1}^{|Dic_t|} c_j \phi(z_j) - \phi(z_t)||^2 < v \qquad (9)$$

$$error = k(z_t, z_t) - \boldsymbol{k}_{Dic_t}^T(z_t)\, \boldsymbol{c}_t, \quad \text{where} \qquad (10)$$

$$\boldsymbol{c}_t = \boldsymbol{K}_{Dic_t}^{-1} \boldsymbol{k}_{Dic_t}(z_t),$$

$$\boldsymbol{k}_{Dic_t}^T = [k(1, z_t), \cdots, k(j, z_t), \cdots, k(|Dic_t|, z_t)]$$

If the error is larger than predefined accuracy v, then z_t will be added to the dictionary elements, i.e. $Dic_{t+1} = Dic_t \cup \{z_t\}$, otherwise $Dic_{t+1} = Dic_t$. After testing all the elements in the data samples set D, the matrix $\boldsymbol{K}_{Dic}^{-1}$ can be computed, this is in the offline learning method. For online learning, the matrix $\boldsymbol{K}_{Dic}^{-1}$ can be updated at each time step [13].

At each time step t, if the error that results from testing the basis functions of z_t is smaller than v, then $Dic_{t+1} = Dic_t$ and $\boldsymbol{K}_{Dic_{t+1}}^{-1} = \boldsymbol{K}_{Dic_t}^{-1}$, otherwise $Dic_{t+1} = Dic_t \cup \{z_t\}$. The matrix $\boldsymbol{K}_{Dic_{t+1}}^{-1}$ is updated as follows:

$$\boldsymbol{K}_{Dic_{t+1}}^{-1} = \frac{1}{error_t} \begin{bmatrix} error_t\, \boldsymbol{K}_{Dic_t}^{-1} & -\boldsymbol{c}_t \\ -\boldsymbol{c}_t^T & 1 \end{bmatrix} \qquad (11)$$

3   Online Kernel-Based LSPI

Online kernel-based LSPI (KBLSPI) is a kernelised version of online-LSPI and the pseudocode is given in Fig. 2. Given the basis function matrix $\boldsymbol{\Phi}$, the initial learned policy π_0, the accuracy parameter v and the initial state s_1. At each time step t, online-KBLSPI uses the KG exploration policy, the algorithm in Fig. 1. to select the action a_t in the state s_t (step: 4) and observes the new state s_{t+1} and reward r_t. The action a_{t+1} in s_{t+1} is chosen by the learning policy π_t. The algorithm in Fig. 2 performs the ALD test, Sect. 2.5 on the basis functions of z_t and z_{t+1} to provide feature selection (steps: 7–14), where z_t is the state-action pair (s_t, a_t) at time step t and z_{t+1} is the state-action pair (s_{t+1}, a_{t+1}) at time step $t + 1$. If the basis functions of a given state-action pair, i.e. z_t and z_{t+1} can not approximated by the basis functions of the elements that stored in the dictionary Dic_t, then the given state-action pair will be added to the dictionary, the inverse kernel matrix $\boldsymbol{K}^{-1}$ will be updated, the number of columns and rows of the matrix $\hat{\boldsymbol{A}}$ will be increased and the number of dimensions of the vector $\hat{\boldsymbol{b}}$ will be increased (step: 11). Otherwise, the given state-action pair will not be added to the dictionary (step: 12). Then, online-KBLSPI updates the matrix $\hat{\boldsymbol{A}}$ and the vector $\hat{\boldsymbol{b}}$ (steps: 15–16). After few samples K_θ obtained from the environment, online-KBLSPI estimates the weight vector $\hat{\boldsymbol{w}}^{\pi_t}$ under the current policy π_t (step: 18) and approximates the corresponding state-action

1. Input: $|S|$;$|A|$;discount factor γ;accuracy v;set of basis functions $\boldsymbol{\Phi} = \{\boldsymbol{\phi}_1, \cdots, \boldsymbol{\phi}_n\}$;initial learned policy π_0;horizon of an experiment L; policy improvement interval K_θ;reward $r \sim N(\mu_a, \sigma_a^2)$;initial state s_1.

2. Intialize: $\hat{A} \leftarrow 0$;$\hat{b} \leftarrow 0$;s_t;$Dic_t = \{\ \}$;
   $K_{|SA| \times |SA|} = <\boldsymbol{\Phi}^T, \boldsymbol{\Phi}>$;$K_{Dic_t}^{-1} = []$;$\hat{Q}_{|SA|} \leftarrow 0$

3. For time step $t = 1, \cdots, L$
4. Select an action $a_t \leftarrow KG$
5. for the current s_t, and a_t;observe: s_{t+1};r_t;$a_{t+1} \leftarrow \pi_t(s_{t+1})$
6. $z_t \leftarrow (s_t) * |A| + a_t$, $z_{t+1} \leftarrow (s_{t+1}) * |A| + a_{t+1}$
7. For $z_i \in \{z_t, z_{t+1}\}$
8. $\boldsymbol{k}^T(., z_i) = [k(1, z_i), \cdots, k(j, z_i), \cdots, k(|Dic_t|, z_i)]$;$\boldsymbol{c}(z_i) = K_{Dic_t}^{-1} * \boldsymbol{k}(., z_i)$
9. $error(z_i) = \boldsymbol{k}(z_i, z_i) - \boldsymbol{k}^T(., z_i) * \boldsymbol{c}(z_i)$
10. If error $error(z_i) > v$
11. $Dic_{t+1} \leftarrow Dic_t \cup z_i$;$K_{Dic_{t+1}}^{-1} \leftarrow \frac{1}{error(z_i)} \begin{pmatrix} error(z_i)K_{Dic_t}^{-1} & -\boldsymbol{c}(z_i) \\ -\boldsymbol{c}(z_i)^T & 1 \end{pmatrix}$;

$$\hat{A}_t \leftarrow \begin{pmatrix} \hat{A}_t & 0 \\ 0 & 0 \end{pmatrix}; \hat{b}_t \leftarrow \begin{pmatrix} \hat{b}_t \\ 0 \end{pmatrix}$$

12. Else $Dic_{t+1} \leftarrow Dic_t$;$K_{Dic_{t+1}}^{-1} \leftarrow K_{Dic_t}^{-1}$
13. End if
14. End for
15. $\hat{A}_{t+1} \leftarrow \hat{A}_t + \boldsymbol{k}(., z_t)[\boldsymbol{k}(., z_t) - \gamma\boldsymbol{k}(., z_{t+1})]^T$
16. $\hat{b}_{t+1} \leftarrow \hat{b}_t + \boldsymbol{k}(., z_t)r_t$, $\boldsymbol{k}(., z_t) = [k(1, z_t), \cdots, k(j, z_t), \cdots, k(|Dic_{t+1}|, z_t)]^T$
17. If $t = (l+1)K_\theta$ then
18. $\hat{w}_l^{\pi_t} \leftarrow \hat{A}_{t+1}^{-1}\hat{b}_{t+1}$
19. for the state-action pair $z = z_1, z_2, \cdots, z_{|SA|}$
   $\boldsymbol{k}(., z) = [k(1, z), \cdots, k(j, z), \cdots, k(|Dic_{t+1}|, z)]^T$;$\hat{Q}_l^{\pi_t}(z) = \hat{w}_l^{\pi_t, T} * \boldsymbol{k}(., z)$
   end
20. $\pi_{t+1} \leftarrow \text{argmax}_a \hat{Q}_l^{\pi_t}(s, a)\ \forall_{s \in} S$;$\pi_t \leftarrow \pi_{t+1}$;$l \leftarrow l + 1$
21. End if
22. $s_t \leftarrow s_{t+1}$
23. End for

24. Output: At each time step t, note down:
   the reward r_t and the learned policy π_t

Fig. 2. Algorithm: (Online-KBLSPI).

value function $\hat{Q}^{\pi_t}$ (step: 19), i.e. *approximate policy evaluation*. Then, online-KBLSPI derives an improved new learned policy π_{t+1} which is a greedy one (step: 20), i.e. *approximated policy improvement*. This procedure is repeated until the end of playing L time steps which is the horizon of an experiment.

4 Experiments

In this section, we describe the test domain, the experimental setup and the experiments where we compare online-LSPI and online-KBLSPI using KG policy. All experiments are implemented in MATLAB.

4.1 Test Domain/Experimental Setup

The *test domain* consists of 5 MDPs as shown in Fig. 3, each with discount factor $\gamma = 0.9$. The first three domains are the 4-, 20-, and 50- chain. The 4-, and 20-domain are also used in [2,3] and the 50-chain is used in [2]. In general, the x-open chain which is originally studied in [14] consists of a sequence of x states, labeled from s_1 to s_x. In each state, the agent has 2 actions, either *GoRight* (R) or *GoLeft* (L). The actions succeed with probability 0.9 changing the state in the intended direction and fail with probability 0.1 changing the state in the opposite direction. The reward structure can vary such as the agent gets reward for visiting the middle states or the end states. For the 4-chain problem, the agent is rewarded 1 in the middle states, i.e. s_2 and s_3, and 0 at the edge states, i.e. s_1 and s_4. The optimal policy is R in states s_1 and s_2 and L in states s_3 and s_4. [14] used a policy iteration method to solve the 4-chain and showed that the resulting suboptimal policies oscillate between R R R R and L L L L. The reason is because of the limited approximation abilities of basis functions in policy evaluation. For the 20-chain, the agent is rewarded 1 in states s_1 and s_{20}, and 0 elsewhere. The optimal policy is L from states s_1 through s_{10} and R from states s_{11} through s_{20}. And, for the 50-chain, The agent gets reward 1 in states s_{10} and s_{41} and 0 elsewhere. The optimal policy is R from state s_1 through state s_{10} and from state s_{26} through state s_{40}, and L from state s_{11} through state s_{25} and from state s_{41} through state s_{50} [2]. The fourth and fifth MDPs, the $grid_1$ and $grid_2$ worlds, are used in [1]. The agent has 4 actions *Go Up, Down, Left* and *Right* and for each of them it transits to the intended state with probability 0.7 and fails with probability 0.1 changing the state to the one of other directions. The agent gets reward 1 if it reaches the goal state, -1 if it hits the wall, and 0 elsewhere.

The *experimental setup* is as follows: For each of the 5 MDPs, we compared online-LSPI and online-KBLSPI using knowledge gradient KG policy as an exploration policy. For number of experiments $EXPs$ equals 1000 for the chain domains, 100 for the $grid_1$ domain and 50 for the $grid_2$ domain, each one with length L. The performance measures are: (1) the average frequency at each time step, i.e. at each time step t for each experiment, we computed the probability that the learned policy (step: 19) in Algorithm 2 reached to the optimal policy, then we took the average of $EXPs$ experiments to give us the average frequency at each time step. (2) the average cumulative frequency at each time step, i.e. the cumulative average frequency at each time step t. [15] used the 50-chain domain with length of trajectories L equals 5000, therefore, we used the same horizon. For other MDP domains we adapted the length of trajectories L according to the number of states, i.e. as the number of states is increased, L will be increased. For instance, L is set to 18800 for the grid world.

Knowledge gradient KG policy, needs estimated standard deviation and estimated mean for each state-action pair. Therefore, we assume that the reward has a normal distribution. For example, for the 50-chain problem, the agent is rewarded 1 if it goes to state 10, therefore, we set the reward in s_{10} to $N(\mu_1, \sigma_a^2)$, where $\mu_1 = 1$. And, the agent is rewarded 0 if it goes to s_1, therefore, we set the

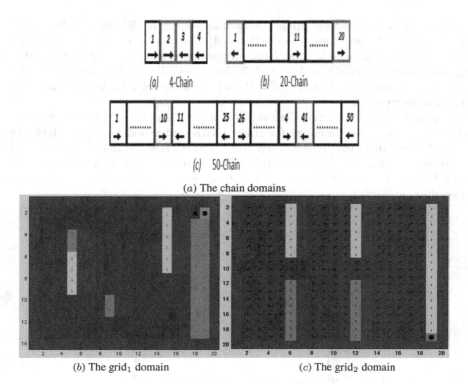

(a) The chain domains

(b) The grid₁ domain   (c) The grid₂ domain

Fig. 3. Subfigure (a) is the chain domains, in the red cells, the agent gets rewards. Subfigure (b) is the grid₁ with 280 states and 188 accessible states. Subfigure (c) is the grid₂ with 400 states and 294 accessible states. The arrows show the optimal actions in each state.

reward to $N(\mu_2, \sigma_a^2)$, where $\mu_2 = 0$. The standard deviation σ_a of the reward is set to a fixed and equal value for each action, i.e. $\sigma_a = 0.01, 0.1, 1$. Moreover, KG exploration policy is a full optimistic policy, therefore, we set the policy improvement interval K_θ to 1. For each run, the initial state s_1 was selected uniformly at random from the state space S. We used the pseudo-inverse when the matrix $\hat{A}$ is non-invertible [15].

For online KBLSPI, we define a kernel function K on state-action pairs, i.e. $K : |SA| \times |SA| \to \mathcal{R}$, we composed K into a state kernel K_s, i.e. $K_s : |S| \times |S| \to \mathcal{R}$ and an action kernel K_a, i.e. $K_a : |A| \times |A| \to \mathcal{R}$ as [13]. Therefore, the kernel function K is $K = K_s \otimes K_a$ where $\otimes$ is the Kronecker product. K is a kernel matrix because K_s and K_a are kernel matrices, we refer to [16] for more details. The kernel state matrix K_s is a Gaussian kernel, i.e. $k(s, s') = \exp^{-||s-s'||^2/(2\sigma_{ks}^2)}$ where σ_{ks} is the standard deviation of the kernel state function, s is the state at time t and s' is the state at time $t + 1$. And, the action kernel is a Gaussian kernel, i.e. $k(a, a') = \exp^{-||a-a'||^2/(2\sigma_{ka}^2)}$ where σ_{ka} is the standard deviation of the kernel action function, a is the action at time t and a' is the action at time

$t + 1$. s and s', and a and a' are normalized as [3], e.g. for 50-chain with number of states $|S| = 50$ and number of actions $|A| = 2$, $s, s' \in \{1/|S|, \cdots, 50/|S|\}$ and $a, a' \in \{0.5, 1\}$. The standard deviation of the kernel state σ_{ks} and kernel action σ_{ka} are tuned empirically and set to 0.55 for the chain domains and 2.25 for the grid world domains (grid$_1$ and grid$_2$). We set the accuracy v in the approximated kernel basis to 0.0001.

For online-LSPI, we used Gaussian basis functions $\phi_s = \exp^{-||s-c_i||^2/(2\sigma_\Phi^2)}$ where ϕ_s is the basis functions for state s with center nodes $(c_i)_{i=1}^n$ which are set with equal distance between each other, and σ_Φ is the standard deviation of the basis functions which is set to 0.55. The number of basis functions n equals 3 for 4-chain, 5 for 20-chain, and 10 for 50-chain as [2] and 40 for the grid$_1$ and grid$_2$ domains as [17].

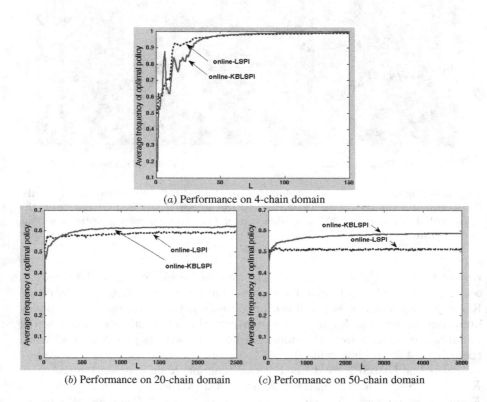

(a) Performance on 4-chain domain

(b) Performance on 20-chain domain   (c) Performance on 50-chain domain

Fig. 4. Performance of the average frequency by the KG policy in online-LSPI in blue and KG in online-KBLSPI in red. Subfigure (a) shows the performance on the 4-chain using standard deviation of reward $\sigma_a = 0.01$. Subfigure (b) shows the performance on the 20-chain using standard deviation of reward $\sigma_a = 1$. Subfigure (c) shows the performance on the 50-chain using $\sigma_a = 0.1$.

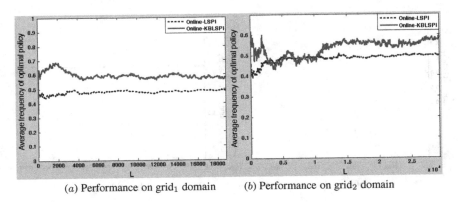

(a) Performance on grid$_1$ domain   (b) Performance on grid$_2$ domain

Fig. 5. Performance of the average frequency by the KG policy in online-LSPI in blue and KG in online-KBLSPI in red. Subfigure (a) shows the performance on the grid$_1$ domain using standard deviation of reward $\sigma_a = 0.01$. Subfigure (b) shows the performance on the grid$_2$ domain using standard deviation of reward $\sigma_a = 1$.

4.2 Experimental Results

Using different values of the standard deviation of reward, i.e. $\sigma_a = 0.01, 0.1$ and 1. We compared the performance of online LSPI and KBLSPI using knowledge gradient policy KG as an exploration policy. The experimental results on the chain domains, i.e. 4-, 20-, and 50-chain show that the online-KBLSPI outperforms the online-LSPI according to the average frequency of optimal policy and cumulative average frequency of optimal policy performances for all values of the standard deviation of reward σ_a i.e. $\sigma_a = 0.01, 0.1$ and 1. Figure 4 gives the average frequency of optimal policy performance. The x-axis is the horizon of each experiment. The y-axis is the average frequency of optimal policy performance which is the average of $EXPs$ experiments. Figure 4 shows how the performance of the learned policy is increased by using online-KBLSPI on the 4-chain, 20-chain and 50-chain.

The experimental results on the grid$_1$ domain show that the online-KBLSPI outperforms the online-LSPI according to the average frequency and cumulative average frequency of optimal policy performances for all values of the standard deviation of reward σ_a i.e. $\sigma_a = 0.01, 0.1$ and 1. And, the experimental results on the grid$_2$ domain show that the online-KBLSPI performs better than the online-LSPI for standard deviation of reward equals 1. Figure 5 shows how the performance of the learned policy is increased by using online-KBLSPI on the grid$_1$ and grid$_2$ domains.

The results clearly show that online-KBLSPI usually converges faster than online-LSPI to the (near) optimal policies, i.e. the performance of the online KBLSPI is increased. Although, the performance of the online-LSPI is better in the beginning and this is because the online-LSPI uses its all basis functions, while online-KBLSPI incrementally constructs its basis functions by the kernel sparsification method.

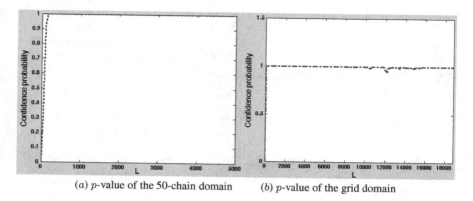

(a) p-value of the 50-chain domain   (b) p-value of the grid domain

Fig. 6. The confidence probability p-value that the average frequency of optimal policy performance of online-KBLSPI performs better than online-LSPI. Subfigure (a) shows the p-value of the 50-chain using standard deviation of reward $\sigma_a = 0.1$. Subfigure (b) shows the p-value of the grid domain using standard deviation of reward $\sigma_a = 0.1$.

4.3 Statistical Methodology

We used a statistical hypothesis test, i.e. student's t-test with significance level $\alpha_{st} = 0.05$ to compare the performance of the average frequency of optimal policy that results from the online-LSPI and the online-KBLSPI at each time step t. The null hypothesis H_0 is the online-KBLSPI average frequency performance (AF_{KBLSPI}) larger than the online-LSPI average frequency performance (AF_{LSPI}) and the alternative hypothesis H_a is AF_{KBLSPI} less or equal AF_{LSPI}. We wanted to calculate the confidence in the null hypothesis, therefore, we computed the confidence probability p-value at each time step t. The p-value is the probability that the null hypothesis is correct. The confidence probability converges to 1 for all standard deviation of reward, i.e. $\sigma_a = 0.01, 0.1$, and 1 and for all domains, i.e. the 4-, 20-, and 50-chain domains and the grid world domains. Figure 6 shows how the p-value converges to 1 using the 50-chain, and the grid$_1$ domain with standard deviation of reward $\sigma_a = 0.1$. The x-axis gives the time steps (the length of trajectories). The y-axis gives the confidence probability, i.e. p-value. Figure 6 shows the confidence in the online kernel-based LSPI performance is very high, where the p-value converged quickly to 1.

5 Conclusions and Future Work

We presented Markov decision process which is a mathematical model for the reinforcement learning. We introduced online and offline least squares policy iteration (LSPI) that find the optimal policy in an unknown environment. We presented knowledge gradient KG policy to be used as an exploration policy in the online learning algorithm. We introduced offline kernel-based LSPI (KBLSPI). We also introduced approximate linear dependency (ALD) method to select feature automatically and get rid of tuning empirically the center nodes. We proposed online-KBLSPI which uses KG exploration policy and ALD method. Finally, we compared online-KBLSPI and online-LSPI and concluded that the average frequency of optimal policy performance is improved by using online-KBLSPI. Future work must compare the performance of online-LSPI and online-KBLSPI using other types of basis functions, e.g. the hybrid shortest path basis functions [18], must compare the performance using continuous MDP domain, e.g. Interval pendulum and must prove a convergence analysis of the online-KBLSPI.

References

1. Sutton, R.S., Barto, A.G.: Reinforcement Learning: An Introduction. MIT Press, Cambridge (1998)
2. Lagoudakis, M.G., Parr, R.: Model-free least squares policy iteration. Technical report, Computer Science Department, Duke University, Durham, North Carolina, United States (2003)
3. Xu, X., Hu, D., Lu, X.: Kernel-based least squares policy iteration for reinforcement learning. J. IEEE Trans. Neural Netw. **18**(4), 973–992 (2007)
4. Vapnik, V.: The Grid: Statistical Learning Theory. Wiley, New York (1998)
5. Engel, Y., Mannor, S., Meir, R.: The kernel recursive least-squares algorithm. J. IEEE Trans. Signal Process. **52**(8), 2275–2285 (2004)
6. Buşoniu, L., Ernst, D., De Schutter, B., Babuška, R.: Online least-squares policy iteration for reinforcement learning control. In: American Control Conference (ACC), pp. 486–491 (2010)
7. Li, L., Littman, M.L., Mansley, C.R.: Online exploration in least-squares policy iteration. J. Comput. (2008)
8. Yahyaa, S., Manderick, B.: Knowledge gradient exploration in online least squares policy iteration. In: 5th International Conference on Agents and Artificial Intelligence (ICAART). Springer-Verlag, Barcelona (2013)
9. Ryzhov, I.O., Powell, W.B., Frazier, P.I.: The knowledge-gradient policy for a general class of online learning problems. J. Oper. Res. **60**, 180–195 (2011)
10. Puterman, M.L.: Markov Decision Processes: Discrete Stochastic Dynamic Programming. Wiley, New York (1994)
11. Powell, W.B., Ryzhov, I.O.: Optimal Learning. Willey, Canada (2012)
12. Engel, Y., Meir, R.: Algorithms and representations for reinforcement learning. Technical report, Computer Science Department, Senate of the Hebrew (2005)
13. Engel, Y., Mannor, S., Meir, R.: Reinforcement learning with Gaussian processes. In: 22nd International Conference on Machine learning (ICML), New York (2005)
14. Koller, D., Parr, R.: Policy iteration for factored MDPs. In: 16th Annual Conference on Uncertainty in Artificial Intelligence American Control Conference (UAI 2000) (2000)
15. Mahadevan, S.: Representation Discovery Using Harmonic Analysis. Morgan and Claypool Publishers, San Rafael (2008)
16. Scholkopf, B., Smola, A.J.: Learning with Kernels: Support Vector Machines, Regularization, Optimization, and Beyond. MIT Press, Cambridge (2002)

17. Sugiyama, M., Hachiya, H., Towell, C., Vijayakumar, S.: Geodesic Gaussian kernels for value function approximation. J. Auton. Robots **25**(3), 287–304 (2008)
18. Yahyaa, S., Manderick, B.: Shortest path Gaussian kernels for state action graphs: an empirical study. In: 24th Benelux Conference on Artificial Intelligence (BNAIC). Maastricht University, The Netherlands (2012)

Statistical Response Method and Learning Data Acquisition using Gamified Crowdsourcing for a Non-task-oriented Dialogue Agent

Michimasa Inaba[1]([✉]), Naoyuki Iwata[2], Fujio Toriumi[3], Takatsugu Hirayama[2],
Yu Enokibori[2], Kenichi Takahashi[1], and Kenji Mase[2]

[1] Graduate School of Information Sciences, Hiroshima City University,
3-4-1 Ozukahigashi, Hiroshima, Asaminami-ku, Japan
{inaba,takahashi}@hiroshima-cu.ac.jp
[2] Graduate School of Information Sciences,
Nagoya University, Furo-cho, Nagoya, Chikusa-ku, Japan
{iwata,enokibori}@cmc.ss.is.nagoya-u.ac.jp,
{hirayama,mase}@is.nagoya-u.ac.jp
[3] School of Engineering, University of Tokyo, 7-3-1 Hongo, Tokyo, Bunkyo-ku, Japan
tori@sys.t.u-tokyo.ac.jp

Abstract. This paper presents a proposal of a construction method for non-task-oriented dialogue agents (chatbots) that are based on the statistical response method. The method prepares candidate utterances in advance. From the data, it learns which utterances are suitable for context. Therefore, a dialogue agent constructed using our method automatically selects a suitable utterance depending on a context from candidate utterances. This paper also proposes a low-cost quality-assured method of learning data acquisition for the proposed response method. The method uses crowdsourcing and brings game mechanics to data acquisition.

Results of an experiment using learning data obtained using the proposed data acquisition method demonstrate that the appropriate utterance is selected with high accuracy.

Keywords: Dialogue agent · Chatbots · Crowdsourcing · Gamification.

1 Introduction

Computerized dialogue agents are used increasingly in different areas. Not only are task-oriented dialogue agents demanded recently. Non-task-oriented dialogue agents are also in high demand. task-oriented dialogue agents are made for accomplishing particular tasks such as reservation services [1] and supplying specific information [2]. Non-task-oriented dialogue agents have no such tasks. They are used for personal communication tasks such as chat.

Non-task-oriented dialogues play a crucially important role in human society because they are important tools for building relationships with humans. Robots

© Springer International Publishing Switzerland 2015
B. Duval et al. (Eds.): ICAART 2014; LNAI 8946, pp. 119–136, 2015.
DOI: 10.1007/978-3-319-25210-0_8

and other anthropomorphic agents are expected to participate in our daily lives to an increasing degree. Therefore, much more investigation must be done to ascertain how non-task-oriented dialogue agents can be designed to develop good relationships with people.

Even a task-oriented dialogue agent can accomplish a task more efficiently using non-task-oriented dialogue. For example, a study by Bickmore showed that when dialogue agents that supported the buying and selling of real estate initially chatted about subjects not pertinent to real estate such as the weather, people were much more motivated to buy real estate through them than through agents who did not engage in non-task-oriented dialogues [3].

As described herein, we propose a method for producing non-task-oriented dialogue agents that are based on the statistical response method. In fact, two major response methods exist for non-task-oriented dialogue agents.

The first of these are rule-based methods that produce utterances in accordance with response rules. Well-known dialogue agents using this strategy are ELIZA [4] and A.L.I.C.E. [5]. Mitsuku [6], which is the Loebner prize contest[1] (non-task-oriented dialogue agent competition) winner of 2013, and Eugene Goostman [7], which is the first agent to pass a Turing test (in 2014), also used this strategy. The problem of this strategy is their substantial cost because the rules are developed manually.

The other is an example-based method [8,9]. A dialogue agent employing this strategy searches a large database of dialogue with user input (a user's utterance) using cosine similarity. It selects an utterance that follows the most similar one as a response. The problem is how it acquires a large quantity of good quality dialogues efficiently because the performance depends on the quality of dialogues in the database. A response method based on statistical machine translation [10] has been proposed. It treats the last user's utterance as an input sentence and translates it into the response utterance. This method, categorized as the example-based method, has the same problem.

A problem shared by both methods is that they cannot use a context (sequence of utterances). They only use a user's last given utterance. According to the rule-based method, necessary rules and costs of creating them are increased extremely. Regarding example-based methods, if it searches the database by a context, in many cases, then it cannot find a similar one because of the diversity of non-task-oriented dialogues. When the method cannot find a similar one, it has no choice but to use random selection.

Our statistical response method belongs to the category of example-based methods because it uses dialogue data. However, our method, which uses no cosine similarity but statistical machine learning, is able to use contexts. Our method prepares candidate utterances in advance. It learns which utterances are suitable for context by the data. Therefore, a dialogue agent that is constructed using our method automatically selects a suitable utterance from candidate utterances, depending on a context. Additionally, we provide a low-cost, quality-assured method of learning data acquisition using crowdsourcing and gamification.

---

[1] http://www.loebner.net/Prizef/loebner-prize.html.

2 Statistical Response Method

2.1 Selection of Candidate Utterances

As described herein, we define "utterance" as a one-time statement and "context" as an ordered set consisting of utterances from the conversation's beginning to a specific point in time. Here, "an utterance is suitable to a context" means that the utterance is a "humanly" and semantically appropriate answer to the context.

First, we define a state of a point of time in a dialogue as context $c = \{u_1, u_2, \ldots, u_l\}$. Each $u_i(i = 1, 2, \ldots, l)$ denotes an utterance appearing in the context. Also, each l denotes a number of utterances. Herein, u_1 is the last utterance; u_l is the first utterance in context c. As a matter of practical convenience, u_0 represents a response utterance to context c.

Second, we define a candidate utterance set $A_c = \{a_1^c, a_2^c, \ldots, a_{|A_c|}^c\}$, where $a_i^c(i = 1, 2, \ldots, |A_c|)$ denotes a candidate utterance. Here, A_c includes suitable and unsuitable utterances to context c. $|A_c|$ represents a number of candidate utterances. We define the correct utterance set $R_c = \{r_1^c, r_2^c, \ldots, r_{|R_c|}^c\} \subseteq A_c$, where $r_i^c(i = 1, 2, \ldots, |R_c|)$ denotes a correct utterance. $|R_c|$ represents a number of correct utterances to context c. The utterance selection means acquisition of a correct utterance set R_c from a candidate utterance set A_c, given a context c. Here, we assume that c and A_c fulfill the following requirements.

Table 1. Example of context c.

No	Speaker	Utterance
u_4	Agent	Are you good at English?
u_3	Human	No I am not. I love Japanese
u_2	Agent	It is said that experience is important to enhance English communication skills
u_1	Human	I see! It might be a good idea to travel abroad during summer vacation
u_0	(Agent)	(Select an utterance from Table 2)

Table 2. Example of candidate utterance set A_c.

No.	Utterance
a_1^c	Are you good at English?
$a_2^c[r_1^c]$	Where do you want to go?
a_3^c	I think dogs are trustworthy and intelligent animals.
$a_4^c[r_2^c]$	That would be nice.
	...
$a_{20}^c[r_3^c]$	Travel can make a person richer inside.
	...
a_{130}^c	A link exists between mental and physical health.

– A_c can be generated by any context c.
– A_c has at least one correct utterance r_i^c for context c.

Tables 1 and 2 present examples of c, A_c, and R_c (R_c is shown by the darker-shaded area). In this example, a suitable utterance to context c shown in Table 1 is selected from the candidate utterance set A_c shown in Table 2. The utterance should be selected from the correct utterance set $R_c = \{a_2^c, a_4^c, a_{20}^c\}$ in this case.

2.2   Ranking Candidate Utterances

We describe our study's method of selecting candidate utterances automatically.

By specifically processing c and $a_i^c (\in A_c)$, we generate n-dimensional feature vector $\Phi(c, a_i^c) = (x_1(c, a_i^c), x_2(c, a_i^c), \ldots, x_n(c, a_i^c))$ that represents relations between the context and the candidate utterance. Each $x_j(c, a_i)(j = 1, 2, \ldots, n)$ is a feature representing a binary value. For instance, when particularly addressing the last utterance u_1 in c and a_i, a feature $x_j(s, a_i^c)$ is represented if it contains a specific word, a word class, or a combination of the two.

We then defined f as a function that will return the evaluated value of a feature vector. In the following passages, we describe expression of the feature vector $\Phi(c, a_i)$ as Φ_i. Here it can be denoted using a linear function, which can be expressed as

$$f(\Phi_j) = \sum_{j=1}^{n} w_j x_j(c, a_i^c). \tag{1}$$

Therein, w_j is a parameter representing the weight of $x_j(c, a_i^c)$.

Using the evaluation function presented above, optimum utterance $\hat{a}$ in response to the context is obtainable by the following equation:

$$\hat{a} = \underset{a \in A_c}{\operatorname{argmax}}\ f(\Phi_j). \tag{2}$$

Therefore, the candidate utterances can be ranked by sorting the value from the evaluation function above.

To estimate the parameter $w = (w_1, w_2, \ldots, w_n)$ in evaluation function f, we use a learning to rank method ListNet [11] algorithm.

2.3   Parameter Estimation

ListNet is constructed for ranking objects. It uses probability distributions for representing the ranking lists of objects. Then, minimizing the distance between learning data and distribution of the model, it learns suitable parameters for ranking.

We define $Y_c = \{y_1^c, y_2^c, \ldots, y_{|A_c|}^c\}$ as a score list to candidate utterance set $A_c = \{a_1^c, a_2^c, \ldots, a_{|A_c|}^c\}$. Each score $y_i^c(i = 1, 2, \ldots, |A_c|)$ denotes the score of a candidate utterance a_i^c with respect to context c. Score y_i^c represents the degree of correctness a_i^c to c and is an evaluated value assigned by humans. For instance, if a candidate utterance is a suitable response to a context, then the score is 10. Alternatively, if an utterance is unsuitable, then the score is 1.

ListNet parameter estimation algorithm uses pairs of $X_c = (\Phi_1, \Phi_2, \ldots, \Phi_{|A_c|})$, which is a list of feature vectors, and Y_c as learning data that are ranked correctly.

Here, for the list of feature vectors X_c, using function f, we obtain a list of scores $Z_c = \big(f(\Phi_1), f(\Phi_2), \ldots, f(\Phi_{|A_c|})\big)$. The objective of learning is to minimize the difference between Y_c and Z_c in terms of their rankings. We then formalize it using a loss function.

$$G(C) = \sum_{\forall c \in C} L(Y_c, Z_c) \tag{3}$$

Therein, C means all contexts in learning data and L is a loss function. In ListNet, the cross entropy is used as a loss function.

$$H(p, q) = - \sum_x p(x) \log q(x) \tag{4}$$

In that equation, $p(x)$ and $q(x)$ are probability distributions. When $p(x)$ and $q(x)$ show an equal distribution, cross entropy $H(p, q)$ takes a minimum value.

Therefore, the lists of scores Y_c and Z_c are converted into probability distributions using the Plackett–Luce model [12,13]. The distribution of Y_c using the Plackett–Luce model for the top rank utterance is expressed as shown below.

$$P_{Y_c}(\Phi_i) = \frac{\mathrm{pow}(\alpha, y_i^c)}{\sum_{j=1}^{|A_c|} \mathrm{pow}(\alpha, y_j^c)} \tag{5}$$

In that equation, $\mathrm{pow}(\alpha, y)$ denotes α to the power of y. This equation represents the probability distribution of a candidate utterance being ranked at the top. The higher the candidate utterance score is, the higher the probability becomes. For instance, when a list of feature vectors X_c is (Φ_1, Φ_2, Φ_3) and a list of scores Y_c is $(1, 0, 3)$, then the probability of Φ_3 being ranked at the top is calculated as $(\alpha = 2)$.

$$
\begin{aligned}
P_{Y_c}(\Phi_3) &= \frac{\mathrm{pow}(2, y_3^c)}{\mathrm{pow}(2, y_1^c) + \mathrm{pow}(2, y_2^c) + \mathrm{pow}(2, y_3^c)} \\
&= \frac{\mathrm{pow}(2, 3)}{\mathrm{pow}(2, 1) + \mathrm{pow}(2, 0) + \mathrm{pow}(2, 3)} \\
&= 0.727
\end{aligned} \tag{6}
$$

Instead, the probability of Φ_1 being ranked at the top is 0.182 and Φ_2 is 0.091, which is the lowest.

Similarly, the distribution of Z_c can be converted into a probability distribution as shown below.

$$P_{Z_c}(\Phi_i) = \frac{\mathrm{pow}\big(\alpha, f(\Phi_i)\big)}{\sum_{j=1}^{|A_c|} \mathrm{pow}\big(\alpha, f(\Phi_j)\big)} \tag{7}$$

Using Eq. (4), (5) and (7), then the loss function $L(Y_c, Z_c)$ becomes the following.

$$L(Y_c, Z_c) = -\sum_{i=1}^{|A_c|} P_{Y_c}(\Phi_i) \log\big(P_{Z_c(f)}(\Phi_i)\big) \tag{8}$$

Optimum parameter w is obtainable using gradient descent. The gradient of the loss function $L(Y_c, Z_c(f_w))$ is expressed as follows.

$$\Delta w = \frac{\partial L\big(Y_c, Z_c(f_w)\big)}{\partial w}$$

$$= \log(\alpha) \left(-\sum_{i=1}^{|A_c|} P_{Y_c}(\Phi_i) \frac{\partial f_w(\Phi_i)}{\partial w} \right.$$

$$\left. + \frac{1}{\sum_{i=1}^{|A_c|} \mathrm{pow}\big(\alpha, f_w(\Phi_i)\big)} \sum_{i=1}^{|A_c|} \mathrm{pow}\big(\alpha, f_w(\Phi_i)\big) \frac{\partial f_w(\Phi_i)}{\partial w} \right) \tag{9}$$

Using Δw, we can obtain optimum w according to the following procedure.

1. Calculate a list of scores $Z_c(f_w)$.
2. Calculate Δw using function (9).
3. Update w. ($w = w - \eta \times \Delta w$)
   here, η indicates the learning rate.
4. Repeat 1. - 3.

The proposed method calculates each score of the candidate utterance using obtained function $f_w(\Phi_i)$ and ranks them based on these scores.

3   Data Acquisition

3.1   Crowdsourcing

The statistical response method requires various learning data. In this section, we propose a low-cost quality-assured method of learning data acquisition. The proposed method uses crowdsourcing and opens a website to acquire the learning data. It enables the gathering of participants irrespective of gender, age, and religion.

The crowdsourcing website shows a context c and candidate utterances a_i^c. The candidate utterances are shown as options and participants select a suitable one to a given context. The number of options is s ($s = 5$ in the opened website). The displayed candidate utterances a_i^c are selected randomly from the candidate utterance set A_c. Context c and candidate utterance set A_c are prepared in advance.

Examples of a context c and options are presented in Fig. 1. In this figure, speakers A and B are talking about communication skills in the context and an option "Having strong communication skills is paramount if you want to be

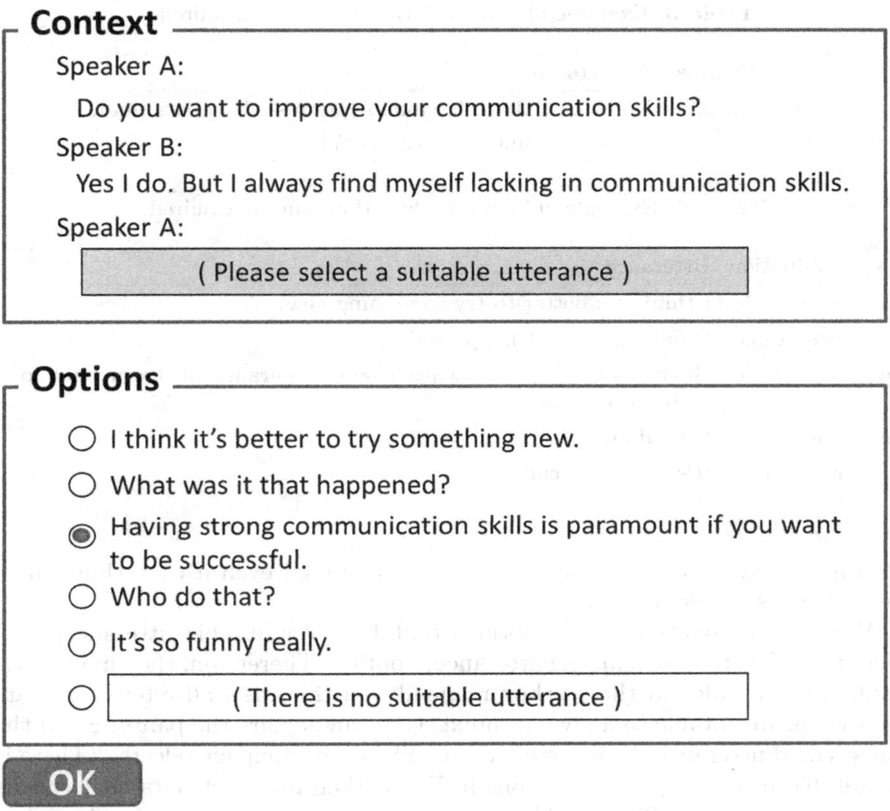

Context

Speaker A:

Do you want to improve your communication skills?

Speaker B:

Yes I do. But I always find myself lacking in communication skills.

Speaker A:

(Please select a suitable utterance)

Options

○ I think it's better to try something new.

○ What was it that happened?

◉ Having strong communication skills is paramount if you want to be successful.

○ Who do that?

○ It's so funny really.

○ (There is no suitable utterance)

OK

Fig. 1. Context and candidate utterances on our crowdsourcing website.

successful" is selected. When an option is selected as suitable and a participant clicks the "OK" button, the evaluation of the displayed dialogue data is finished. The website then displays the next one. In our method, a selection of suitable candidate utterances is a one trial and participants work through N trials in all. The bottom option in Fig. 1 indicating "(there is no suitable utterance)" is a free description area. If a participant thinks that there is no suitable utterance, then the participant selects it and must write a suitable utterance manually in the textbox. This way, we can acquire new candidate utterances.

Participants select the most suitable utterance from given options. We can acquire the pair of c and the selected utterance as correct data. When a participant selects "Having strong communication skills is paramount if you want to be successful" from options in Fig. 1, the correct data that can be acquired are shown in Table 3. The selected utterance "Having strong communication skills is paramount if you want to be successful" is a correct datum that can be acquired. The proposed method uses no non-selected options as learning data because

Table 3. Example of correct data that can be acquired.

No	Evaluation	Utterance
$a_3^c[r_1^c]$	suitable	Having strong communication skills is paramount if you want to be successful

Table 4. Example of incorrect data that can be acquired.

No	Evaluation	Utterance
a_1^c	unsuitable	I think it's better to try something new.
a_2^c	unsuitable	What was it that happened?
a_3^c	unsuitable	Having strong communication skills is paramount if you want to be successful
a_4^c	unsuitable	Who do that?
a_5^c	unsuitable	It's so funny really.

participants can select only one utterance in one trial even if they think there are multiple suitable alternatives.

When participants consider options that have no suitable utterance, they select the "(There is no suitable utterance)" option. Thereupon, they must write a suitable utterance in the textbox manually. In this case, all alternatives are regarded as unsuitable to a given context. One can acquire the pair of c and the non-selected utterances as incorrect data. When a participant selects "(There is no suitable utterance)" from options in Fig. 1, then incorrect data that can be acquired are shown in Table 4. All displayed utterances are acquired as incorrect data.

Additionally, when participants select "(There is no suitable utterance)" , the website shows another s options only once for the same c to acquire more correct data. Updated options are candidate utterances that are selected as suitable one or more times by other participants. The shortfalls are selected randomly from A_c if the number of such utterances is less than s.

Each pair of context and an utterance are evaluated repeatedly by participants. Here, we define the number of evaluations to a candidate utterance a_i^c as $n_{a_i^c}$, the number of evaluations for which a_i^c is suitable as $n_{\hat{a_i^c}}$, and the number of evaluations that are unsuitable as $n_{\acute{a_i^c}}$. The eventual evaluation to a_i^c is determined according to $n_{\hat{a_i^c}}$ and $n_{\acute{a_i^c}}$. Adequate evaluations can be acquired from consideration of multiple participants' results.

3.2 Confidence Estimation

When we use crowdsourcing, quality control of acquired data is necessary [14,15]. We offer this to the general public. Therefore, quality gaps are unavoidable. Slapdash workers are not uncommon. Solving the problem of collecting high-quality data is an extremely important challenge in using crowdsourcing.

Table 5. Example of degree of confidence estimation.

| Participant | Evaluated data | | | | | Confidence |
	1	2	3	4	5	
A	Suitable	Suitable	Suitable	Suitable	Suitable	5
B	Suitable	Suitable	Unsuitable	Unsuitable	Suitable	3
C	Unsuitable	Unsuitable	Unsuitable	Suitable	Unsuitable	1

Suitable: Suitable utterance is selected
Unsuitable: Unsuitable utterance is selected

An example of data quality control is that the crowdsourcing system itself has a mechanism to control quality. This mechanism estimates the worker quality by the number of tasks that the worker has done, by quality evaluations of tasks from clients, and so on. Alternatively, the quality can be estimated by comparing other workers' results. Dekel et al. estimated by assumption that a worker whose results are well matched with the correct answer creates high-quality data [16]. This method constructed a prediction model of a correct answer and eliminated workers whose results depart radically from prediction of the model.

In this study, we prepare several evaluated data to estimate the degree of confidence of participants p. Participants undertake trials of two kinds at our website: trials to estimate p and to acquire the data. Letting N_p and N_a respectively denote the number of trials to estimate the degree of confidence p and to acquire the data, then the total number of trials N is

$$N = N_p + N_a. \tag{10}$$

For adequate estimation, the website does not tell participants of the existence of such trials.

The degree of confidence p is estimated by whether or not the participant can select obviously suitable utterance r_i^c to a given context c from options. The obviously suitable utterances r_i^c satisfy the following two conditions.

– Candidate utterances that are judged as suitable by multiple evaluators.
– All other options, except for r_i^c, that are judged as unsuitable by multiple evaluators.

The degree of confidence p is a the number of times that a participant selects r_i^c correctly in the number of trials N_p. Consequently, the range of p is $0 \leq p \leq N_p$.

Table 5 shows the degree of confidence estimation result under the condition of $N_p = 5$. In this example, because the participant A selected suitable utterances in all trials, the degree of confidence is the highest 5. However, participant C selected suitable utterances only once; the degree of confidence is 1. As described in this paper, because participants whose degree of confidence is lower than threshold T_p have the potential to be careless, we do not use acquired data that were evaluated by them.

3.3   Gamification

An important consideration with crowdsourcing is the rewards given to participants. If we set high rewards, then we can gather many participants and acquire many data. To construct a better non-task-oriented dialogue agent that can accommodate topics of many kinds, it is desirable to acquire new data continuously. Although the agent requires much new data, setting high rewards increases the cost of construction and the participation of non-serious users who do not address the task properly.

In this study, we bring game mechanics to data acquisition to gather participants with no rewards. When using game mechanics, participants come to enjoy the task as game play. Such a method brings game mechanics to accomplish an objective called "gamification" [17,18].

3.4   Gamified Data Acquisition Environment

We opened a website "The diagnosis game of dialogue skills"[2] (Japanese text only) as a gamified crowdsourcing data acquisition environment.

At this site, participants work through 10 ($N = 10$) times trials. It finally shows a score for dialogue skills. The score increases to 100 points. By scoring the selection results, we expect participants' repetitive challenges to produce a higher score.

The score is sum of following scores of two kinds. Selections of suitable utterance and similarity of other participants' selection make one's score high.

- **Scores Based on the Degree of Confidence (Maximum 50 Points)**
  Using the degree of confidence p and number of trials for confidence estimation N_p, this score is calculated as follows.

$$p * 50/N_p \tag{11}$$

  When $N_p = 5$ (same as our website) and $p = 4$, this score becomes $4 \times 50 \div 5 = 40$ points.
- **Scores Based on Other Participants' Selections (Maximum 50 Points)** Using $n_{a_i^c}$ (number of evaluation to a candidate utterance a_i^c) and $n_{\hat{a}_i^c}$ (number of evaluation that a_i^c is suitable) , The rate of the selected option of total $n_{a_i^c}$ times evaluations is $n_{\hat{a}_i^c}/n_{a_i^c}$. Using this and the number of total trials N, the desired score is calculated as follows.

$$\sum \frac{n_{\hat{a}_i^c}}{n_{a_i^c}} * 50/N_a \tag{12}$$

  For example, when $N_a = 5$, the rates of the selected option in each trial are $0.7, 0.4, 0.6, 0.6, 0.9$ in a sequence, the score becomes $(0.7 + 0.4 + 0.6 + 0.6 + 0.9) \times 50 \div 5 = 32$ points.

---

[2] http://beta.cm.info.hiroshima-cu.ac.jp/DialogCheck/.

Additionally, the website records all scores by participants and shows a graph of score distribution for comparison with other participants. Figure 2 portrays an example of a game result, a calculated score and a graph of the score distribution. In this graph, participants' scores are separated by 10 points and are classified into 11 bars. In Fig. 2, 70–79 points are the largest number of participants with subsequent 60–69 points and 80–69 points. The white bar shows the position of a current participant's score.

Scoring the results of selection and comparison with those of the other participants stimulates participants' retrial motivation, by which they want to obtain a higher score. Additionally, by posting the score on SNS or micro blogs by themselves, we expect advertising effects for other people (the website has a tweet button to tweet their score easily).

3.5   Candidate Utterance Acquisition

If participants think there is no suitable utterance, then the participant selects the "(there is no suitable utterance)" option and must write a suitable utterance manually in our website. This handwritten utterance is considered as a lack utterance in a candidate set A_c. However such utterances might include inappropriate ones created in a sloppy manner. To resolve this problem, we evaluate acquired utterances by the degree of confidence p. The website rejects the

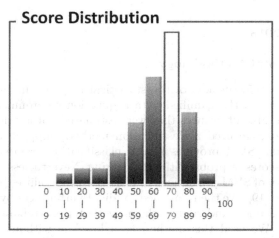

Fig. 2. Diagnosis game result.

utterances created by participants whose degree of confidence p is less than the threshold T_p. Additionally, by including a new utterance in options of a trial to other participants, the website conducts a third person evaluation. It is judged as an appropriate candidate utterance if other multiple participants evaluate it as suitable to a given context.

Another problem of acquired utterances is semantic overlap. For example "Today is sunny" and "The weather is sunny today" are different expressions, but they are semantically similar. The efficiency of the learning data acquisition becomes worse because evaluations to utterances are dispersed if a number of similar candidate utterances exist. Therefore, we delete similar utterances using the cosine similarity.

Cosine similarity is represented using vector x, y as follows.

$$\frac{x \bullet y}{|x| * |y|} \tag{13}$$

Therein, $x \bullet y$ denotes the inner product of x and y. If two vectors are similar, then the similarity becomes closer to 1. If different, then the similarity becomes closer to 0. In our case, when we regard an utterance as a word vector, we can calculate the similarity of two utterances a_1 and a_2 as follows.

$$\frac{\text{number of common words between } a_1 \text{ and } a_2}{\sqrt{\text{number of words in } a_1} * \sqrt{\text{number of words in } a_2}} \tag{14}$$

Using the example given above, let a_1 and a_2 be "Today is sunny" and "The weather is sunny today". Common words between a_1 and a_2 are three words, "today", "is", and "sunny". The number of words of a_1 is 3 and a_2 is 5. Therefore, the cosine similarity between a_1 and a_2 becomes $3 \div (\sqrt{3} \times \sqrt{5}) = 0.77$. One of the utterances is removed if the similarity between two utterances is higher than threshold T_c.

4   Experiments

4.1   Experimental Methodology

To underscore the effectiveness of the statistical response method that learns data acquired through the gamified data acquisition environment, we checked the ranking of suitable utterances that were estimated automatically.

For comparison, we used a classification method, support vector machine (SVM). In general, SVM provides binary classification results and no direct means to obtain scores or probabilities for ranking. Nevertheless, Piatt proposed the transformation of SVM predictions to posterior probabilities by passing them through a sigmoid [19]. We then classified candidate utterances by SVM, selected correctly classified ones, and ranked them by posterior probabilities using the sigmoid method. We used this method as a baseline without the use of the learning to rank method.

4.2   Features

To rank the utterances, we converted pairs of a context and a candidate utterance into a feature vector. We used features of 11 types to represent relations between a context and an utterance. Here, we describe one of these, the noun feature, as the most basic one.

In the noun feature, we use a combination of a noun in a context and an utterance. Using this feature, we expect that a candidate utterance that includes words related to words in a context ranks higher. We only use u_1, u_2, u_3, and u_4 in a context for this feature because it is often the case that semantic relations between old utterances in a context and suitable candidate utterances are few. The usage range of utterances in a context differs according to the type of feature. In the noun feature, whether a particular noun pair exists between utterances represents a binary feature value. We use noun pairs that appear three or more times in learning data.

Table 6 presents an example. The upper table presents an example of the context and candidate utterances. The lower shows part of a feature vector generated from them. As the table shows, we distinguish noun pairs by the number of utterances in the context. For instance, the vector value of "u_2 : travel & u_0 : overseas" is 1 because u_2 includes the word "travel" and u_0 includes "overseas". Similarly, the vector value of "u_1 : summer & u_0 : trip" is 0 because u_0 includes "trip" but u_1 does not include "summer".

The features should be designed to represent various aspects of relations between contexts and utterances such as sentence structures, discourse structures, semantics, and topics.

Table 6. Feature vector generation (noun feature).

No.	Speaker	Utterance
u_4	Agent	Enjoy this season fully because it's long-awaited summer vacation.
u_3	Human	Yes, I will.
u_2	Agent	Do you plan to travel?
u_1	Human	No. However, I would like to go.
u_0	(Agent)	Why don't you go on a trip overseas?

Noun pair	Vector value
u_1 : travel & u_0 : Europe	0
u_1 : summer & u_0 : trip	0
u_2 : part-timer & u_0 : overseas	0
u_2 : travel & u_0 : trip	1
u_2 : travel & u_0 : overseas	1
u_3 : friends & u_0 : trip	0
u_4 : summer & u_0 : overseas	1
u_4 : vacation & u_0 : trip	1

Table 7. Data acquisition result.

Number of participants	460
Number of evaluated contexts	320
Number of evaluated utterances	4694
Average of the confidence p	4.215

4.3   Data Set

Candidate Utterances. We made 980 utterances by hand for crowdsourcing and the experiment. The topics of utterances were selected to interest as many people as possible: healthcare, marriage, travel, sport, etc. We also produced versatile utterances such as "I think so." and "It's wonderful!".

Learning Data. To acquire the data, we opened the gamified website for crowd-sourcing. Table 7 presents data acquisition results.

We used 4520 evaluated utterances for which confidence p is $p > 3.0$ for the experiment.

Additionally, we used other data produced by 50 part-time participants intended to compensate for data deficiency. The methods of producing data are nearly identical, with the exception of using the game mechanics. Results show that we obtained 239,897 evaluated utterances to 14,900 contexts. We used these data all together as learning data.

The scores of utterances are given depending on the evaluation. If an utterance is suitable to a context, then the score is 30. The score is 1 if unsuitable. The values of the score are decided on an empirical basis.

Test Data. We prepared 500 contexts as test data. The ranked utterances using the proposed method and SVM were evaluated manually. Each utterance was evaluated by three evaluators who judged whether each utterance was semantically suitable or unsuitable to the context. The eventual judgment was decided by majority. Therefore, when two evaluators judge an utterance as suitable and one evaluator judged it as unsuitable, the utterance was determined to be suitable.

4.4   Results

Figure 3 shows the experiment result and 95 % confidence intervals obtained using the proposed method and SVM.

The x-axis represents the rank of the first appearance of a suitable utterance. The y-axis shows the cumulative frequency. In other words, the figure shows the rate of the contexts that include at least one appropriate utterance within each rank.

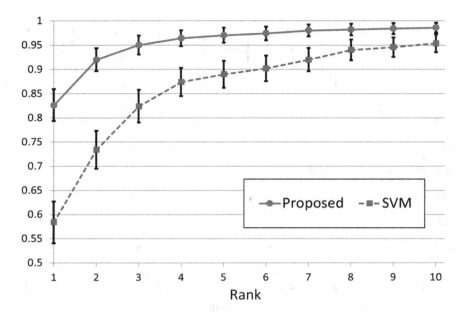

Fig. 3. Rate of appropriate candidate utterance.

In the figure, the proposed method ranked a suitable utterance on the top at 82.6 %, within the top 3 at 95.0 %, and the top 10 at 98.6 %. However, SVM was ranked on the top at 58.4 %, within the top 3 at 82.4 %, and at the top 10 at 95.4 %. As shown in the result, the proposed method outperformed SVM overall. The discussion above demonstrates that the proposed method is effective for the selection of candidate utterances.

When we implement the proposed method to dialogue agents, the rate of reply to a suitable utterance (82.6 %) is inadequate for smooth communication. The set of candidate utterances has at least one correct utterance for each context (test data). This might not always be the case. The rate might drop when the agent actually talks to humans. However, the proposed method produced rankings within the top 3 at over 90 % to use new effective features. To improve the ranking algorithm, apparently it is possible to improve the statistical response method performance further.

4.5   Discussion

A great benefit of the proposed method is that it can use contexts for responses. To demonstrate that effectiveness, we created feature vectors using the last user's utterance (u_1) only and conducted an experiment.

Figure 4 portrays the results. The rate of the top 1 was 69.2 %, 13.4 % lower. All results in the figure are lower by at least 1.6 % than that using contexts (Fig. 3). This result is the natural result because a context has more hints than an utterance for selecting a suitable utterance.

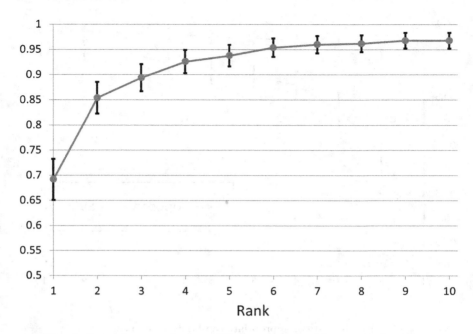

Fig. 4. Rate of appropriate candidate utterance without use of contexts.

However, as described at the beginning of this paper, existing response methods cannot use contexts for response generation. Various problems persist because of such information loss. For instance, a dialogue agent broaches a topic that was discussed previously or makes contradictory comments to what it had said before. In fact, this experimentally obtained result indicates that using the last utterance in addition to contexts is necessary to realize superior non-task-oriented dialogue agents. Therefore, in terms of the availability of contexts, the effectiveness of the statistical response method was clarified.

5   Conclusions

As described in this paper, we proposed a statistical response method that automatically ranks previously prepared candidate utterances in order of suitability to the context by the application of a machine learning algorithm. Non-task-oriented dialogue agents that applied the method use the top utterance from the ranking result for carrying out their dialogues. To collect learning data for ranking, we used crowdsourcing and gamification. We opened a gamified crowdsourcing website and collected learning data through it. Thereby, we achieved low-cost and continuous learning data acquisition. To prove the performance of the proposed method, we checked the ranked utterances to contexts and conclude that the method is effective because a suitable utterance is ranked at the top at 82.6 % and within the top 10 at 98.6 %.

The non-task-oriented dialogue agents are fundamentally evaluated manually, which requires a tremendous amount of time and effort. Using the proposed gamified crowdsourcing platform, we can evaluate the performance of non-task-oriented dialogue agents at low cost. We prepare several types of agents for evaluation. Each agent generates a response to the given context. The platform shows the context and the generated responses to participants in the same way as our website does. The responses generated by a high-performance agent should be selected more than others.

The candidate utterances are created manually. Future work includes automatic candidate utterance generation. Our crowdsourcing website has a function that collects new utterances. However, these utterances present some problems such as spelling errors and phraseology because they are written by users in free description format. We must fix them to use the new utterances. As an alternative utterance generation method, using microblog data is promising. Using microblog data, it can be expected to generate a new utterances set that includes numerous or newest topics.

We also intend to improve the feature vector. It is important to devise new effective features because the performance of our method depends heavily on the features. The features used in the experiment (not illustrated in detail here) did not deeply consider the semantics of context and utterances. Realizing appropriate responses requires semantical features. We are now deliberating about such features.

References

1. Zue, V., Seneff, S., Polifroni, J., Phillips, M., Pao, C., Goodine, D., Goddeau, D., Glass, J.: Pegasus: a spoken dialogue interface for on-line air travel planning. Speech Commun. 15(3–4), 331–340 (1994)
2. Chu-Carroll, J., Nickerson, J.S.: Evaluating automatic dialogue strategy adaptation for a spoken dialogue system. In: Proceedings of the 1st North American chapter of the Association for Computational Linguistics conference, pp. 202–209 (2000)
3. Bickmore, T., Cassell, J.: Relational agents: a model and implementation of building user trust. In: Proceedings of the SIGCHI conference on Human factors in computing systems, pp. 396–403 (2001)
4. Weizenbaum, J.: ELIZA-a computer program for the study of natural language communication between man and machine. Commun. ACM 9(1), 36–45 (1966)
5. Wallace, R.S.: The anatomy of alice, pp. 181–210. Parsing the Turing, Test (2009)
6. Worswick, S.: Mitsuku Chatbot (2013). http://www.mitsuku.com/
7. Veselov, V., Demchenko, E., Ulasen, S.: Eugene Goostman (2014). http://www.princetonai.com/
8. Murao, H., Kawaguchi, N., Matsubara, S., Yamaguchi, Y., Inagaki, Y.: Example-based spoken dialogue system using woz system log. In: SIGdial Workshop on Discourse and Dialogue, pp. 140–148 (2003)
9. Banchs, R.E., Li, H.: Iris: a chat-oriented dialogue system based on the vector space model. In: Proceedings of the ACL 2012 System Demonstrations, pp. 37–42. Association for Computational Linguistics (2012)

10. Ritter, A., Cherry, C., Dolan, W.B.: Data-driven response generation in social media. In: Proceedings of the conference on empirical methods in natural language processing, pp. 583–593. Association for Computational Linguistics (2011)
11. Cao, Z., Qin, T., Liu, Y., Tsai, M.F., Li, H.: Learning to rank: from pairwise approach to listwise approach. In: Proceedings of the 24th international conference on Machine learning, pp. 129–136 (2007)
12. Plackett, R.L.: The analysis of permutations. Applied Statistics, pp. 193–202 (1975)
13. Luce, R.D.: Individual Choice Behavior: A Theoretical Analysis. Wiley, New York (1959)
14. Ipeirotis, P.G., Provost, F., Wang, J.: Quality management on amazon mechanical turk. In: Proceedings of the ACM SIGKDD workshop on human computation, pp. 64–67. ACM (2010)
15. Lease, M.: On quality control and machine learning in crowdsourcing. In: Proceedings of the AAAI workshop on human computation, pp. 97–102 (2011)
16. Dekel, O., Shamir, O.: Vox populi: collecting high-quality labels from a crowd. In: Proceedings of the 22nd Annual Conference on Learning Theory (COLT) (2009)
17. Von Ahn, L., Dabbish, L.: Labeling images with a computer game. In: Proceedings of the SIGCHI conference on Human factors in computing systems, pp. 319–326. ACM (2004)
18. Deterding, S., Sicart, M., Nacke, L., O'Hara, K., Dixon, D.: Gamification. using game-design elements in non-gaming contexts. In: Proceedings of the 2011 annual conference extended abstracts on Human factors in computing systems, pp. 2425–2428. ACM (2011)
19. Platt, J., et al.: Probabilistic outputs for support vector machines and comparisons to regularized likelihood methods. Adv. Large Margin Classifiers $10(3)$, 61–74 (1999)

A Method for Binarization of Document Images from a Live Camera Stream

Mattias Wahde[✉]

Department of Applied Mechanics, Chalmers University of Technology,
41296 Göteborg, Sweden
mattias.wahde@chalmers.se

Abstract. This paper describes a method for binarization of document images from a live camera stream. The method is based on histogram matching over partial images (referred to as *tiles*). A method developed previously has been applied successfully to images with artificially added noise. Here, an improved method is presented, in which the user has more direct control over the specification of the binarizer. The resulting system is then taken a step further, by considering the more difficult case of binarization of live camera images. It is demonstrated that the improved method works well for this case, even when the image stream is obtained using a (slightly modified) low-cost web camera with low resolution. For typical images obtained this way, a standard OCR reader is capable of reading the binarized images, detecting around 87.5 % of all words without any error, and with mostly minor, correctable errors for the remaining words.

Keywords: Document image binarization · Image processing

1 Introduction

Reliably identifying text in a live image stream from (for example) a web camera is a difficult task, due to variable lighting, image noise, skewing, misalignment, and focusing problems. In general, the problem of identifying text from images has been considered by many authors, and a large number of applications have been identified [1,2], for example reading the license plates of vehicles, identifying labels on packaging, helping the visually impaired to read signs and other texts etc.

An important special case is that of detecting, extracting, and reading text in document images, for example letters, bank statements etc. This application is slightly less complex than reading, say, all the text available in an arbitrary image, but is nevertheless far from trivial, and has been the subject of much research; see e.g. [3–6].

Of course, nowadays, many letters are received in electronic form. However, particularly for elderly users with visual impairment, a system for automatic reading of a document held in front of a camera, would potentially be very

© Springer International Publishing Switzerland 2015
B. Duval et al. (Eds.): ICAART 2014; LNAI 8946, pp. 137–150, 2015.
DOI: 10.1007/978-3-319-25210-0_9

useful. Such a system could then be integrated in an intelligent agent intended for helping the visually impaired to manage everyday tasks. The method presented below is intended to form a part of such a system. Once completed, the agent, represented as a face on a screen, and running on a computer equipped with a microphone, a camera, and loudspeakers, will interact with a user to aid in a variety of tasks, including (but not limited to) the one just mentioned.

In order to read the text in a document image using, for example, an already available optical character recognition (OCR) system, a common first step is to binarize the image, i.e. taking an often noisy image with varying illumination levels and converting it to a black-and-white image, ideally containing easily identifiable black characters on a white background. In the binarization of an image (or a part thereof) any pixel with gray value at or above a threshold will then be set white (gray level 255) and any pixel with gray value below the threshold will be set black (gray level 0). In many cases, different binarization thresholds are needed for different parts of the image. In general, the main difficulties in binarization concern brightness variations (for example, due to spotlights or bad lighting altogether), stains, misaligned text (due to bending or tilting, or a combination thereof), and other noise sources.

In recent years, several binarization methods have been suggested for document images; see e.g. [3,4,6–8]. In an earlier paper [6], we described a method for text image binarization using histogram matching. While this method did very well on noisy text images without bending or misalignment, outperforming benchmark methods such as Otsu's method [9], Niblack's method [10], and Sauvola's method [11], it did less well on images captured from a live video stream. In addition to being noisy, such images are also typically misaligned and skewed. Furthermore, it is common that some of the text in the image is not exactly in focus. In order to cope with such images, which of course is a necessity for a system of the kind described above to be useful, a modified method has been developed, and it will be the subject of this paper.

The paper is organized as follows: In Sect. 2 the method is introduced and described. The results are presented in Sect. 3 and are followed by a discussion and some conclusions in Sect. 4.

2   Method

As mentioned above, an earlier version of the proposed method was introduced in [6]. Both the previous method and the modified method introduced here generate a system for document binarization (henceforth referred to as a *binarizer*) consisting of (i) a set of histograms, denoted $\mathcal{H}$, obtained during training, (ii) a set of binarization thresholds $\mathcal{T}$, one for each histogram in $\mathcal{H}$, also obtained during training, and (iii) a set of user-specified parameters, which will be further explained below.

The method operates in two phases, a *training phase* and a *usage phase*. In the first version of the method, described in detail in [6], a set of image pairs is used, in which each pair consists of a noisy version and a clean, ground truth version. In that version, the training images were generated artificially, by starting from a perfect image, and then adding suitable amounts of noise, without bending or misaligning the text. However, in subsequent testing using a live camera stream, it turned out that the performance was less good than one might have hoped. In other words, the artifically generated noise was not sufficiently similar to the distortions generated when using a live camera. This is so, since the use of a clean ground truth version of each image is not very realistic: In reality, when a text document is held in front of a web camera, the resulting image is not only noisy but is also, in most cases, skewed and misaligned (due to bending of the paper, for example). Even though one *can* generate a ground truth version (i.e. a perfectly noise-free image, but with the same amount of skewing and misalignment as in the image captured by the camera) even in such a case or, alternatively, attempt to remove the skewing and misalignment in the captured image, it would be a rather time-consuming procedure that would, moreover, have to be repeated for every new image considered.

By contrast, in the modified method presented here, the user manually selects the binarization thresholds for different parts of the image, thus avoiding the problem of generating a perfectly clean ground truth image with which to compare the binarization results. Instead, the assessment of binarization quality is based on simple inspection of the binarized image. Now, this might seem as a step backwards given that the first version of the method automatically found suitable binarization thresholds for each part of the image. However, it turns out that setting suitable thresholds is a rather quick procedure: A binarizer can be built manually in the space of an hour. More importantly, unlike the binarizers obtained with the previous version of the method, it will be able to handle live camera streams. The procedure for generating a binarizer, in the modified method, will now be described.

2.1 Training a Binarizer

The first part of the training phase is to generate a set $I_{tr} = \{I_1, I_2, \ldots, I_k\}$ of k training images. Using a live video stream (obtained from a web camera), snapshots are taken of k different documents. Next, the user considers the first image, which is divided into a mosaic consisting of $N \times M$ tiles, denoted $\tau_{n,m}$. An example is shown in Fig. 1. The top panel shows the original camera image, and in the bottom panel the grid defining the tiles has been superposed. The fact that some tiles are highlighted with thick squares will be explained in Sect. 3 below.

As can be seen in Fig. 1, the image quality is quite poor. Most of the noise comes from the fact that light can shine through a single sheet of paper held in front of a camera. Of course, one can always fine-tune the settings, for any given lighting situation, in order to obtain a much better image quality. However, one should obviously not need to fine-tune the settings for the camera every time that it is used; if this turned out to be necessary, the binarization system would not be very useful for, say, a visually impaired person. Instead, the binarizer must be able to cope with noise of the kind present in Fig. 1.

He had risen from his chair and was standing between the parted
blinds gazing down into the dull neutral-tinted London street. Looking
over his shoulder, I saw that on the pavement opposite there stood a
large woman with a heavy fur boa round her neck, and a large curling
red feather in a broad-brimmed hat which was tilted in a coquettish
Duchess of Devonshire fashion over her ear. From under this great
panoply she peeped up in a nervous, hesitating fashion at our
windows, while her body oscillated backward and forward, and her
fingers fidgeted with her glove buttons. Suddenly, with a plunge, as of
the swimmer who leaves the bank, she hurried across the road, and we
heard the sharp clang of the bell.

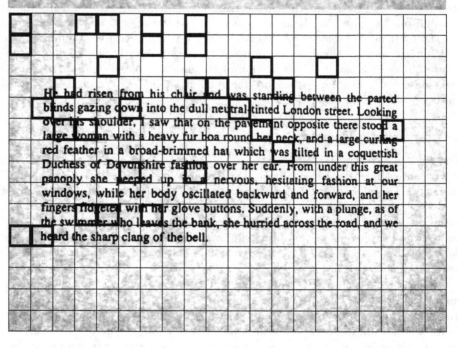

Fig. 1. Upper panel: An example of a typical image captured from the live video
stream. Lower panel: The same image, after grayscale conversion and sharpening, and
with the tiles indicated. The reason for highlighting some of the tiles is described in
Sect. 3 below. The size of the image is 640×480 pixels, and each tile is 32×32 pixels.

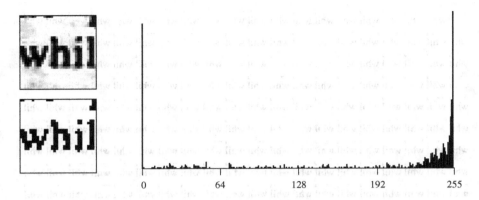

Fig. 2. Top left panel: A blowup of one tile from the image shown in Fig. 1, namely the tile covering the first part of the word *while* on the eighth line. Right panel: the histogram obtained from the tile in the top left panel. The horizontal axis runs from 0 to 255 (the gray levels). For each gray level the corresponding vertical bar indicates the fraction of pixels at that gray level. This particular histogram has two peaks, the left-most peak (at 0) corresponding to the text in the tile, and the right-most peak (from around 220 to 255) corresponding to the background. The bottom left panel shows the result of binarizing the tile, using a threshold of 86.

As an aid to the user, a computer program has been written in connection with the development of the method. This program is responsible for presenting tiles to the user, computing histogram distances etc. As will be seen below, the user's sole tasks are (i) to select tiles, and (ii) to choose a suitable binarization threshold for each selected tile. This procedure is repeated until the binarizer contains a sufficient number of histograms for reliable binarization.

From the first presented image (I_1), the user considers the tiles one by one, starting (for example) from the upper left corner. Now, for each tile a histogram H is generated, showing the fraction of pixels at each grayscale value $(0, 1, \ldots, 255)$ in the tile. The histogram is then normalized so that $\sum_i H(i) = 1$, where $H(i)$ denotes the contents of bin i of the histogram and the sum extends over all 256 bins. As usual, a gray scale value of 0 corresponds to a black pixel, a value of 255 represents a white pixel, and intermediate values correspond to different shades of gray. An example of a tile histogram (taken from one of the tiles in the image shown in Fig. 1) is shown in Fig. 2. As can be seen, the histogram has two distinct peaks, one narrow peak at gray level 0, corresponding to the text in the tile, and one broader peak at high gray levels, corresponding to the background.

Initially, the binarizer is empty, i.e. the sets $\mathcal{H}$ and $\mathcal{T}$ contain no elements. The user then selects the threshold (in the range [0,255]) that provides the best binarization for the first tile. In order to help the user, the computer program written in connection with the development of the method presents a grid of 256 images, each binarized with a different threshold $0, 1, \ldots$ etc. up to 255. An example is shown in Fig. 3. By visual inspection of this grid, it is usually very

whil whil whil whil whil whil whil whil whil whil whil whil whil whil whil whil whil whil whil whil
whil whil whil whil whil whil whil whil whil whil whil whil whil whil whil whil whil whil whil whil
whil whil whil whil whil whil whil whil whil whil whil whil whil whil whil whil whil whil whil whil
whil whil whil whil whil whil whil whil whil whil whil whil whil whil whil whil whil whil whil whil
whil whil whil whil whil whil whil whil whil whil whil whil whil whil whil whil whil whil whil whil
whil whil whil whil whil whil whil whil whil whil whil whil whil whil whil whil whil whil whil whil
whil whil whil whil whil whil whil whil whil whil whil whil whil whil whil whil whil whil whil whil
whil whil whil whil whil whil whil whil whil whil whil whil whil whil whil whil whil whil whil whil
whil whil whil whil whil whil whil whil whil whil whil whil whil whil whil whil whil whil whil whil
whil whil whil whil whil whil whil whil whil whil whil whil whil whil whil whil whil whil whil whil
whil whil whil whil whil whil whil whil whil whil whil whil whil whil whil whil whil whil whil whil
whil whil whil whil whil whil whil whil whil whil whil whil whil whil whil whil whil whil whil whil

Fig. 3. The grid showing binarization results (for a given tile, namely the same tile as in Fig. 2), for all possible binarization thresholds, from 0 to 255. The upper row shows binarizations with threshold values of $0, 1, \ldots, 19$, the second row shows binarizations with threshold values of $20, 21, \ldots, 39$ etc. As can be seen, for this particular tile, suitable thresholds can be found roughly in the range 1 to 120. Note that the range of suitable thresholds is rather wide, making the choice of a specific threshold quite simple.

simple to select a suitable binarization threshold T. Typically, many different thresholds will give roughly the same result, meaning that the resulting binarizer will not be very sensitive to the exact threshold chosen for a given tile.

The histogram of the first tile is always added to $\mathcal{H}$, while the corresponding chosen binarization threshold is added to $\mathcal{T}$. Thus, at this point, the two sets contain one element each. The user then proceeds to the second tile, for which the histogram is also computed. Next, the program computes the minimum distance between any histogram in $\mathcal{H}$ (at this stage, there is only one histogram available) and the histogram of the current (second) tile. The distance between two histograms H_p and H_q is computed using the chi-square histogram distance measure [12], defined as

$$\chi^2(H_p, H_q) = \frac{1}{2} \sum_i \frac{(H_p(i) - H_q(i))^2}{(H_p(i) + H_q(i))}, \tag{1}$$

where $H_s(i)$ denotes the contents of bin i of histogram H_s, $s = p, q$, and the sum extends over all bins.

```
set H = ∅
set T = ∅
for each training image I do
    for each tile τ_{n,m} ∈ I do
        generate the tile histogram H
        set d_min = ∞
        for each histogram H_j ∈ H do
            compute d = dist(H, H_j) ≡ χ²(H, H_j)
            if (d < d_min) then
                d_min ← d
            end if
        end do
        if (d_min > D) or (H = ∅) then
            (1) Select a suitable binarization threshold T_b for the current tile.
            (2) Add H to H and T_b to T.
        end if
    end do
end do
```

Fig. 4. The training algorithm for the binarizer. See the main text for a description of the algorithm.

If, for all histograms in $\mathcal{H}$, the chi-square distance to the histogram of the current tile exceeds a limit D (typically set to around 0.10–0.12; see below), i.e. if

$$\chi^2(H_j, H) > D \ \forall \ H_j \in \mathcal{H} \tag{2}$$

the user selects a suitable binarization threshold for the current tile, by considering the grid of all possible tile binarizations, as explained above and as shown in Fig. 3. Next, the histogram of the current tile is added to $\mathcal{H}$ and the corresponding selected binarization threshold is added to $\mathcal{T}$. If instead there is some histogram for which the chi-square distance to the current tile histogram is below the limit D, the current tile histogram is discarded, and the user proceeds to the next tile etc.

The rationale behind the method is to make sure that the binarizer will contain a representative set of histograms, along with the corresponding binarization thresholds, to correctly binarize any tile content. The limit D is introduced in order to keep the number of such histograms to a minimum: If $\mathcal{H}$ already contains a histogram that is so similar to the histogram of the current tile that their distance is below the threshold, there is no need to add the histogram of the current tile to $\mathcal{H}$. The training method is summarized in Fig. 4.

Now, for an image size of 640×480 pixels, and using a tile size of 32×32 pixels, there will be $20 \times 15 = 300$ tiles for each image. However, because of the rather slow variation in lighting, focus, and noise levels, typically only a rather small fraction of those histograms must be included in $\mathcal{H}$. Moreover, as the user proceeds to the 300 tiles of the *second* image in the training set (having completed the first image), there will normally be fewer histograms fulfilling the distance criterion than for the first image, since $\mathcal{H}$ will then already contain several representative histograms. In fact, after going through a few images, it becomes difficult to find *any* histogram fulfilling the distance criterion in Eq. (2). The training phase is then complete.

2.2   Using a Binarizer

As mentioned above, in addition to the set of histograms $\mathcal{H}$ and the set of binarization thresholds $\mathcal{T}$, a binarizer also contains a set of parameters. The first parameter is the distance threshold D used for determining which histograms to add, during training (see above). However, during use, another (higher) threshold D_{use} is employed: When the binarizer is applied to an image, it starts by dividing the image into tiles, exactly as during training. Next, for each tile, the corresponding histogram H is computed. Then, the binarizer runs through all the stored histograms H_j in the set $\mathcal{H}$, computing the distance between H and H_j, and keeping track of the index (denoted j_{use}) that yields the smallest distance ($d_{\min}$). If, after running through all the histograms, the smallest distance $d_{\min}$ is smaller than the parameter D_{use}, the tile is binarized using the threshold T (from the stored set $\mathcal{T}$) associated with the corresponding histogram, i.e.

$$j_{\text{use}} = \operatorname{argmin}_j \ \text{dist}(H, H_j) \tag{3}$$

and

$$T = T_{j_{\text{use}}} \tag{4}$$

If instead $d_{\min}$ exceeds D_{use}, the corresponding tile is set completely white. Originally, it was thought that the threshold used for training (D) would also be used when applying the binarizer. However, some experimentation showed that a larger threshold gave a better result when applying the binarizer. Thus, the binarizer contains a total of two parameters, D, which is typically set to around 0.10–0.13, and D_{use}, which is typically set to around 0.40–0.55.

3   Results

A binarizer was trained, as described in Subsect. 2.1 above, using a set of text images taken as snapshots of a live video stream, obtained from a web camera with fixed focus (see also Sect. 4 below). A total of five documents were photographed, thus generating a training set containing five images. Now, this might seem like a rather small set of images. However, it should be noted that each image provided 300 tiles (again using an image size of 640×480 pixels, and a tile size of 32×32 pixels), so that the total number of considered tiles equalled $5 \times 300 = 1500$. Moreover, when the fifth image was reached, only a few histograms could be added, keeping in mind the distance criterion in Eq. (2). The limit D was set to 0.13. The (manual) training took around one hour to complete, with the aid of the computer program (described above) that generates histograms, computes histogram distances, and also generates grids showing, for any given tile, the result of binarizing the tile with all 256 possible thresholds.

Two preprocessing steps, which are not really part of the binarizer, were applied, namely (i) conversion to grayscale, and (ii) a sharpening of the image, using a standard convolution sharpening method, i.e. convolving the image with the matrix

$$S = \begin{pmatrix} -\frac{\alpha}{8} & -\frac{\alpha}{8} & -\frac{\alpha}{8} \\ -\frac{\alpha}{8} & 1+\alpha & -\frac{\alpha}{8} \\ -\frac{\alpha}{8} & -\frac{\alpha}{8} & -\frac{\alpha}{8} \end{pmatrix}, \tag{5}$$

with α set to 0.75, after some experimentation. These preprocessing steps were applied to all images (both during training and use), before generating tiles, histograms etc. The first image used during training can be seen in Fig. 1. The highlighted tiles are those whose histograms were added to $\mathcal{H}$ during the training phase. For example, the first added histogram (and the corresponding binarization threshold), i.e. the one obtained from the upper left tile, was sufficient for binarizing the subsequent two tiles in the first row. When the fourth tile was reached, however, the minimum histogram distance (see Eq. (2)) exceeded D. Thus, the histogram for this tile was added, along with the corresponding selected binarization threshold etc. In all, 30 histograms were added for the first training image. Next, the procedure was repeated for the second image etc. until all five images had been considered. At this stage, $\mathcal{H}$ contained a total of 93 histograms, of which only very few were added in connection with the last image of the training set.

The resulting binarizer was stored, and was then used for binarizing images from a live camera stream, with the parameter D_{use} set to 0.55. To this end, five additional documents (different from the ones used during training) were generated, and were then held, one by one, in front of the live camera stream. A snapshot was then taken, and the preprocessing steps (grayscale conversion and sharpening) described above were applied. Next, the image was passed to the binarizer. The binarization results obtained for two of the five test images can be seen in Fig. 5. As is evident from the figure, the binarizer achieved very good results over these images (and also achieved results of similar quality over the three test images *not* shown in the figure), removing almost all noise while keeping, and enhancing, almost all text.

In the previous version of the method, as described above, it was assumed that a clean, ground truth image was available, so that the resulting binarized image could be compared to that image, and a quality measure could be computed. However, in the method presented here, no such clean, ground truth images are available. Thus, in order to judge the quality of the binarization, a different approach was used: For each of the five test images, both the original version and the binarized version of the image were passed through a freely available online OCR engine [13], and the quality of the text obtained was measured. The results are summarized in Table 1. As can be seen, the results for the non-binarized, original images, i.e. images of the kind shown in the left panels of

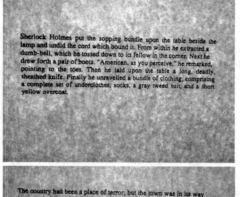

Fig. 5. Two examples of the results obtained when applying the binarizer to (previously unseen) test images, taken as snapshots from the live camera stream. For each of the two rows, the left panel shows the original, non-binarized image, whereas the right panel shows the binarized image.

Fig. 5, are more or less disastrous: The OCR engine was simply not able to make sense of these noisy images, except for a few rare words here and there. By contrast, once the images had been binarized using the binarizer described above, the results were greatly improved: Around 87.5 % of the words were completely correctly identified (i.e. without any error at all).

For those words that are incorrectly identified, i.e. words for which there is any error at all, errors of two kinds are defined: Minor errors, for which the identified word has *at least* two correctly identified characters and *at most* two incorrectly identified ones, and major errors, for which the conditions just mentioned are not fulfilled. Thus, for example, any error in a word with at most two characters is considered major. An example of a minor error is the misidentification of the word *remarked* as *nemarked*; see also Fig. 6. In fact, in many cases, the errors are minor and could easily be corrected using a dictionary and possibly some information regarding context.

In Fig. 6, a blowup is shown of the first four lines of text in the first image from Fig. 5, both from the original image (top panel), and the binarized image (middle panel). The OCR result obtained before binarizing, i.e. when passing

Table 1. Results obtained when passing the original test images $(O_i, i = 1, 2, \ldots, 5)$ and the corresponding binarized images $(B_i, i = 1, 2, \ldots, 5)$ through an OCR engine. The first column identifies the image index (i), and the second column shows the number of words (N_w) in the corresponding image. The third column shows the number of correctly identified words $(N_{correct})$ for the binarized images, whereas the fourth column shows the number of words for which minor errors were obtained (N_{minor}), and the fifth column shows the number of words for which major errors were obtained (N_{major}), in both cases for the binarized images. The difference between minor and major errors is described in the main text. Finally, the sixth to eighth columns show the same information for the original (non-binarized) images. The results show that the text in the binarized images was essentially correctly identified, whereas the text in the original images was impossible for the OCR engine to read.

		Binarized image			Original image		
i	N_w	$N_{correct}$	N_{minor}	N_{major}	$N_{correct}$	N_{minor}	N_{major}
1	87	81	5	1	0	0	87
2	80	68	6	6	0	0	80
3	99	87	5	7	0	0	99
4	88	79	4	5	1	1	86
5	78	64	4	10	4	2	72

the original image through the OCR engine, is not shown: It consisted of an essentially random sequence of letters and symbols, bearing no resemblance to

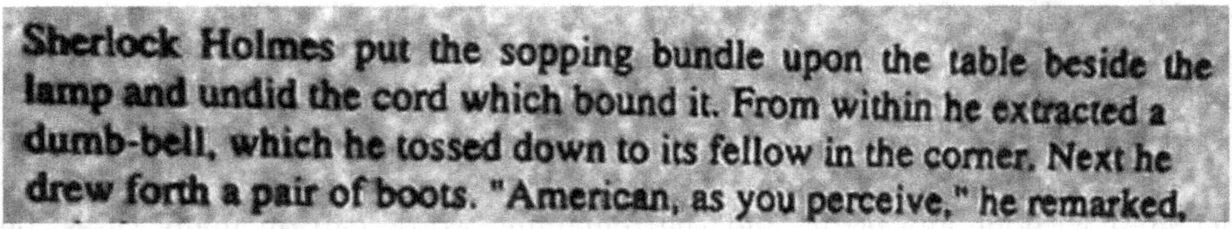

Sherlock Holmes put the sopping bundle upon the table beside the
lamp and undid the cord which bound it. From within he extracted a
dumb-bell, which he tossed down to its fellow in the corner. Next he
drew forth a pair of boots. "American, as you perceive," he remarked.

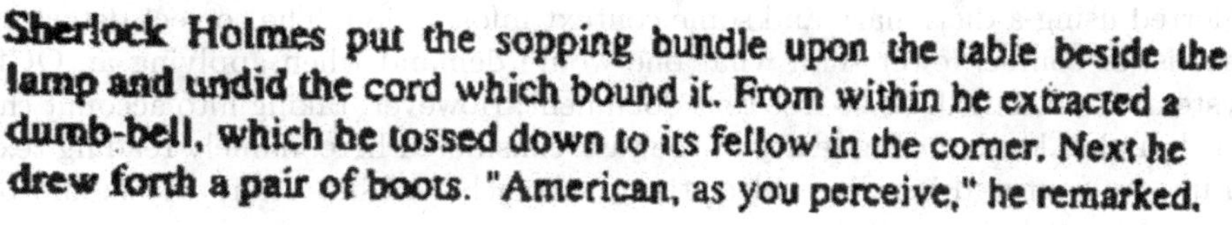

Sherlock Holmes put the sopping bundle upon the table beside the
lamp and undid the cord which bound it. From within he extracted a
dumb-bell, which he tossed down to its fellow in the corner. Next he
drew forth a pair of boots. "American, as you perceive," he remarked.

```
Sherlock Holmes put the sopping bundle upon the table beside lhe
lump and undid the cord which bound it. From within he extracted a
dumb,-bell. which he tossed down to its fellow in the comer. Next he
drew forth a pair of boots. "American. as you perceive." he nemarked.
```

Fig. 6. A blowup of the first four lines of the first image in Fig. 5. The upper panel shows the original image, the middle panel shows the binarized image, and the bottom panel shows the text extracted by the OCR engine. As can be seen, the extracted text was correct except for a few minor errors.

the actual text. By contrast, the text obtained from the OCR engine applied to the binarized image is shown at the bottom of the figure. Here, there are only a few minor errors: The word *the* at the end of the first line was interpreted as *lhe*, the word *lamp* on the second line was interpreted as *lump*, the word *corner* on the third line was interpreted as *comer*, and the final word, *remarked*, was interpreted as *nemarked*. Also, an extra comma was erroneously inserted in the word *dumb-bell*, and two commas were misinterpreted as full stops. Clearly, most, if not all, of these errors could easily be corrected.

4   Discussion and Conclusion

The most important result of this work is that the proposed method is able to generate a binarizer that can reliably binarize text images from a live video stream, using a rather simple web camera (the rationale being that, if reliable binarization can be obtained with such a camera, it should be possible to obtain equally good, or better, results with a more advanced camera). Here, a low cost fixed-focus camera (purchased at a price of around 15 euro) was used. Now, a typical fixed-focus web camera is ill-suited for the type of application considered here: Such cameras are normally optimized to provide a clear and focused image of, say, a person sitting at a typical distance from a computer, rather than a text image which would normally be held closer to the camera, in order to obtain sufficient resolution of the individual characters in the text. In order to overcome this problem, the camera was disassembled, and the focal ring, which originally was glued in place, was made adjustable by simply removing the glue. The focal ring was then manually adjusted (while running the camera) until a suitable setting was found for text images held at around 0.25–0.50 m from the camera.

The test reported above comprised only five images. However, as is evident from Table 1, the results obtained differ very little between images. In fact, some additional tests were carried out, with other images, and the results were more or less the same as those reported in the table: Around 87.5 % of all words are detected without any error, and the remaining words can, in many cases, be inferred using a dictionary and some context information. The correct detection rate is, of course, lower than what one would demand when applying an OCR system to a scanned, noise-free text document. However, taking into account the much higher level of complexity of the task considered here, namely reading text in images captured in a live video stream, the achieved detection rate should be sufficient.

Comparing with the previous method, introduced in [6], the method presented here has even lower complexity, since it does not include the sequence of repeated enhancements that was an integral part (albeit seldom used, in practice) in the previous method. As mentioned above, the modified method does require that the user, rather than the method itself, should find suitable histograms. However, doing so is not very difficult, with the help of the computer program developed together with the method. Typically, a complete binarizer can be generated in around one hour. Moreover, when using the method, if one finds that some new test image is not binarized correctly, one can easily resume the training to include a few more histograms. It should be noted, however, that this was *not* necessary in order to binarize the five test images above, using the binarizer described in the beginning of Sect. 3. Another important advantage of the method presented here (over the previous method) is that it does not require that one should define a set of noise-free, ground truth images for the training phase. Instead, it is sufficient just to have the images captured by the camera.

The resolution of the images used here was deliberately set to a rather low value (640×480 pixels), again in order to provide a stringent test of the method: If it is capable of correct binarization of such an image, it should be even easier for it to binarize an image with, say, four times as many pixels (1280×960). As for the time requirements, with a tile size of 32×32, using a computer with an Intel Core i7-2600 CPU (3.40 GHz), the binarizer (with 93 histograms, as described in Sect. 3 above) runs at a speed of 2.3 ms per tile so that a full image takes around 0.75 s to binarize. While this is too slow, by around a factor 10, for real-time operation, it is certainly sufficient for the application at hand, namely reading (aloud) the text in a document held in front of the camera: The actual reading will, of course, take several seconds, and a slight delay in the beginning is hardly noticeable. Note also that a speedup can be obtained by, for example, just binarizing a fraction of the image, and then binarizing the rest (in a separate thread) while reading the text extracted from the first fraction.

The next steps in the project will be to define and integrate a custom OCR engine, which will be optimized to work together with the binarizer. Furthermore, the OCR engine will be equipped with error-correcting methods, which will be applied to the raw text extracted from the image. Another important topic for future work is to speed up the method, so that it can operate in real time. Finally, the resulting system will be integrated into the intelligent agent mentioned in Sect. 1.

To conclude, a method has been presented which allows a user, assisted by a computer program, to manually define (in the space of an hour, or less) a binarizer capable of generating readable text from a document image obtained as a snapshot from a live camera stream. The method works well even with a low-cost camera with rather low image resolution, provided that the focus point has first been manually adjusted.

Acknowledgements. The author gratefully acknowledges financial support from *De blindas vänner*.

References

1. Neumann, L., Matas, J.: A method for text localization and recognition in real-world images. In: Kimmel, R., Klette, R., Sugimoto, A. (eds.) ACCV 2010, Part III. LNCS, vol. 6494, pp. 770–783. Springer, Heidelberg (2011)
2. González, A., Bergasa, L.: A text reading algorithm for natural images. Image vis. comput. **31**, 255–274 (2013)

3. Stathis, P., Kavallieratou, E., Papamarkos, N.: An evaluation technique for binarization algorithms. J. Univ. Comput. Sci. **14**(18), 3011–3030 (2008)
4. Shi, J., Ray, N., Zhang, H.: Shape based local thresholding for binarization of document images. Pattern Recogn. Lett. **33**, 24–32 (2012)
5. Valizadeh, M., Kabir, E.: An adaptive water flow model for binarization of degraded document images. Int. J. Doc. Anal. Recogn. **16**(2), 165–176 (2013)
6. Wahde, M.: A method for document image binarization based on histogram matching and repeated contrast enhancement. In: Duval, B., van der Herik, J., Loiseau, S., Filipe, J. (eds.) Proceedings of the 6th International Conference on Agents and Artificial Intelligence (ICAART 2014), pp. 34–41 (2014)
7. Chen, K.-N., Chen, C.-H., Chang, C.-C.: Efficient illumination compensation techniques for text images. Digit. Signal Process. **22**, 726–733 (2012)
8. Lu, S., Su, B., Tan, C.: Document image binarization using background estimation and stroke edges. Int. J. Doc. Anal. Recogn. **13**(4), 303–314 (2010)
9. Otsu, N.: A threshold selection method from gray-level histograms. IEEE Trans. Syst. Man. Cybern. **9**, 62–66 (1979)
10. Niblack, W.: An Introduction to Image Processing. Prentice-Hall, Englewood Cliffs (1986)
11. Sauvola, J., Pietikäinen, M.: Adaptive document image binarization. Pattern Recogn. **33**, 225–236 (2010)
12. Pele, O., Werman, M.: The Quadratic-Chi Histogram Distance Family. In: Daniilidis, K., Maragos, P., Paragios, N. (eds.) ECCV 2010, Part II. LNCS, vol. 6312, pp. 749–762. Springer, Heidelberg (2010)
13. FreeOCR, accessed 20140722. www.free-ocr.com

A Probabilistic Semantics for Cognitive Maps

Aymeric Le Dorze[1]([✉]), Béatrice Duval[1], Laurent Garcia[1], David Genest[1],
Philippe Leray[2], and Stéphane Loiseau[1]

[1] Laboratoire d'Étude et de Recherche en Informatique d'Angers,
Université d'Angers, 2 Boulevard Lavoisier, 49045 Angers Cedex 01, France
{ledorze,bd,garcia,genest,loiseau}@info.univ-angers.fr
[2] Laboratoire d'Informatique de Nantes Atlantique, École Polytechnique,
Université de Nantes, La Chantrerie - Rue Christian Pauc,
44306 Nantes Cedex 3, France
philippe.leray@univ-nantes.fr

Abstract. Cognitive maps are a graphical knowledge representation
model that describes influences between concepts, each influence being
quantified by a value. Most cognitive map models use values the semantics of which is not formally defined. This paper introduces the probabilistic cognitive maps, a new cognitive map model where the influence
values are assumed to be probabilities. We formally define this model and
redefine the propagated influence, an operation that computes the global
influence between two concepts in the map, to be in accordance with this
semantics. To prove the soundness of our model, we propose a method
to represent any probabilistic cognitive map as a Bayesian network.

Keywords: Cognitive map · Probabilities · Causality · Bayesian
network

1 Introduction

Graphical models for knowledge representation help to easily organize and understand information. A *cognitive map* [1] is a graph that represents influences
between concepts. A *concept* is a short textual description of an idea of the real
world such as an action or an event and is represented by a labeled node in
the graph. An *influence* is an arc between two of these concepts. A cognitive
map provides an easy visual communication medium for humans, especially for
the analysis of a complex system. It can be used for instance to take a decision
in a brainstorming meeting. These maps are used in several domains such as
biology [2], ecology [3], or politics [4].

In a cognitive map, each influence is labeled with a value that quantifies
it. This value describes the strength of the influence. It belongs to a previously defined set, called a *value set*. A cognitive map can be defined on several
kinds of value sets. These value sets can be sets of symbolic values such as
$\{+, -\}$ [1] or $\{none, some, much, a\ lot\}$ [5], or an interval of numeric values such

© Springer International Publishing Switzerland 2015
B. Duval et al. (Eds.): ICAART 2014; LNAI 8946, pp. 151–169, 2015.
DOI: 10.1007/978-3-319-25210-0_10

as $[-1; 1]$ [6, 7]. Thanks to these values, we are able to compute the global influence of any concept of the map on any other one. Such an operation is called the *propagated influence*. To compute it, the values of the influences that compose the paths linking the two concepts are aggregated according to their semantics. The propagated influence is what makes cognitive maps useful for decision-making since it provides an overview of the consequences of a decision.

The main advantage of cognitive maps is that they are simple to use; people who are not familiar with formal frameworks need this simplicity. Consequently, the semantics of the values is sometimes not clearly defined. The drawback is that it is often hard to interpret the real meaning of the values associated to the influences and to verify the soundness of the computed propagated influence.

Some approaches exist to formally define the semantics of cognitive maps. The fuzzy cognitive maps links the cognitive maps to the fuzzy set framework [6, 8]. They consider that the concepts are fuzzy sets and that the values represent the degrees of causality between these concepts. These maps are generally easy to use but the inference is sometimes quite obscure for a layman since fuzzy sets are not a very popular framework.

There exist other knowledge representation models that represent both a graph and values associated to a strong semantics. The graphical structure of a cognitive map and the values given by a concept influencing another one remind us of the Bayesian network framework [9, 10]. Bayesian networks express dependency relations between variables. These relations are quantified with conditional probabilities. They are more expressive than cognitive maps but their building and their use are more complex. It is then interesting to improve the formal aspect of cognitive maps when dealing with values assumed to be probabilities since probabilities are generally a popular framework. Such a model would keep the simplicity of cognitive maps while tending to be as formal as Bayesian networks.

This paper introduces a new cognitive map model, the *probabilistic cognitive maps*. This model keeps the simplicity of cognitive maps while improving the formal representation of the values by providing a probabilistic interpretation for the influence values. Such an interpretation is formal enough without being restrictive to users but needs to adapt the semantics of the concepts and the influences. Therefore, the propagated influence has to be redefined to fit the semantics. To show the validity of our model, we propose a procedure to represent a cognitive map as a Bayesian network and show that the propagated influence in the probabilistic cognitive map corresponds to a specific probability in the Bayesian network. The studied model is the causal Bayesian network model [10] because, as shown in this paper, it is more closely related to cognitive maps.

There exist other works that link cognitive maps to probabilities. For example, [11] defines the fuzzy probabilistic cognitive map model, which is based on the fuzzy cognitive map model. However, in this model, the probabilities are only expressed on the concepts since they are used to compute whether a concept can or cannot influence other concepts. The probabilistic cognitive map model that we define must not be confused with the Incident Response Probabilistic Cognitive Map model (IRPCM) [12]. In this model, the links between the concepts are not necessarily causal, therefore what they call a "cognitive map" is not the same model as the one we define here. IRPCM is mostly used for diagnosis whereas our model proposes a framework that studies influences between concepts.

Qualitative Probabilistic Networks (QPN) [13] are a probabilistic model that acts as a bridge between cognitive maps and Bayesian networks. Indeed, according to their inventor M. Wellman, QPNs generalize cognitive maps defined on $\{+, -\}$. They also allow to express new relations such as synergies that describe the effect of two combined concepts. However, the values labeling the arcs are not really influence values, but rather constraints expressed on the probability distribution associated to the QPN. Hence, QPNs are *qualitative* rather than *quantitative*. Nevertheless, we prefer to keep using cognitive maps rather than a different model since cognitive maps come with useful operations that we would like to be able to apply.

In this article, we present in Sect. 2 the cognitive map model and a simple introduction to Bayesian networks. We then define the probabilistic cognitive map model in two parts. First, we focus on the semantics of the model in Sect. 3. Then, we define the propagated influence for this model in Sect. 4. In Sect. 5, we ensure the soundness of our model by encoding a cognitive map into a causal Bayesian network. Finally, we present in Sect. 6 a software we developed that implements the probabilistic cognitive map model.

2   State of the Art

In this section, we first present the cognitive map model in Sect. 2.1. Then, we introduce the Bayesian network model in Sect. 2.2. Finally, we outline the causal Bayesian network model in Sect. 2.3.

2.1   Cognitive Maps

A cognitive map is a knowledge representation model that represents influences between concepts with a graph. An influence is a causal relation between two concepts labeled with a value that quantifies it. It expresses how much a concept influences another one regardless of the other concepts. This value belongs to a predefined set, called the *value set*.

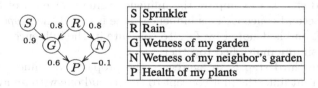

Fig. 1. *CM1*, a cognitive map defined on the value set $[-1; 1]$.

Definition 1 (Cognitive Map). *Let C be a concept set and I a value set. A cognitive map CM defined on I is a directed graph $CM = (C, A, \text{label})$ where:*

- *the concepts of C are the nodes of the graph;*
- *$A \subseteq C \times C$ is a set of arcs, called* influences;
- *label: $A \to I$ is a function labeling each influence with a value of I.*

Example 1. The cognitive map *CM1* (Fig. 1) represents the influences of some concepts on the health of my plants. It is defined on the value set $[-1; 1]$. An influence between two concepts labeled with a positive value means that the first concept positively influences the second one. A negative value means on the contrary that the first concept negatively influences the second one. A value of 1 means that the influence is total. A value of 0 means that there is no direct influence between two concepts whereas the absence of an influence between two concepts means that the designer of the map does not know if there is such a relation between these concepts. The classic cognitive map model does not define more precisely the semantics of this notion of influence and how to interpret it.

If we consider the concepts R and G, the rain influences the wetness of my garden by 0.8. On the contrary, if we consider the concepts N and P, the wetness of my neighbor's garden influences the health of my plants by -0.1 because his growing trees shade my garden.

Thanks to the influence values, the global influence of a concept on another one can be computed. This global influence is called the *propagated influence* and is computed by aggregating the values on the influences that belong to any path linking these two concepts. Many operators to compute the propagated influence exist. We will only present the most common one for the value set $[-1; 1]$ [14]. It is composed of three steps.

The first step is to list the different paths that link the first concept to the second one. Since a cognitive map may be cyclic, there is potentially an infinite number of paths between the two concepts. To avoid an infinite computation, only the most meaningful paths are considered, which are the paths that does not contain any cycle. Indeed, if a path contains a cycle, it means that a concept influences itself. Because the effect of this influence cannot have immediate consequences, it occurs in fact at a future time frame. Therefore, since the influences of a path should belong to the same time frame, the paths that contain a cycle are not considered. A path that contains no cycle is called a *minimal path*.

The second step is to compute the influence value that each of these paths brings to the second concept. This influence value is called the *propagated influence on a path* and is denoted by $\mathcal{IP}$. To compute it, the influence values of the said path are simply multiplied together.

Finally, the third step is to aggregate the propagated influences on every minimal path that links the first concept to the second one with an average. The propagated influence $\mathcal{I}$ of a concept on another one is thus defined as the sum of the propagated influences on every minimal path between the two concepts divided by the number of minimal paths.

Definition 2 (Propagated Influence). *Let c_1 and c_2 be two concepts.*

1. *An* influence path *P from c_1 to c_2 is a sequence of length $k \geq 1$ of influences $(u_i, u_{i+1}) \in A$ with $i \in [0; k-1]$ such that $u_0 = c_1$ and $u_k = c_2$. P is said minimal iff $\forall i, j \in [0; k-1], i \neq j \Rightarrow u_i \neq u_j \wedge u_{i+1} \neq u_{j+1}$; we denote by $\mathcal{P}_{c_1,c_2}$ the set of all minimal paths from c_1 to c_2.*

2. *The* propagated influence on *P is $\mathcal{IP}(P) = \prod_{i=0}^{k-1} \text{label}\big((u_i, u_{i+1})\big)$.*

3. *The* propagated influence of c_1 on c_2 *is:*

$$\mathcal{I}(c_1, c_2) = \begin{cases} 0 & \text{if } \mathcal{P}_{c_1,c_2} = \emptyset \\ \frac{1}{|\mathcal{P}_{c_1,c_2}|} \times \sum_{P \in \mathcal{P}_{c_1,c_2}} \mathcal{IP}(P) & \text{otherwise} \end{cases} .$$

Example 2. In *CM1*, we want to compute the propagated influence of R on P.

1. there are two minimal paths between R and P:
   $\mathcal{P}_{R,P} = \{p_1, p_2\}$ with $p_1 = \{R \rightarrow G \rightarrow P\}$ and $p_2 = \{R \rightarrow N \rightarrow P\}$
2. the propagated influences on p_1 and p_2 are:
   $\mathcal{IP}(p_1) = 0.8 \times 0.6 = 0.48$ $\mathcal{IP}(p_2) = 0.8 \times -0.1 = -0.08$
3. the propagated influence of R on P is:
   $\mathcal{I}(R, P) = \frac{1}{|\mathcal{P}_{R,P}|} \times \big(\mathcal{IP}(p_1) + \mathcal{IP}(p_2)\big) = \frac{1}{2} \times (0.48 - 0.08) = 0.2$

Note that the complexity of the computation of the propagated influence depends on the chosen operator. The complexity is in the worst case at least factorial, as all paths between two concepts must be considered. However, some operators are expressible as a matrix multiplication and are therefore computable in polynomial time [15]. The complexity of some of them may even be linear.

2.2   Bayesian Networks

Bayesian networks [9, 10] are graphical models that represent probabilistic dependency relations between discrete variables as conditional probabilities. Each variable takes its value from many predefined states. In such a graph, each variable is assimilated to a node and an arc represents a probabilistic dependency relation between two variables. This graph is acyclic. Each variable is associated to a table of conditional probabilities. Each entry of this table provides the probability that a variable has some value given the state of each parent of this variable in the graph.

A Bayesian network allows to compute the probabilities of the states of the variables according to the observation of some other variables in the network. The structure of the graph is used to simplify the computations by using the independence relations between the variables. However, these computations are generally NP-hard [16].

Example 3. The Bayesian network *BN1* (Fig. 2) represents dependency relations between variables related to the wetness of my garden. These variables are binary events. We denote the state $A = \top$ by A and $A = \bot$ by $\overline{A}$ for any event A. Each

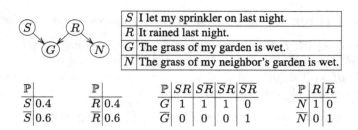

S	I let my sprinkler on last night.
R	It rained last night.
G	The grass of my garden is wet.
N	The grass of my neighbor's garden is wet.

$\mathbb{P}$	
S	0.4
$\overline{S}$	0.6

$\mathbb{P}$	
R	0.4
$\overline{R}$	0.6

$\mathbb{P}$	SR	$\overline{S}R$	$S\overline{R}$	$\overline{S}\,\overline{R}$
G	1	1	1	0
$\overline{G}$	0	0	0	1

$\mathbb{P}$	R	$\overline{R}$
N	1	0
$\overline{N}$	0	1

Fig. 2. The Bayesian network *BN1*.

node is associated to a probability table. The first row of the first table means that the probability that I let my sprinkler on last night is $\mathbb{P}(S) = 0.4$. The values in the table of the variable G means that I am sure that my garden is wet either if I let my sprinkler on last night, or if it rained last night, or both. Otherwise, I am sure that my garden is not wet. From this network, some information can be deduced, like the probability of the states of each node or the independence of two nodes. We can also compute conditional probabilities.

For example, as I am leaving my home, I notice that the grass of my garden is wet. The grass can only be wetted by the rain or my sprinkler. So, I ask myself if I have let my sprinkler on. Thanks to this network, we compute $\mathbb{P}(S|G) = 0.625$. This value is greater than $\mathbb{P}(S)$. This means that knowing that my garden is wet increases the probability that I let my sprinkler on. However, we also compute $\mathbb{P}(R|G) = 0.625$. Thus, we are unable to know what wetted my garden between my sprinkler and the rain because these events are equiprobable given that my garden is wet. Then, I notice that the grass of my neighbor's garden is not wet. If it rained last night, then both our gardens should be wet. We need so to compute the probability that my sprinkler is on given that my grass is wet, contrary to my neighbor's. We compute $\mathbb{P}(S|G\overline{N}) = 1$. Therefore, I am now sure that I let my sprinkler on.

2.3   Causal Bayesian Networks

The causal Bayesian network model [10] extends the classic Bayesian network model. The main difference is the fact that the arcs of a classic Bayesian network can represent any kind of probabilistic dependency relation whereas they have to be causal in a causal Bayesian network. Contrary to classic Bayesian networks, causal Bayesian networks also distinguishes *observation* and *intervention*. When an observation is made on a variable, the information is propagated to the nodes linked to this variable regardless of the direction of the arcs. When an intervention is made on a variable, the information is propagated only to its children, following the direction of the arcs. Thus, with intervention, only the descendants of the variable are influenced by it.

For example, if I *observe* that my garden is wet and I want to compute the probability that it rained last night, I compute $\mathbb{P}(R|G)$, as discussed earlier.

This kind of reasoning can be both deductive and abductive [17]. Now, if I *make* my garden wet, I *intervene* on the wetness of my garden. To represent that intervention, the causal Bayesian network model defines a new operator, called do($\cdot$) [10]. Here, if I want to compute the probability that it rained given the fact that I made my garden wet, I compute $\mathbb{P}(R|\,\mathrm{do}(G))$. Applying do($G$) is thus equivalent to remove the arcs ending on G in the Bayesian network and separate it from its parents [18]. Intuitively, the fact that I made my garden wet has no consequence whatsoever on the fact that it rained and therefore $\mathbb{P}(R|\,\mathrm{do}(G)) = \mathbb{P}(R)$. That kind of reasoning is strictly deductive and only affects the descendants of G.

3   The Probabilistic Cognitive Map Model

We now present our new probabilistic cognitive map model. In such a cognitive map, the influence values are interpreted as probability values. The semantics of the concepts and the influences must be defined according to this interpretation. For the same reason, the propagated influence of a concept on another one must be redefined according to this semantics. In this section, we focus on the semantics of the model by first presenting the semantics of the concepts in Sect. 3.1 and then the semantics of the influences in Sect. 3.2.

3.1   Semantics of the Concepts

To better understand the idea between the semantics of a probabilistic cognitive map, let us consider a simple cognitive map made of concepts A and B linked by a unique influence from A to B with a value α. Note that in the general case, the relationships between the influences, the values and the probabilities are more complex but this basic example helps to get the basic idea behind our approach. Such a map means that A influences B at a level α. Since α is a probability, the concepts A and B must be associated to random variables.

A random variable is defined over a set of values covering its possible states. We would like this set to be as small as possible and to be the same for every variable associated to a concept, in order to keep the simplicity of the model. These values need to represent an information of the real world.

In a cognitive map, a concept is often associated to a piece of information of the real world which is quantifiable. For example, if we consider the concept S in Example 3, it can be seen as the strength of the sprinkler or as the quantity of water it delivers. We define the possible values of the random variable associated to the concept using this quantity. However, we cannot use directly the possible values of this quantity since it may be a continuous scale.

In order to have the same set of values for every random variable, we define two values, inspired by [19]. The value + means that the concept is increasing. The value − means that the concept is decreasing.

Example 4. We consider the concept S that represents a sprinkler from Example 1. The quantity associated to S is the quantity of water that the sprinkler is delivering. We define the random variable X_S associated to S. The increase state $X_S = +$ means that S is increasing, that is the sprinkler is delivering more and more water. The decrease state $X_S = -$ means that S is decreasing, that is the sprinkler is delivering less and less water.

Note that we do not provide a state that represents the fact that a concept is stagnating. This implies that the quantity associated to the concept cannot remain unchanged and has to either increase or decrease. However, we consider that this should not have strong consequences since we consider that cognitive maps aim to study only influences between increasing or decreasing concepts. Hence, we cannot study if a concept stagnates due to an influence by another concept but we can know if a concept is not influenced by another concept if the propagated influence is 0.

Note also that in [19], the state $X_S = +$ means that the *causal effect* of S is positive whereas $X_S = -$ means that the effect is negative. This representation is close to ours but the semantics of the causal effect is stronger with our approach.

Now that the states of the random variables associated to the concepts are defined, we have to define a probability distribution on these states. To compute the probabilistic propagated influence, we need the a priori probability of the states of every random variable of the map. The a priori probability of a state is given when we have no information about the states of any concept. Since there is no information in a cognitive map providing the a priori probability of any state of any concept, we assume that the states of every random variable of the map are equiprobable. Since the random variable associated to each concept has only two states, for every concept A of the map, $\mathbb{P}(X_A = +) = \mathbb{P}(X_A = -) = 0.5$.

3.2   Semantics of the Influences

We focus now on the semantics of the influences and especially the influence values, to define them more precisely than the values presented in Example 1, using probabilities. To evaluate the influence of a concept on another one, the idea is to study how the influenced concept reacts relatively to the different states of the influencing concept. In our case, this leads to study the probabilities of the states of the influenced concept given that the influencing concept is increasing or decreasing. Therefore, if we consider the previous simple map from Sect. 3.1, the influence between A and B is linked to the probabilities of X_B when $X_A = +$ and when $X_A = -$. The value α of an influence should represent how the influenced concept reacts and is thus tied to these conditional probabilities.

A has two ways to influence B: either when A is increasing or when A is decreasing. Thus, the influence should have two values: one for the state $X_A = +$, and one for the state $X_A = -$. To consider this fact, [20] allows to label each influence with two values. However, we want only one value for each influence in the cognitive map, in order to keep the simplicity of the model. Therefore, we need to express a relation between the two values. According to [6], we assume that; an influence being a causal relation, the effect of the increase of A on the increase of B equals the effect of the decrease of A on the decrease of B. Thus, the probability of X_B when $X_A = +$ should be the complement of the probability of X_B when $X_A = -$. In our model, we consider that the influence value α represents the influence of A on B when they are both increasing.

Giving a value α to the direct influence between A and B would lead to answer questions such as "Given that A is increasing, how the probability that B is increasing is modified?". The influence value α quantifies the modification of the a priori probability of B caused by A, in other words, the difference between the conditional probability of B given that A is increasing and the a priori probability of B. Thus, α is linked to the difference between $\mathbb{P}(X_B = +|X_A = +)$ and $\mathbb{P}(X_B = +)$.

This relation between the notion of influence and a conditional probability has consequences on the structure of the cognitive map. Indeed, to compute the global influence of a concept on another one, we aggregate influences. Thus, when we compute the global influence, we manipulate in fact conditional probabilities. Therefore, the global influence of a concept on itself is linked to the conditional probability of a variable given that variable. In such a case, the value of the conditional probability must check certain properties: for example, it has to be equal to either 0 or 1 according to the different values of the variable. Thus, if there are influences that link a concept to itself, the values of these influences should respect this property. As we consider this constraint too strong for the designer of a cognitive map, we forbid cycles in a probabilistic cognitive map.

Now, we express formally the link between α and the difference between $\mathbb{P}(X_B = +|X_A = +)$ and $\mathbb{P}(X_B = +)$. Since $\mathbb{P}(X_B = +|X_A = +)$ is a probability that therefore belongs to $[0; 1]$ and $\mathbb{P}(X_B = +) = 0.5$, α should belong to $[-0.5; 0.5]$. However, in the cognitive map of Example 1, it is obviously not the case since this map is defined on $[-1; 1]$. The idea is to convert α into a value of $[-0.5; 0.5]$. Therefore, a *conversion function* $\mathcal{F}$ must be defined such that whatever the value set I the cognitive map is defined on, its values are converted into values of $[-0.5, 0.5]$. Moreover, a *reverse conversion function* $\mathcal{F}^{-1}$ is defined to get back an influence value that belongs to I when the computation of the propagated influence is done. This reverse conversion function is defined such that $\mathcal{F}^{-1}(\mathcal{F}(\alpha)) = \alpha$. If the conversion function is bijective, then the reverse conversion function is simply its reciprocal function. The conversion function allows us to say that we have $\mathcal{F}(\alpha) = \mathbb{P}(X_B = +|X_A = +) - \mathbb{P}(X_B = +)$. Note that this relation is more complex when B has more than one parent.

Example 5. Since the cognitive map *CM1* is defined on $[-1; 1]$, we define the conversion function $\mathcal{F} \colon [-1; 1] \to [-0.5; 0.5]$ as $\mathcal{F}(\alpha) = \frac{\alpha}{2}$. We define the reverse conversion function $\mathcal{F}^{-1} \colon [-0.5; 0.5] \to [-1; 1]$ as $\mathcal{F}^{-1}(\alpha) = \alpha \times 2$.

4   Probabilistic Propagated Influence

The semantics of a direct influence being established, we define how to combine influences to compute the propagated influence in a probabilistic cognitive map.

We call the operation of propagated influence in a probabilistic cognitive map *the probabilistic propagated influence*. We consider that such an influence should take its values in the same value set as the one the cognitive map is defined on. However, we have stated that the value of a direct influence is linked to the difference between a conditional probability and an a priori probability and that this difference belongs to $[-0.5; 0.5]$. The propagated influence being the combination of many direct influences, its value should also belong to $[-0.5; 0.5]$. Before computing the probabilistic propagated influence, we compute what we call the *partial probabilistic propagated influence* $\mathcal{IP}_\mathbb{P}'$ that represents this difference. Since it takes its values in $[-0.5; 0.5]$, we use the reverse conversion function to compute the probabilistic propagated influence and get back a value of the original value set.

The computation of the partial probabilistic propagated influence of a concept on another one is based on that of the propagated influence described in Definition 2. First, we list the paths between the two concepts. Then we compute the influence value of each path. Finally, we aggregate these influence values.

Since a probabilistic cognitive map is acyclic, the set of paths between two concepts is necessarily finite.

We need then to compute the influence value of each of these paths. The *probabilistic propagated influence on a path* $\mathcal{IP}_\mathbb{P}$ represents the influence value of the said path. To compute this value, we cannot simply multiply the converted values in the same way we did for the values of $[-1; 1]$ in the previous section as the result of such a product would belong to something like $[-(0.5^n); 0.5^n]$. A better way to aggregate the values is to multiply the converted values by 2 before the product and then divides the final result by 2. Thus, we get a value that belongs to $[-0.5; 0.5]$.

Definition 3 (Probabilistic Propagated Influence on a Path). *Let $\mathcal{F}$ be a conversion function. Let P be a path of length k between two concepts of CM and made of influences (u_i, u_{i+1}) with $i \in [0; k-1]$. The probabilistic propagated influence on P is* $\mathcal{IP}_\mathbb{P}(P) = \frac{1}{2} \times \prod_{i=0}^{k-1} 2 \times \mathcal{F}\big(\text{label}\big((u_i, u_{i+1})\big)\big).$

Example 6. We consider the path $p_1 = R \to G \to P$ in *CM6* (Example 2). We use the conversion function defined in Example 5. The probabilistic propagated influence on p_1 is $\mathcal{IP}_\mathbb{P}(p_1) = \frac{1}{2} \times \big(2 \times \mathcal{F}(0.6)\big) \times \big(2 \times \mathcal{F}(0.8)\big) = 0.24$.

To compute the probabilistic propagated influence, we aggregate the values of the probabilistic propagated influences on the paths between two concepts. This aggregation is also different from the one defined in the previous section. Before the aggregation, we need to weight each propagated influence on a path. This weight is called the *part of a path*. The idea is to consider that the influence values of the parents of each concept are of equal importance during the computation of the probabilistic propagated influence.

Following that reasoning on paths, the part of a path is simply 1 divided by the product of the number of parents of every concept crossed by this path, except the first one.

Definition 4 (Part of a Path). *Let P be a path of length k between two concepts of CM and made of influences (u_i, u_{i+1}) with $i \in [0; k-1]$. Let $\mathcal{C}(c)$ denote the parents of any concept c. The part of P is* $\mathrm{part}(P) = \prod\limits_{i=1}^{k} \frac{1}{|\mathcal{C}(u_i)|}$.

Example 7. We consider again the path $p_1 = R \rightarrow G \rightarrow P$ from Example 2. The part of p_1 is $\mathrm{part}(p_1) = \frac{1}{|\mathcal{C}(G)|} \times \frac{1}{|\mathcal{C}(P)|} = \frac{1}{2} \times \frac{1}{2} = \frac{1}{4}$.

Using the part and the probabilistic propagated influence on a path, we are able to compute the partial probabilistic propagated influence of a concept on another one. It is defined as the sum of the products of the part and the probabilistic propagated influence on each path between the two concepts. With such a definition, when there is no path from a concept to another one, the probabilistic propagated influence is 0, which is what we would expect since there is no way any of the first concept may influence the second one.

However, there is an exception to this definition when we want to compute the probabilistic propagated influence of a concept on itself. Since, for any random variable X and any one of its possible values x, we have $\mathbb{P}(X = x | X = x) = 1$, we should have, for any concept A, $\mathbb{P}(X_A = + | X_A = +) = 1$. Since we defined the partial probabilistic propagated influence of a concept on another one as the difference between a conditional probability and the a priori probability, the partial probabilistic propagated influence of a concept on itself should be 0.5.

Definition 5 (Partial Probabilistic Propagated Influence). *Let $\mathcal{F}$ be a conversion function. Let c_1 and c_2 be two concepts. The partial probabilistic propagated influence of c_1 on c_2 is:*

$$\mathcal{I}_\mathbb{P}'(c_1, c_2) = \begin{cases} 0.5 & \text{if } c_1 = c_2 \\ \sum\limits_{P \in \mathcal{P}_{c_1, c_2}} \mathrm{part}(P) \times \mathcal{IP}_\mathbb{P}(P) & \text{otherwise} \end{cases}$$

Example 8. We want to compute the partial probabilistic propagated influence of R on P in *CM1*. We already stated in Example 2 that there is two paths between R and P: $p_1 = R \rightarrow G \rightarrow P$ and $p_2 = R \rightarrow N \rightarrow P$. We have also already computed $\mathcal{IP}_\mathbb{P}(p_1) = 0.24$ and $\mathrm{part}(p_1) = \frac{1}{4}$ in Examples 6 and 7. We compute in the same way $\mathcal{IP}_\mathbb{P}(p_2) = -0.04$ and $\mathrm{part}(p_2) = \frac{1}{2}$. The partial probabilistic propagated influence of R on P is:
$\mathcal{I}_\mathbb{P}'(R, P) = \mathrm{part}(p_1) \times \mathcal{IP}_\mathbb{P}(p_1) + \mathrm{part}(p_2) \times \mathcal{IP}_\mathbb{P}(p_2) = \frac{1}{4} \times 0.24 + \frac{1}{2} \times -0.04 = 0.04$
The partial probabilistic propagated influence of N on S is $\mathcal{I}_\mathbb{P}'(N, S) = 0$, as there is no path linking the two concepts.
The partial probabilistic propagated influence of S on itself is $\mathcal{I}_\mathbb{P}'(S, S) = 0.5$.

We said earlier that the probabilistic propagated influence is defined as the value of the partial probabilistic propagated influence converted using the reverse conversion function. Looking closely at the definition of the partial probabilistic propagated influence, we notice that this definition looks like a weighted average of the probabilistic propagated influence on the paths. The weights are given by the respective parts of these paths. However, the sum of these weights does not equal 1. Normalizing the partial probabilistic propagated influence by the sum of the parts of the paths before converting the value has two advantages. First, we compute a real weighted average. Second, it ensures that, if two concepts are linked by a single direct influence, the probabilistic propagated influence of the first concept on the second one equals the value of the direct influence.

After this normalization is done, we can convert the value using the reverse conversion function to get our probabilistic propagated influence. Note that to avoid a division by 0 when there is no path between the two concepts, we simply convert the partial probabilistic propagated influence without any normalization.

Definition 6 (Probabilistic Propagated Influence). *Let $\mathcal{F}$ be a conversion function and $\mathcal{F}^{-1}$ be its reverse conversion function. Let c_1 and c_2 be two concepts. The probabilistic propagated influence of c_1 on c_2 is:*

$$\mathcal{I}_{\mathbb{P}}(c_1, c_2) = \begin{cases} \mathcal{F}^{-1}\left(\; \mathcal{I}_{\mathbb{P}}'(c_1, c_2) \; \right) & \text{if } \mathcal{P}_{c_1, c_2} = \emptyset \\ \mathcal{F}^{-1}\left(\dfrac{\mathcal{I}_{\mathbb{P}}'(c_1, c_2)}{\sum\limits_{P \in \mathcal{P}_{c_1, c_2}} \text{part}(P)} \right) & \text{otherwise} \end{cases}$$

Example 9. As in Example 8, we compute this time the probabilistic propagated influence of R on P. We use the reverse conversion function defined in Example 5. The probabilistic propagated influence of R on P is:

$$\mathcal{I}_{\mathbb{P}}(R, P) = \mathcal{F}^{-1}\left(\frac{\mathcal{I}_{\mathbb{P}}'(R, P)}{\text{part}(p_1) + \text{part}(p_2)} \right) = \left(\frac{0.04}{\frac{1}{4} + \frac{1}{2}} \right) \times 2 = 0.1067$$

As there is no path between N and S, the probabilistic propagated influence is 0. For the same reason, the probabilistic propagated influence of S on itself is 1.

5   Relations with the Bayesian Network Model

In order to prove the validity of the probabilistic cognitive map model and the definition of the probabilistic propagated influence associated to it, we define a procedure to encode any probabilistic cognitive map into a Bayesian network. We demonstrate also that, in such a cognitive map, the computation of the probabilistic propagated influence equals the computation of a specific conditional probability in the related Bayesian network.

We give first the idea of the encoding in Sect. 5.1. We then show more clearly the relation between the probabilistic propagated influence and a conditional probability in the associated Bayesian network in Sect. 5.2.

5.1   Encoding a Cognitive Map as a Bayesian Network

The Bayesian network is built from the cognitive map such that each node of the map (concept) is encoded as a node in the network. Each influence between two concepts of the map is also encoded as an arc between the two nodes in the network that represent these concepts. So, the network has the same graphical structure as the map. Thus, we give the same name to the cognitive map nodes and to the Bayesian network nodes.

Having the same structure as the Bayesian network and the network being acyclic, the cognitive map has also to be acyclic. To remove the cycles of a cognitive map, [21,22] describe how to obtain a map structure suitable for a Bayesian network. One way to prevent cycles is to discuss with the map designer to explain what is the meaning of the links to avoid redundancy or inconsistency. Another way is to disaggregate a concept of the cycle into two time frames. That is why we consider only acyclic cognitive maps in this paper.

Each node of the Bayesian network is associated to a random variable that corresponds to the random variable the concept of the cognitive map is associated to. The probability table associated to each variable is computed from the values of the influences that end to its associated concept in the cognitive map.

We consider first the nodes that have no parent. With such nodes, the only probability values to provide are a priori probabilities. We already know these values as we stated earlier that the different states of a concept are equiprobable.

Example 10. The probability table of the node S from Example 1 gives:
$$\mathbb{P}(X_S = +) = 0.5 \qquad \mathbb{P}(X_S = -) = 0.5.$$

For the nodes that have several parents, we have to provide the conditional probabilities for every possible configuration of the states of their parents. Thus, we have to merge the values from the arcs that end to one of these nodes to express these probabilities. There are several methods to compute such probability values with only few values given by an expert. We outline briefly three of them.

Some of these methods are dedicated to the representation of a cognitive map as a Bayesian network. [19] provides a procedure that works only for cognitive maps defined on $[-1; 1]$. However, it leads to obtain a probability of 1 in each probability table. The combined influence of several parents may thus be total even if the values of each influence is low. This problem is obvious when we consider only two concepts linked by an influence: if the influence has either a value of 0.1 or 0.9, these values would be represented by the same value of 1 in the probability table. Thus, the original influence value is lost. Note that [20] uses a similar method, but with two values on each influence.

The noisy-OR model [23] leads to compute the table from individual conditional probabilities. In this model, the variables must be binary and the combined influence of several parents does not matter, as in cognitive maps. However, it is necessary to suppose that the given probabilities correspond to the case where only one parent is set to a specific value and all the others are set to the opposite value. This means that we have to give probabilities such as $\mathbb{P}(X_B = + | X_{A_1} = -, \ldots, X_{A_{i-1}} = -, X_{A_i} = +, X_{A_{i+1}} = -, \ldots, X_{A_n} = -)$. This is not consistent with the fact that the notion of influence is independent from the other parents.

[24] uses a weighted average on many values. These values and the weights are given by an expert. Each expert value represents the probability of a node considering only one of its parents. The weights represent the relative strengths of the influence of the parents. This method is suitable for cognitive maps. The

question asked to the expert is indeed: "Given that the value of the parent Y is y, compatible with the values of the other parents, what should be the probability distribution over the states of the child X?". A parent Y_i with a value y_i is said *compatible* with another parent Y_j with a value y_j if, according to the expert's mind, the state $Y_i = y_i$ is most likely to coexist with the state $Y_j = y_j$ [24]. This configuration helps the expert to focus only on the state $Y_i = y_i$. We use this method in our encoding of a cognitive map as a Bayesian network to fill the probability table of a node with many parents.

In a cognitive map, the expert values are given by the influence values, provided by the map designer fulfilling the role of the expert. In the previous section, we stated that an influence value is linked to the difference between a conditional probability and an a priori probability. The expert values being considered as conditional probabilities, we define the expert value associated to an influence as the sum of the a priori probability and the converted influence value. Let us consider a concept B with n parents A_i, each of them bringing an influence value α_i. With our example, the expert value of $X_B = +$ when $X_{A_i} = +$ is therefore $0.5 + \mathcal{F}(\alpha_i)$. Thus, the question to ask to the map designer to get an influence value is: "Given that A is increasing, this increase being compatible with the states of the other parents of B, how much the probability that B is increasing should increase?". We also stated in the previous section that the probability of X_B when $X_{A_i} = +$ is the complement of the probability of X_B when $X_{A_i} = -$. Therefore, the expert value of $X_B = +$ when $X_{A_i} = -$ is $0.5 - \mathcal{F}(\alpha_i)$.

Besides the values given by the expert, we also need to provide a weight for each value. However, in a cognitive map, it is not possible to indicate that the influence of a concept is more important than the influence of another one. Thus, the values of the influences are considered to be evenly important and we give the same weight for each value.

Definition 7 (Probability Table of a Concept). *Let $\mathcal{F}$ be a conversion function. Let B be a concept and let X_B be the random variable associated to B. Let $A_i \in \mathcal{C}(B)$ be the parents of B, each of them being associated to a random variable X_{A_i}. We note, for each A_i, $\alpha_i = \text{label}\big((A_i, B)\big)$ and a_i the value of X_{A_i}. The probability table of X_B is:*

$\mathbb{P}$	$\ldots$	$X_{A_1} = a_1, \ldots, X_{A_n} = a_n$	$\ldots$
$X_B = +$		$0.5 + \frac{1}{n}\sum_{i=1}^{n} c(a_i)$	
$X_B = -$		$0.5 - \frac{1}{n}\sum_{i=1}^{n} c(a_i)$	

where $c(a_i) = \begin{cases} \mathcal{F}(\alpha_i) & \text{if } a_i = + \\ -\mathcal{F}(\alpha_i) & \text{if } a_i = - \end{cases}$

Example 11. Let us consider the node G of *CM1* (Example 1). We give just one example of a computation of a conditional probability: the conditional probability that G is increasing given that S is decreasing and R is increasing:

$$\mathbb{P}(X_G = + \mid X_S = -, X_R = +) = 0.5 + \tfrac{1}{2}\big(-\mathcal{F}(0.9) + \mathcal{F}(0.8)\big) = 0.475$$

The full probability table of the variable X_G is:

$\mathbb{P}$	$X_S = +, X_R = +$	$X_S = +, X_R = -$	$X_S = -, X_R = +$	$X_S = -, X_R = -$
$X_G = +$	0.925	0.525	0.475	0.075
$X_G = -$	0.075	0.475	0.525	0.925

5.2 Relation Between the Probabilistic Propagated Influence and a Conditional Probability

To ensure that the probabilistic cognitive map model is valid, we still need to show that our probabilistic propagated influence corresponds to some inference in the associated causal Bayesian network. Let us consider two concepts A and B. We want to express the link between the probabilistic propagated influence and a probability expressed on A and B. Being causal, the reasoning in a cognitive map is only deductive. The notion of intervention in a causal Bayesian network leads also to a strictly deductive reasoning. That's why this model is closer to the cognitive maps than the classic one: studying the influence of a concept is indeed similar to intervene on the value of a variable. Therefore, the propagated influence of A on B is linked to $\mathbb{P}(X_B = + | \mathrm{do}(X_A = +))$. We stated in Sect. 3.2 that the partial probabilistic propagated influence between A and B is based on the difference between the conditional probability of B given A and the a priori probability of B. This conditional probability is thus $\mathbb{P}(X_B = + | \mathrm{do}(X_A = +))$ and the a priori probability of B is $\mathbb{P}(X_B = +) = 0.5$.

Theorem 1 formally expresses the link between the partial probabilistic propagated influence of concept on another one and a conditional probability on the random variables associated to these concepts.

Theorem 1. *Let CM be a probabilistic cognitive map. Let A and B be two concepts of CM. We have $\mathcal{I}_{\mathbb{P}}{}'(A, B) = \mathbb{P}(X_B = + | \mathrm{do}(X_A = +)) - 0.5$.*

Due to a lack of space, the whole proof is not shown here but it is available in a technical report [25]. The idea is first to define the partial probabilistic propagated influence as a recursive operator, given by the following lemma.

Lemma 1. *The Definition 5 of the partial probabilistic propagated influence is equivalent to:*

$$\mathcal{I}_{\mathbb{P}}{}'(A, B) = \begin{cases} 0.5 & \text{if } A = B \\ 0 & \text{if } \mathcal{P}_{A,B} = \emptyset \\ \frac{2}{|\mathcal{C}(B)|} \times \sum_{B' \in \mathcal{C}(B)} \Big(\mathcal{F}(\mathrm{label}(B', B)) \times \mathcal{I}_{\mathbb{P}}{}'(A, B') \Big) & \text{otherwise} \end{cases}$$

The equivalence is proven by considering each case separately. For the general case, each definition (Definitions 3, 4 and 5) is unfolded into a single sum on the influences of each path between the two concepts, and the recursion is deduced by extracting the terms linked to the common last influence of all paths.

Then, we prove the fact that, in a causal Bayesian network that represents a cognitive map, any $\mathbb{P}(X_B = + | \mathrm{do}(X_A = +)) - 0.5$ can also be written as a recursive operator equivalent to the one of Lemma 1. First, we consider a Bayesian network that represents the cognitive map. We apply the $\mathrm{do}(\cdot)$ operator by removing from this network the arcs that ends on A, as stated in Sect. 2.3. Then, different cases are evaluated separately: $A = B$, A is not an ancestor of B, A is an ancestor and a parent of B, and A is an ancestor but not a parent of B. The first two cases are easily proven by applying basic probability relations. The two other ones are more difficult since they are both recursive. The idea is develop the computation of the conditional probability on the parents of B as a sum, and then to use d-separation in order to simplify the expression. By using Definition 7, and by considering the sign of each influence value, the value of the last influence to B can be extracted from the sum, in order to deduce the recursion. The equivalence between the two recursive operators is then obvious once it is proven that $\mathcal{P}_{A,B} = \emptyset$ iff A is not an ancestor of B, and that the two last cases of the Bayesian network are equal to the general case of Lemma 1.

We do not currently known the complexity of the computation of the probabilistic propagated influence but it seems that the computation can be expressed as a simple value propagation over the nodes in the topological order. Therefore, its complexity should be polynomial.

6   System

We implemented our model in a software called VSPCC[1]. This software allows to build cognitive maps and to perform diverse operations on them, such as the computation of the propagated influence. Figure 3 (left) shows how to build a cognitive map with our software. The top part is used to build the cognitive map by selecting the concepts of the bottom part and then by adding influences between them. The value of each influence must belong to the value set that was chosen beforehand.

Once the cognitive map is built, the propagated influence between any concept of the map on any other one can be computed, as shown in Fig. 3 (right).

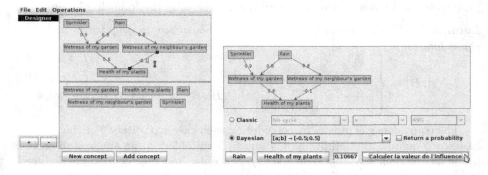

Fig. 3. Screen captures of the building of a cognitive map and of the computation of the propagated influence in VSPCC.

---

[1] Available at: http://forge.info.univ-angers.fr/~ledorze/vspcc/.

First, we choose if we want to use the "classic" propagated influence (Definition 2) or the probabilistic propagated influence defined in Sect. 4. If the later is chosen, then we select a conversion function with the drop-down list. This list proposes every registered function that is compatible with the value set of the map. Finally, we select the influencing concept and the influenced concept. The software then computes automatically the propagated influence between the two concepts. Notice the "Return a probability" checkbox: it allows to return a probability instead of an influence value. To do so, 0.5 is simply added to the partial probabilistic propagated influence, as shown by Theorem 1.

7 Conclusions

In this paper, we introduced the new probabilistic cognitive map model where the influence values are interpreted as probabilities. We defined consequently the semantics of the concepts and the influences and how to compute the propagated influence of a concept on another one in such a map. This model gives thus a stronger semantics to cognitive maps and provides a better usability. It also helps to clarify the links between cognitive maps and Bayesian networks.

This new model can be quite difficult to master, especially for laymen that are not familiar with probabilities. One way to help them in their building task is to validate their maps in order to ensure that they correctly built it [26].

As said in the introduction, the QPNs are another approach to link Bayesian networks and cognitive maps. It would be interesting to know if they could be related to probabilistic cognitive maps. To do so, we could consider some extensions of the QPN model [27,28] that quantify the constraints to express the strength of the relations between variables, as cognitive maps. Regarding causality, since QPNs are based on Bayesian network, representing causality in a QPN is mostly the same as in a causal Bayesian network.

Last, even if it was not the initial goal, we can see the work presented in this paper as a first step about learning Bayesian networks when the information is expressed with a cognitive map, a cognitive map being an easy model to capture informal knowledge. Conversely, representing a Bayesian Network as a cognitive map could help an expert to better understand the network he has built.

References

1. Axelrod, R.M.: Structure of Decision: The Cognitive Maps of Political Elites. Princeton University Press, Princeton (1976)
2. Tolman, E.C.: Cognitive maps in rats and men. Psychol. Rev. **55**(4), 189–208 (1948)
3. Celik, F.D., Ozesmi, U., Akdogan, A.: Participatory ecosystem management planning at Tuzla Lake (Turkey) using Fuzzy cognitive mapping (2005). eprint arXiv:q-bio/0510015
4. Levi, A., Tetlock, P.E.: A cognitive analysis of Japan's 1941 decision for war. J. Confl. Resolut. **24**, 195–211 (1980)

5. Zhou, S., Zhang, J.Y., Liu, Z.Q.: Quotient FCMs - a decomposition theory for Fuzzy cognitive maps. IEEE Trans. Fuzzy Syst. **11**, 593–604 (2003)
6. Kosko, B.: Fuzzy cognitive maps. Int. J. Man-Mach. Stud. **24**, 65–75 (1986)
7. Satur, R., Liu, Z.Q.: A contextual Fuzzy cognitive map framework for geographic information systems. IEEE Trans. Fuzzy Syst. **7**, 481–494 (1999)
8. Aguilar, J.: A survey about Fuzzy cognitive maps papers. Int. J. Comput. Cogn. **3**, 27–33 (2005)
9. Pearl, J.: Probabilistic Reasoning in Intelligent Systems: Networks of Plausible Inference. Morgan Kaufmann Publishers Inc., San Francisco (1988)
10. Pearl, J.: Causality: Models, Reasoning and Inference, 2nd edn. Cambridge University Press, New York (2009)
11. Song, H.J., Shen, Z.Q., Miao, C.Y., Liu, Z.Q., Miao, Y.: Probabilistic Fuzzy cognitive map. In: FUZZ-IEEE 2006, pp. 1221–1228. IEEE (2006)
12. Krichéne, J., Boudriga, N.: Incident response probabilistic cognitive maps. In: Proceedings of the IEEE International Symposium on Parallel and Distributed Processing with Applications (ISPA 2008), pp. 689–694. IEEE, Los Alamitos (2008)
13. Wellman, M.P.: Fundamental concepts of qualitative probabilistic networks. Artif. Intell. **44**, 257–303 (1990)
14. Lionel, C., David, G., Aymeric, L.D., Stéphane, L.: User centered cognitive maps. In: Guillet, F., Pinaud, B., Venturini, G., Zighed, D.A. (eds.) Advances in Knowledge Discovery and Management. SCI, vol. 471, pp. 203–220. Springer, Heidelberg (2013)
15. Genest, D., Loiseau, S.: Modélisation, classification et propagation dans des réseaux d'influence. Technique et Science Informatiques **26**, 471–496 (2007)
16. Cooper, G.F.: The computational complexity of probabilistic inference using Bayesian belief networks. Artif. Intell. **42**, 393–405 (1990)
17. Charniak, E., McDermott, D.: Introduction to Artificial Intelligence. Addison-Wesley, Reading (1985)
18. Spirtes, P., Glymour, C., Scheines, R.: Causation, Prediction, and Search, vol. 1. MIT Press, Cambridge (2001)
19. Cheah, W.P., Kim, K.-Y., Yang, H.-J., Choi, S.-Y., Lee, H.-J.: A manufacturing-environmental model using Bayesian belief networks for assembly design decision support. In: Okuno, H.G., Ali, M. (eds.) IEA/AIE 2007. LNCS (LNAI), vol. 4570, pp. 374–383. Springer, Heidelberg (2007)
20. Sedki, K., Bonneau de Beaufort, L.: Cognitive maps and Bayesian networks for knowledge representation and reasoning. In: ICTAI 2012, pp. 1035–1040. IEEE (2012)
21. Nadkarni, S., Shenoy, P.P.: A Bayesian network approach to making inferences in causal maps. Eur. J. Oper. Res. **128**, 479–498 (2001)
22. Nadkarni, S., Shenoy, P.P.: A causal mapping approach to constructing Bayesian networks. Decis. Support Syst. **38**, 259–281 (2004)
23. Lemmer, J.F., Gossink, D.E.: Recursive noisy or - a rule for estimating complex probabilistic interactions. Trans. Syst. Man Cybern. Part B **34**, 2252–2261 (2004)
24. Das, B.: Generating conditional probabilities for Bayesian networks: easing the knowledge acquisition problem. CoRR cs.AI/0411034 (2004)
25. Le Dorze, A., Duval, B., Garcia, L., Genest, D., Leray, P., Loiseau, S.: Probabilistic cognitive maps. Technical report, LERIA, Université d'Angers (2013)
26. Le Dorze, A., Garcia, L., Genest, D., Loiseau, S.: Validation of a cognitive map. In: Duval, B., van den Herik, J., Loiseau, S., Filipe, J. (eds.) ICAART 2014, vol. 1 of Science and Technology Publications, pp. 320–327. SCITEPRESS (2014)

27. Renooij, S., van der Gaag, L. C.: From qualitative to quantitative probabilistic networks. In: Darwiche, A., Friedman, N. (eds.) UAI 2002, pp. 422–429. Morgan Kaufmann (2002)
28. Renooij, S., Parsons, S., Pardieck, P.: Using kappas as indicators of strength in qualitative probabilistic networks. In: Nielsen, T.D., Zhang, N.L. (eds.) ECSQARU 2003. LNCS (LNAI), vol. 2711, pp. 87–99. Springer, Heidelberg (2003)

New Techniques for Checking Dynamic Controllability of Simple Temporal Networks with Uncertainty

Luke Hunsberger$^{(\boxtimes)}$

Computer Science Department, Vassar College, Poughkeepsie, NY 12604-0444, USA
hunsberg@cs.vassar.edu

Abstract. A Simple Temporal Network with Uncertainty (STNU) is a structure for representing time-points, temporal constraints, and temporal intervals with uncertain—but bounded—durations. The most important property of an STNU is whether it is dynamically controllable (DC)—that is, whether there exists a strategy for executing its time-points such that all constraints will necessarily be satisfied no matter how the uncertain durations turn out. Algorithms for checking from scratch whether STNUs are dynamically controllable are called (full) DC-checking algorithms. Algorithms for checking whether the insertion of one new constraint into an STNU *preserves* its dynamic controllability are called *incremental* DC-checking algorithms. This paper introduces novel techniques for speeding up both full and incremental DC checking. The first technique, called *rotating Dijkstra,* enables constraints generated by propagation to be immediately incorporated into the network. The second uses novel heuristics that exploit the nesting structure of certain paths in STNU graphs to determine good orders in which to propagate constraints. The third technique, which only applies to incremental DC checking, maintains information acquired from previous invocations to reduce redundant computation. The most important contribution of the paper is the incremental algorithm, called *Inky,* that results from using these techniques. Like its fastest known competitors, *Inky* is a cubic-time algorithm. However, a comparative empirical evaluation of the top incremental algorithms, all of which have only very recently appeared in the literature, must be left to future work.

Keywords: Temporal Networks · Uncertainty · Dynamic controllability

1 Introduction

An intelligent agent must be able to plan, schedule and manage the execution of its activities. Invariably, those activities are subject to a variety of temporal constraints, such as release times, deadlines and precedence constraints. In addition, in some domains, the agent may control the starting times for actions, but not their durations [1,7]. For a simple example, I may control the starting time for my taxi ride to the airport, but not its duration. Although I may know that

© Springer International Publishing Switzerland 2015
B. Duval et al. (Eds.): ICAART 2014; LNAI 8946, pp. 170–193, 2015.
DOI: 10.1007/978-3-319-25210-0_11

the ride will last between 15 and 30 minutes, I only *discover* the actual duration in real time, when I arrive at the airport. Therefore, if I need to ensure that I arrive at the airport no later than 10:00, I must start my taxi ride no later than 9:30 to guard against the possibility that the ride might last 30 minutes. In more complicated examples involving large numbers of actions with uncertain durations, generating a succesful *execution strategy* becomes more challenging.

A *Simple Temporal Network with Uncertainty* (STNU) is a data structure that an agent can use to support the planning, scheduling and executing of its activities, some of which may have uncertain durations [10]. The most important property of an STNU is whether it is *dynamically controllable* (DC)—that is, whether there exists a dynamic strategy for executing the constituent actions such that all temporal constraints are guaranteed to be satisfied no matter how the uncertain action durations happen to turn out in real time. The strategy is *dynamic* in that its execution decisions may depend on past execution events, but not on advance knowledge of future events. Algorithms for determining from scratch whether an STNU is DC are called (full) DC-checking algorithms. The fastest DC-checking algorithm reported so far is the $O(N^3)$-time algorithm presented by Morris in 2014 [9], where N is the number of time-points in the network.

In most applications, an STNU is not populated with constraints all at once, but instead one constraint—or a few constraints—at a time. As each new constraint is added to the network, it is important to know whether the dynamic controllability property has been preserved. An *incremental* DC-checking algorithm is an algorithm for determining whether the insertion of a single new (or tighter) constraint into a DC STNU preserves the DC property. Several related incremental DC-checking algorithms have been reported in the literature [12–16]. The latest algorithm in this sequence [14], which is quite similar to Morris' full DC-checking algorithm, also runs in $O(N^3)$ time.

This paper introduces several new techniques for speeding up both full and incremental DC checking. The first technique, called *rotating Dijkstra*, enables constraints generated by propagation to be immediately incorporated into the network. The second uses novel heuristics that exploit the nesting structure of certain paths in STNU graphs to determine good orders in which to propagate constraints. The third, which applies only to incremental DC checking, maintains information from prior invocations to significantly reduce redundant computation.

The full DC-checking algorithm that results from using these techniques is called *Speedy*. A preliminary version of this paper [6] showed that *Speedy* achieves a significant improvement over the $O(N^4)$-time DC-checking algorithm of Morris [8] which, at the time, was the fastest known DC-checking algorithm. Morris subsequently presented his $O(N^3)$-time DC-checking algorithm [9]. It is not known whether *Speedy* will be competitive with this new algorithm, which follows a completely different approach to DC checking. *Speedy* is most likely to be competitive in scenarios involving relatively small numbers of contingent links, which may make it a practical alternative.

The main contribution of this paper is the incremental algorithm, called *Inky*, that results from the new techniques. The worst-case performance of *Inky* is

Fig. 1. The graphs for the STNs discussed in the text.

$O(N^3)$, which matches that of the fastest known alternatives [9,14]. Furthermore, given its reduction of redundant computations, it is expected not only to be competitive, but perhaps much faster than these algorithms. Unfortunately, since the fastest known alternatives have only very recently been published, a full comparative evaluation of the top competitors must be left to future work.

2   Background

This section presents relevant background about Simple Temporal Networks (STNs) and Simple Temporal Networks with Uncertainty (STNUs). The presentation highlights the strong analogies between STNs and STNUs, culminating in the analogous Fundamental Theorems that explicate the relationships between an STN/STNU, its associated graph, and its associated shortest-paths matrix.

2.1   Simple Temporal Networks

A Simple Temporal Network is a pair, $(\mathcal{T}, \mathcal{C})$, where $\mathcal{T}$ is a set of real-valued variables called *time-points,* and $\mathcal{C}$ is a set of binary constraints of the form, $Y - X \leq \delta$, where $X, Y \in \mathcal{T}$ and $\delta \in \mathbb{R}$ [3]. An STN is called *consistent* if it has a solution (i.e., a set of values for the time-points that jointly satisfy the constraints). Consider the STN where:

$$\mathcal{T} = \{A, C, X, Y\}; \quad \mathcal{C} = \{(C - A \leq 10), (A - C \leq -5), (C - Y \leq 3), (X - C \leq -2)\}.$$

It is consistent. One of its solutions is: $\{(A = 0), (C = 6), (X = 3), (Y = 4)\}$.

Each STN, $\mathcal{S} = (\mathcal{T}, \mathcal{C})$, has an associated graph, $\mathcal{G} = \langle \mathcal{T}, \mathcal{E} \rangle$, where the time-points in $\mathcal{T}$ serve as the nodes for the graph, and the constraints in $\mathcal{C}$ correspond one-to-one to its edges. In particular, each constraint, $Y - X \leq \delta$, in $\mathcal{C}$ corresponds to an edge, $X \xrightarrow{\delta} Y$, in $\mathcal{E}$. The graph for the STN above is shown on the lefthand side of Fig. 1. For convenience, the constraints and edges associated with an STN are called *ordinary* constraints and *ordinary* edges.

Each path in an STN graph, $\mathcal{G}$, corresponds to a constraint that must be satisfied by any solution for the associated STN, $\mathcal{S}$. In particular, if $\mathcal{P}$ is a path from X to Y of length $|\mathcal{P}|$ in $\mathcal{G}$, then the constraint, $Y - X \leq |\mathcal{P}|$, must be satisfied by any solution to $\mathcal{S}$. For example, in the STN from Fig. 1, the path from Y to C to A of length -2 represents the constraint, $A - Y \leq -2$ (i.e., $Y \geq A + 2$). The righthand graph in Fig. 1 includes a dashed edge from Y to A that makes this constraint explicit. Note that this *derived* constraint is satisfied by the solution given earlier. Similar remarks apply to the edge from A to X.

Due to these sorts of connections, the all-pairs, shortest-paths (APSP) matrix—called the *distance matrix*, $\mathcal{D}$—plays an important role in the theory of STNs. In fact, the *Fundamental Theorem of STNs* states that the following are equivalent: (1) $\mathcal{S}$ is consistent; (2) each *loop* in $\mathcal{G}$ has non-negative length; and (3) $\mathcal{D}$ has only non-negative entries down its main diagonal [3,4].

2.2   Simple Temporal Networks with Uncertainty

A Simple Temporal Network with Uncertainty augments an STN to include a set, $\mathcal{L}$, of *contingent links*, each of which represents a temporal interval whose duration is bounded but uncontrollable [10]. Each contingent link has the form, (A, x, y, C), where $A, C \in \mathcal{T}$ and $0 < x < y < \infty$. A is called the *activation* time-point; C is called the *contingent* time-point. Although the link's duration, $C - A$, is uncontrollable, it is guaranteed to lie within the interval, $[x, y]$. When an agent uses an STNU to manage its activities, contingent links typically represent actions with uncertain durations. The agent may control the action's starting time (i.e., when A executes), but only *observes,* in real time, the action's ending time (i.e., when C executes)[1]. For example, consider the STNU:

$$\mathcal{T} = \{A, C, X, Y\}; \quad \mathcal{C} = \{(C - Y \leq 3), (X - C \leq -2)\}; \quad \mathcal{L} = \{(A, 5, 10, C)\}.$$

It is similar to the STN seen earlier, except for one important difference. In the STN, the duration, $C - A$, was constrained to lie within the interval $[5, 10]$, but the agent was free to choose any values for A and C that satisfied that constraint. In contrast, in the STNU, $C - A$ is the duration of a contingent link. This duration is guaranteed to lie within $[5, 10]$, but the agent does not get to choose this value. For example, if A is executed at 0, then the agent only gets to *observe* the execution of C when it happens, sometime between 5 and 10. In this sense, the contingent duration is uncontrollable, but bounded.

Dynamic Controllability. For an STNU, $(\mathcal{T}, \mathcal{C}, \mathcal{L})$, the most important property is whether it is *dynamically controllable* (DC)—that is, whether there exists a *strategy* for executing the controllable (i.e., non-contingent) time-points in $\mathcal{T}$ such that all constraints in $\mathcal{C}$ are guaranteed to be satisfied no matter how the durations of the contingent links in $\mathcal{L}$ turn out in real time—within their specified bounds [10]. Such strategies, if they exist, are called *dynamic execution strategies*—*dynamic* in that their execution decisions may depend on the observation of past execution events, but not on advance knowledge of future events. It is not hard to verify that the following strategy is an example of a dynamic execution strategy for the sample STNU:

- Execute A at 0, and X at 3.

---

[1] Agents are not part of the semantics of STNUs. They are used here for expository convenience.

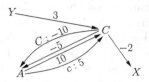 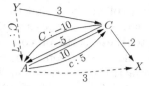

Fig. 2. The graph for the sample STNU before (left) and after (right) generating new edges.

Table 1. Morris and Muscettola's edge-generation rules for STNUs.

(No Case)	$A \xleftarrow{x} C \xleftarrow{y} D$	adds:	$A \xleftarrow{x+y} D$
(Lower Case)	$A \xleftarrow{x} C \xleftarrow{c:y} D$	adds:	$A \xleftarrow{x+y} D$
(Upper Case)	$A \xleftarrow{B:x} C \xleftarrow{y} D$	adds:	$A \xleftarrow{B:x+y} D$
(Cross Case)	$A \xleftarrow{B:x} C \xleftarrow{c:y} D$	adds:	$A \xleftarrow{B:x+y} D$
(Label Removal)	$B \xleftarrow{b:x} A \xleftarrow{B:z} C$	adds:	$A \xleftarrow{z} C$

- If C executes *before* time 7, then execute Y at time $C + 1$; otherwise, execute Y at 7.

Thus, the sample STNU is dynamically controllable. The given strategy is dynamic in that the decision to execute Y depends on observations about C.

STNU Graphs. Each STNU, $(\mathcal{T}, \mathcal{C}, \mathcal{L})$, has an associated graph, $\langle \mathcal{T}, \mathcal{E}^+ \rangle$, where the time-points in $\mathcal{T}$ serve as the nodes in the graph; and the constraints in $\mathcal{C}$ and the contingent links in $\mathcal{L}$ together give rise to the edges in $\mathcal{E}^+$ [11]. To capture the difference between constraints and contingent links, the edges in $\mathcal{E}^+$ come in two varieties: ordinary and labeled. As with an STN, each constraint, $Y - X \le \delta$, in $\mathcal{C}$ corresponds to an ordinary edge, $X \xrightarrow{\delta} Y$, in $\mathcal{E}^+$. In addition, each contingent link, (A, x, y, C), in $\mathcal{L}$ gives rise to two ordinary edges that together represent the constraint, $C - A \in [x, y]$. Finally, each contingent link, (A, x, y, C), also gives rise to the following *labeled* edges: a *lower-case* edge, $A \xrightarrow{c:x} C$, and an *upper-case* edge, $A \xleftarrow{C:-y} C$. The lower-case (LC) edge represents the *uncontrollable possibility* that the duration, $C - A$, might assume its lower bound, x. The upper-case (UC) edge represents the *uncontrollable possibility* that $C - A$ might assume its upper-bound, y. The graph for the sample STNU is shown on the lefthand side of Fig. 2.

Edge Generation for STNUs. Because the labeled edges in an STNU graph represent uncontrollable possibilities, edge generation (equiv., constraint propagation) for STNUs is more complex than for STNs. In particular, a variety of rules are required to handle the interactions between different kinds of edges. Table 1 lists the edge-generation rules for STNUs given by Morris and Muscettola

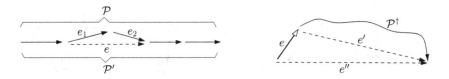

Fig. 3. (a) A generic path transformation; (b) reducing away a lower-case edge.

$(2005)^2$. The *No Case* rule encodes ordinary STN constraint propagation. The *Lower Case* rule generates edges/constraints that guard against the possibility of a contingent link taking on its minimum duration. The *Upper Case* rule generates edges/constraints that guard against the possibility of a contingent link taking on its maximum duration. The *Cross Case* rule addresses the interaction of LC and UC edges from different contingent links.

The edge-generation rules are *sound* in the sense that the edges they generate correspond to constraints that must be satisfied by any dynamic execution strategy. For example, consider the righthand graph in Fig. 2. For ease of exposition, suppose that A has been executed at 0. The edge, $C \xrightarrow{-2} X$, represents the constraint, $X - C \leq -2$ (i.e., $X \leq C - 2$), which requires X to be executed *before* the contingent time-point C. To ensure that this constraint will be satisfied even if C eventually happens to take on its minimum value of 5, X must be executed no later than 3 units after A, whence the dashed edge from A to X. This dashed edge can be generated by applying the *Lower Case* rule to the path from A to C to X.

Next, consider the edge, $Y \xrightarrow{3} C$, which represents the constraint, $C - Y \leq 3$ (i.e., $Y \geq C - 3$). To ensure that this constraint is satisfied, the following *conditional* constraint must be satisfied: *While C remains unexecuted, Y must occur 7 or more units after A.* This conditional constraint, represented by the upper-case edge, $Y \xrightarrow{C:-7} A$, effectively guards against the possibility of C taking on its *maximum* value, 10. The UC edge can be generated by applying the *Upper Case* rule to the path from Y to C to A.

It is not hard to verify that the constraints corresponding to these generated edges are satisfied by the sample dynamic execution strategy given earlier.

Semi-reducible Paths. Recall that, for an STN, each path in its graph corresponds to a constraint that must be satisfied by any solution to that STN. In contrast, for an STNU, it is the *semi-reducible* paths—defined below—that

---

[2] The rules are shown using Morris and Muscettola's notation. Note that: the x's and y's here are not necessarily bounds for contingent links; C is only required to be contingent in the *Lower Case* and *Cross Case* rules, where its activation time-point is D and its *lower* bound is y; and in the *Upper Case* and *Cross Case* rules, B is contingent, with activation time-point A. The *Lower Case* rule only applies when $x \leq 0$ and $A \neq C$; the *Cross Case* rule only applies when $x \leq 0$ and $B \neq C$; and the *Label Removal* rule only applies when $z \geq -x$.

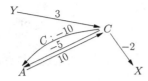
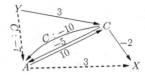

Fig. 4. The OU-graphs, $\mathcal{G}^{ou}$, for the corresponding graphs from Fig. 2.

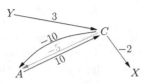
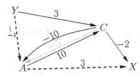

Fig. 5. The *AllMax* graphs, $\mathcal{G}_x$, for the corresponding graphs from Fig. 2.

correspond to the (possibly conditional) constraints that must be satisfied by any dynamic execution strategy for that STNU. Whereas an STN is consistent if and only if its graph has no negative-length loops, an STNU is dynamically controllable if and only if its graph has no *semi-reducible* negative-length loops [8].

Before defining semi-reducible paths, it is useful to view the edge-generation rules from Table 1 as *path-transformation* rules, as follows. Suppose e_1 and e_2 are consecutive edges in a path $\mathcal{P}$, and that one of the first four rules can be applied to e_1 and e_2 to generate a new edge e, as illustrated in Fig. 3(a), where $\mathcal{P}'$ is the path obtained from $\mathcal{P}$ by replacing the edges, e_1 and e_2, with e. We say that $\mathcal{P}$ has been transformed into $\mathcal{P}'$. Similar remarks apply to the *Label Removal* rule, which operates on a single edge.

A path in an STNU graph is called *semi-reducible* if it can be transformed into a path that has only ordinary or upper-case edges [8]. Thus, for any semi-reducible path $\mathcal{P}$, there must be some transformation of $\mathcal{P}$ whereby each lower-case edge e in $\mathcal{P}$ is eventually "reduced away" by either the *Lower Case* or *Cross Case* rule. In other words, as illustrated in Fig. 3(b), for each lower-case edge e in $\mathcal{P}$, there must be some sub-path, $\mathcal{P}^\dagger$, following e in $\mathcal{P}$, such that $\mathcal{P}^\dagger$ can be transformed into a single edge, e', using the rules from Table 1. Depending on whether e' is ordinary or upper-case, either the *Lower Case* or *Cross Case* rule can then be used to transform e and e' into a single edge, e'', effectively reducing away the lower-case edge e.

The soundness of the edge-generation rules ensures that the constraints represented by semi-reducible paths must be satisfied by any dynamic execution strategy. Since shorter paths correspond to stronger constraints, the all-pairs, shortest-*semi-reducible*-paths (APSSRP) matrix, $\mathcal{D}^*$, plays an important role in the theory of STNUs. The *Fundamental Theorem of STNUs* states that the following are equivalent for any STNU $\mathcal{S}$: (1) $\mathcal{S}$ is dynamically controllable; (2) every *semi-reducible* loop in its associated graph has non-negative length;

and (3) its APSSRP matrix, $\mathcal{D}^*$, has only non-negative entries along its main diagonal [5,8].

2.3   Morris' $O(N^4)$-time DC-checking Algorithm

In 2006, Morris presented an $O(N^4)$-time DC-checking algorithm—hereinafter called the *Morris-N^4* algorithm—that uses the rules from Table 1 to generate new edges. Each newly generated edge is added not only to $\mathcal{G}$, but also, in a stripped down form, to a related *STN* graph, called the *AllMax* graph. If the *AllMax* graph ever exhibits a negative-length loop, the original STNU is declared to be non-DC. The algorithm achieves its efficiency by focusing its edge-generation activity on reducing away lower-case edges. Although Morris has since presented a faster, $O(N^3)$-time algorithm [9], the Morris-N^4 algorithm lays a foundation for the rest of this paper and, so, is summarized below.

Let $\mathcal{S}$ be an STNU and $\mathcal{G}$ its associated graph. The lengths of all shortest *semi-reducible* paths in $\mathcal{G}$ can be determined as follows. First, for convenience, any *ordinary* or *upper-case* edge may be called an *OU-edge,* and the graph consisting of all of the OU-edges from $\mathcal{G}$ shall be called the *OU-graph* for $\mathcal{G}$, denoted by $\mathcal{G}^{\mathrm{ou}}$. Since the edges in $\mathcal{G}^{\mathrm{ou}}$ are drawn from $\mathcal{G}$, any path in $\mathcal{G}^{\mathrm{ou}}$ also appears in $\mathcal{G}$. In addition, since each path in $\mathcal{G}^{\mathrm{ou}}$ contains only OU-edges, it is trivially semi-reducible. Thus, the paths in $\mathcal{G}^{\mathrm{ou}}$ are a subset of the semi-reducible paths in $\mathcal{G}$. Furthermore, since the rules from Table 1 generate only OU-edges, inserting any such edges into *both* $\mathcal{G}^{\mathrm{ou}}$ and $\mathcal{G}$ necessarily preserves the property that the paths in $\mathcal{G}^{\mathrm{ou}}$ are a subset of the semi-reducible paths in $\mathcal{G}$. For example, Fig. 4 shows the OU-graph for the sample STNU from Fig. 2 before (left) and after (right) the insertion of two newly generated edges.

Next, with the goal of computing the *lengths* of the paths in $\mathcal{G}^{\mathrm{ou}}$, let $\mathcal{G}_x$ be the graph obtained by removing the *alphabetic labels* from all upper-case edges in $\mathcal{G}^{\mathrm{ou}}$. $\mathcal{G}_x$ is called the *AllMax* graph because it can be obtained from the original STNU by forcing each contingent link to take on its maximum value [11]. The *AllMax* graphs corresponding to the graphs from Fig. 2 are shown in Fig. 5. In the figure, the ordinary edge, $C \xrightarrow{-5} A$, is drawn in light gray because it represents a weaker constraint than the edge, $C \xrightarrow{-10} A$, and thus can be ignored. Note that if the edge-generation rules produce an upper-case edge (e.g., the UC edge from Y to A in Fig. 4), then that edge is stripped of its alphabetic label before being added to the *AllMax* graph, $\mathcal{G}_x$.

Since the *AllMax* graph contains only ordinary edges, it is an *STN* graph. Its associated distance matrix, $\mathcal{D}_x$, is called the *AllMax* matrix. For any X and Y, $\mathcal{D}_x(X, Y)$ equals the length of a shortest path from X to Y in $\mathcal{G}_x$. $\mathcal{D}_x(X, Y)$ also equals the length of a shortest *semi-reducible* path from X to Y in the OU-graph, $\mathcal{G}^{\mathrm{ou}}$. Because $\mathcal{G}^{\mathrm{ou}}$ may contain only a subset of the semi-reducible paths from $\mathcal{G}$, $\mathcal{D}_x(X, Y)$ only provides an *upper bound* on the length of the shortest semi-reducible path from X to Y in $\mathcal{G}$. However, as newly generated edges are inserted into the appropriate graphs, the upper bounds on the lengths of shortest semi-reducible paths provided by $\mathcal{D}_x$ typically tighten.

The Morris-N^4 algorithm focuses its attention on finding paths in the graph that can be used to reduce away lower-case edges. However, it need not find all such paths; instead, it suffices to follow shortest *allowable* paths (defined below), seeking to find *extension sub-paths* (defined below). The following sequence of definitions is provided for expository convenience. First, let e be any lower-case edge, $A \xrightarrow{c:x} C$. A *quasi-allowable* path for e is any path, $\mathcal{P}_e$, such that:

- $\mathcal{P}_e$ is a loopless path emanating from C;
- $\mathcal{P}_e$ contains only OU-edges (i.e., edges in $\mathcal{G}^{ou}$); and
- $\mathcal{P}_e$ is *breach-free* (i.e., does *not* contain any upper-case edges labeled by C)[3].

Next, the path $\mathcal{P}_e$ is said to have the *positive proper prefix* (PPP) property if every *proper* prefix of $\mathcal{P}_e$ has *positive* length. A quasi-allowable path that has the PPP property is called an *allowable* path. If, further still, $\mathcal{P}_e$ itself has non-positive length, then $\mathcal{P}_e$ is called an *extension sub-path* for e. It is not hard to show that any extension sub-path for e can be transformed into a single edge, e', which can then be used to reduce away e, as described earlier (cf. Fig. 3(b)). The resulting edge, e'', is then inserted into $\mathcal{G}^{ou}$ and—after removing any alphabetic label—$\mathcal{G}_x$. Furthermore, it is not necessary to search through paths that are quasi-allowable but not allowable, since any such path must have a proper prefix that is an extension sub-path. Of course, reducing away a lower-case edge e_1 might generate a new edge E_1 that could subsequently be used as part of an extension sub-path that reduces away another lower-case edge e_2, to generate another new edge, E_2. In such a case, the extension sub-path that generated E_1 is said to be *nested* inside the extension sub-path that generated E_2. However, Morris proved that, for an STNU with K contingent links, it suffices to consider at most K levels of such nesting. As a result, the Morris-N^4 algorithm performs at most K rounds of searching through *shortest* allowable paths to determine whether the original graph, $\mathcal{G}$, contains any semi-reducible negative loops. Pseudo-code for the Morris-N^4 DC-checking algorithm is given in Table 2. Its most important features are:

- The outer loop (Lines 1–12) runs at most K times.
- Each outer iteration begins (Line 2) by applying the Bellman-Ford single-source, shortest-paths (SSSP) algorithm [2] to the *AllMax* graph $\mathcal{G}_x$. This serves two purposes. First, if Bellman-Ford determines that the *AllMax* graph is inconsistent, then the algorithm immediately returns **False**. However if $\mathcal{G}_x$ *is* consistent, then the shortest-path information generated by Bellman-Ford can be used to create a *potential function* (Line 4) to transform the lengths of all edges in $\mathcal{G}_x$—and hence all edges in $\mathcal{G}^{ou}$—into non-negative values, as in Johnson's algorithm [2].
- During each iteration of the outer loop, the inner loop (Lines 6–9) runs exactly K times, once per contingent link.
- The jth iteration of the inner loop (Lines 7–8) focuses on C_j, the contingent time-point for the jth contingent link. The algorithm uses the potential func-

---

[3] A breach edge could prevent application of the *Cross Case* rule.

Table 2. Pseudo-code for the Morris-N^4 DC-checking algorithm.

```
-1. Gou := OU-graph for G.
 0. Gx  := AllMax graph for G.
 1. for i = 1, K:                                        (Outer Loop)
 2.   result := Bellman_Ford_SSSP(Gx).
 3.   if (result == inconsistent) return False.
 4.   else generate_potential_function(result).
 5.   newEdges := {}.
 6.   for j = 1, K:                                      (Inner Loop)
 7.     Let Cⱼ be the jᵗʰ contingent time-point.
 8.     Traverse shortest allowable paths in 𝒢ᵒᵘ emanating from Cⱼ, searching for
        extension sub-paths that generate new edges. Add new edges to newEdges.
 9.   end for j = 1, K.
10.   if newEdges empty, return True.
11.   else insert newEdges into Gou andGx.
12. end for i = 1, K.
13. result := Bellman_Ford_SSSP(Gx).
14. if (result == inconsistent) return False.
15. else return True.
```

tion generated in Line 4 to enable a Dijkstra-like traversal of shortest allowable paths emanating from C_j in the graph $\mathcal{G}^{ou}$.

- New edges generated by K iterations of the *inner* loop are accumulated in a set, **newEdges** (Line 8). Afterward, if it is discovered that no new edges have been generated, then the algorithm immediately returns **True** (Line 10). On the other hand, if some new edges were generated by the inner loop, then they are inserted into the graphs (Line 11) in preparation for the running of Bellman-Ford at the beginning of the next iteration of the outer loop (Line 2).

- If, after completing K iterations of the *outer* loop, the *AllMax* graph remains consistent (Lines 14–15), then the network must be DC[4].

The complexity of the algorithm is dominated by the $O(N^3)$-time complexity of the Bellman-Ford algorithm (Line 2), as well as the Dijkstra-like traversals of shortest allowable paths (Line 8). Since Bellman-Ford is run a maximum of K times, and $O(K) = O(N)$, the overall complexity due to the use of Bellman-Ford is $O(N^4)$. Each Dijkstra-like traversal of shortest allowable paths (Line 8) is $O(N^2)$ in the worst case. Since these traversals are run at most K^2 times, the overall contribution is again $O(N^4)$.

---

[4] This conclusion is justified by Morris' Theorem 3 that an STNU contains a semi-reducible negative loop if and only if it contains a *breach-free semi-reducible negative loop* in which the extension sub-paths are nested to a depth of at most K [8].

3   *Speedy*: Speeding Up Full DC Checking

As discussed above, the Morris-N^4 algorithm uses the Bellman-Ford algorithm to compute a potential function at the beginning of each iteration of the outer loop (Lines 2–4). This same potential function is then used for all K iterations of the inner loop (Lines 6–9). For this reason, any new edges discovered during the K iterations of the inner loop cannot be inserted into $\mathcal{G}^{ou}$ or $\mathcal{G}_x$ until preparing for the next iteration of the *outer* loop (Line 11). To see this, consider that the Dijkstra-like traversal of shortest allowable paths (Line 8) depends on all edge-lengths having been converted into non-negative values by the potential function. Incorporating new edges into this traversal without recomputing the potential function could introduce negative-length edges, violating the conditions of a Dijkstra-like traversal. A second important consequence of delaying the integration of new edges until the next *outer* iteration, is that the *order* in which the contingent links are processed by the inner loop cannot make any difference to the Morris-N^4 algorithm.

Given these observations, Hunsberger [6] made two inter-related modifications to the Morris-N^4 algorithm to significantly improve its performance. For convenience, the modified algorithm will hereinafter be called *Speedy*. The first modification used by *Speedy* is called *rotating Dijkstra*. It enables the edges generated by one iteration of the *inner* loop to be *immediately* inserted into the network for use during the very next iteration of the *inner* loop, instead of waiting until the beginning of the next *outer* iteration. Next, because each iteration of the inner loop can use the edges generated by any prior iteration, *Speedy* uses a heuristic function to choose a "good" order in which to process the lower-case edges during the K iterations of the inner loop. The heuristic is inspired by the nesting structure of so-called *magic loops* analyzed in prior work [5]. In some networks, these two changes combine to produce an order-of-magnitude speed-up in DC checking [6]. Although Morris has, in the interim, presented an $O(N^3)$-time DC-checking algorithm that *might* be faster than *Speedy,* the techniques used by *Speedy* also have novel applications to *incremental* DC checking, to be discussed in Sect. 4; therefore, these new techniques are briefly summarized below.

Recalling Johnson's Algorithm. Johnson's algorithm [2] is an all-pairs, shortest-paths algorithm that can be used on graphs whose edges have any numerical lengths: positive, negative or zero. It begins by using the Bellman-Ford single-*source*, shortest-paths algorithm to generate a potential function, h. In particular, for any node X, $h(X)$ is defined to be the length of the shortest path from some source node S to X. Johnson's algorithm then uses that potential function to convert edge lengths to non-negative values, as follows. For any edge, $U \xrightarrow{\delta} V$, the converted length is $h(U) + \delta - h(V)$. This is guaranteed to be non-negative since the path from S to V via U cannot be shorter than the shortest path from S to V. Then, for each time-point X in the graph, Johnson's algorithm runs Dijkstra's single-*source*, shortest-paths algorithm on the re-weighted edges using X as the source. This works because shortest paths in the re-weighted graph correspond to shortest paths in the original graph.

In particular, for any X and Y, the length of the shortest path from X to Y in the original graph is $h(Y) + D(X, Y) - h(X)$, where $D(X, Y)$ is the length of the shortest path from X to Y in the re-weighted graph.

Rotating Dijkstra. The rotating Dijkstra technique is based on several observations. First, just as single-*source*, shortest-paths information can be used to generate a potential function to support the conversion of edge-lengths to non-negative values, so too can single-*sink*, shortest-paths information be used in this way [4]. Furthermore, whether the re-weighting of edges is done using a source-based or sink-based potential function, Dijkstra's algorithm can be run on the re-weighted graph following edges forward from a single source or following edges backward from a single sink. This paper refers to the different combinations as *sinkPot/srcDijk, sinkPot/sinkDijk*, and so on.

Second, when a contingent link, (A, x, y, C), is being processed during one iteration of the inner loop of the Morris-N^4 algorithm, any new edge generated during that iteration must have that link's activation time-point, A, as its source[5]. However, adding edges whose *source* time-point is A cannot cause changes to the lengths of shortest paths *terminating* in A [4]. Thus, adding new edges whose source is A cannot cause any changes to entries of the form, $\mathcal{D}_x(T, A)$, for any time-point T. As a result, if the potential function used to re-weight the edges for this Dijkstra-like traversal is a sink-based potential function with A as its sink, then adding new edges generated by that traversal cannot cause any changes to that potential function. Thus, that same potential function can be used along with Dijkstra's single-*sink*, shortest-paths algorithm to re-compute the values, $\mathcal{D}_x(T, A')$, for any time-point T, in preparation for the next iteration of the inner loop, where A' is the activation time-point for the next contingent link to be processed.

Given these observations, the rotating Dijkstra technique takes the following steps to support the Dijkstra-like traversal of shortest allowable paths emanating from the contingent time-point C associated with the contingent link (A, x, y, C).

(1) Given: All entries, $\mathcal{D}_x(T, A)$, for all time-points T. This collection of entries provides a *sink*-based potential function, h_A, where A is the sink.
(2) Use h_A to convert all edge-lengths in $\mathcal{G}^{ou}$ to non-negative values in preparation for a *source*-Dijkstra traversal of shortest allowable paths emanating from C, as in the Morris-N^4 algorithm (Line 8).
(3) Since any edges generated by this traversal must have A as their source, the new edges cannot cause any changes to the sink-based potential function, h_A. Thus, h_A can subsequently be used to support a *sinkDijkstra* computation of all entries of the form $\mathcal{D}_x(T, A')$, for any T, where A' is the activation time-point for the next contingent link to be processed. This computation

---

[5] This follows immediately from how new edges are generated [8]. In particular, each new edge is generated by reducing the path consisting of the lower-case edge, $A \xrightarrow{c:x} C$, and some extension sub-path into a single new edge. Since such a reduction preserves the endpoints of the path, the generated edge must have A as its source.

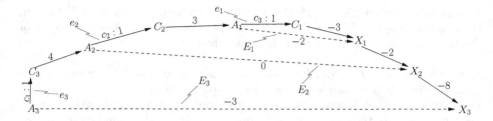

Fig. 6. A path with nested extension sub-paths.

is abbreviated as *sinkPot/sinkDijk*(A, A'), since A is the sink for the potential function, and A' is the sink for the Dijkstra traversal. It generates the potential function, $h_{A'}$, for the next iteration.

For the very first iteration of the inner loop, the entries, $\mathcal{D}_x(T, A)$, needed in Step 1 are provided by an initial run of Johnson's algorithm. For every subsequent iteration, the information needed in Step 1 is obtained from Step 3 of the preceding iteration.

Choosing an Order in which to Process the Contingent Links. Because the rotating Dijkstra technique enables newly generated edges to be inserted into the network immediately, rather than waiting for the next iteration of the *outer* loop, edges generated by one iteration of the inner loop can be used by the very next iteration. Thus, subsequent iterations of the inner loop may generate new edges sooner than they would in the Morris-N^4 algorithm, which can significantly improve performance.

Consider the path shown in Fig. 6. The innermost sub-path, from A_1 to X_1, reduces to (i.e., can be transformed into) a new edge, E_1, from A_1 to X_1. In turn, that enables the next innermost sub-path, from A_2 to X_2, to be reduced to a new edge, E_2, from A_2 to X_2. Finally, that then enables the outermost path, from A_3 to X_3, to be reduced to a single new edge, E_3, from A_3 to X_3. Thus, in this example, if the contingent links are processed in the order, C_1, C_2, C_3, then all three edges, E_1, E_2 and E_3, will be generated in *one* iteration of the outer loop—involving *three* iterations of the inner loop. However, if the contingent links are processed in the opposite order, then *three* iterations of the outer loop—involving *nine* iterations of the inner loop—will be required to generate E_1, E_2 and E_3. To see this, notice that if C_3 is processed first, then the edges E_1 and E_2 will not have been generated yet. And, since allowable paths do not include lower-case edges, the initial search through allowable paths emanating from C_3 will not yield any new edges. Similarly, the initial search through allowable paths emanating from C_2 will not yield any new edges. Only the processing of C_1 will yield a new edge—namely, E_1—during the first iteration of the outer loop. During the second iteration of the outer loop, the processing of C_2 will yield the edge E_2. Finally, during the third iteration of the outer loop, the processing of C_3 will yield the edge E_3. Crucially, since the Morris-N^4 algorithm does not insert new edges until the beginning of the next iteration of the outer loop, that algorithm would exhibit the same behavior for this path. In general, the Morris-N^4 algorithm requires d iterations of the outer loop to generate new edges arising from semi-reducible paths in which extension sub-paths are nested to a depth d.

Nesting Order. Prior work [5] has defined a *nesting order* for semi-reducible paths as follows. Suppose $e_1, e_2, \ldots, e_n$ are the lower-case edges that appear in a semi-reducible path $\mathcal{P}$. Then that ordering of those edges constitutes a *nesting order* for $\mathcal{P}$ if $i < j$ implies that no extension sub-path for e_j is nested within an extension sub-path for e_i in $\mathcal{P}$. For example, the path shown in Fig. 6 has a nesting order e_1, e_2, e_3. The relevance of a nesting order to DC checking is that if a semi-reducible path $\mathcal{P}$ has extension sub-paths nested to a depth d, with a nesting order, $e_1, e_2, \ldots, e_d$, and the lower-case edges (i.e., the contingent links) are processed in that order using the rotating Dijkstra technique, then only *one* iteration of the outer loop will be necessary to generate all edges derivable from $\mathcal{P}$. For the purposes of this paper, it is not necessary to prove this result— although it follows quite easily from the definitions involved—because it is not claimed that for any STNU graph there is a single nesting order that applies to all semi-reducible paths in that graph. However, it does suggest that it might be worthwhile to spend some modest computational effort to find a "good" order in which to process the contingent links in the inner loop of the algorithm.

Toward that end, suppose that e_1 and e_2 are lower-case edges corresponding to the contingent links, (A_1, x_1, y_1, C_1) and (A_2, x_2, y_2, C_2). Suppose further that $\mathcal{P}$ is a shortest allowable path emanating from e_2; and that $\mathcal{P}$ contains e_1. That is, e_1 is nested inside e_2 (e.g., as illustrated in Fig. 6). Although allowable paths for e_2 cannot include any upper-case edges labeled by C_2, the OU-graph invariably includes at least one upper-case edge labeled by C_2. Thus, the *All-Max* matrix entry, $\mathcal{D}_x(C_2, A_1)$, is not a perfect substitute for the length of the shortest *allowable* path from C_2 to A_1. Instead, $\mathcal{D}_x(C_2, A_1)$ is a *lower* bound on that length. Nonetheless, *Speedy*'s heuristic uses it as an imperfect substitute.

The Heuristic, H. Let $\mathcal{G}$ be the graph for an STNU with K contingent links, and $\mathcal{G}_x$ the corresponding *AllMax* graph. Run Johnson's algorithm on $\mathcal{G}_x$ to generate the *AllMax* matrix $\mathcal{D}_x$. For each i, let $Q(i)$ be the number of entries of the form $\mathcal{D}_x(C_i, A_j)$ that are non-positive. Let $H(\mathcal{G})$ be a permutation, $\sigma_1, \sigma_2, \ldots, \sigma_K$, such that $r < s$ implies $Q(\sigma_r) \leq Q(\sigma_s)$. In other words, $H(\mathcal{G})$ is obtained by sorting the numbers $1, 2, \ldots, K$ according to the corresponding Q values.

The Speedy Algorithm. Pseudo-code for the *Speedy* DC-checking algorithm is given in Table 3. The algorithm first constructs the OU-graph, $\mathcal{G}^{ou}$, and the *All-Max* graph, $\mathcal{G}_x$ (Lines -1 and 0). It then uses Johnson's algorithm to compute the *AllMax* matrix, $\mathcal{D}_x$ (Line 1). As discussed above, $\mathcal{D}_x$ is used during the computation of the heuristic function, H (Line 2), which determines the processing order for the contingent links. These $\mathcal{D}_x$ entries also provide the potential function in Line 9 during the very first iteration of the inner loop. GlobalIters, initially 0 (Line 3), counts the total number of iterations of the inner loop. If this

Table 3. Pseudo-code for the *Speedy* DC-checking algorithm.

```
-1. Gou  := OU-graph for G;
 0. Gx   := AllMax graph for G;
 1. Dx   := Johnson(Gx);
 2. Order := Heuristic(Dx);
 3. GlobalIters := 0;
 4. LocalIters := 0;
 5. for i = 1, K:                                              (Outer Loop)
 6.   for j = 1, K:                                            (Inner Loop)
 7.     newEdges := {};
 8.     Let (Aⱼ,xⱼ,yⱼ,Cⱼ) be the jᵗʰ contingent link according to Order.
 9.     Use the values, 𝒟ₓ(T,Aⱼ), as a sink-based potential function, hₐⱼ, to transform the edge-
        lengths in Gx and Gou to non-negative values.
10.     Traverse shortest allowable paths in 𝒢ᵒᵘ emanating from Cⱼ, searching for extension sub-
        paths that generate new edges. Add such edges to newEdges.
11.     for each edge Aⱼ ──δ──▸ X in newEdges: if (δ < −𝒟ₓ(X,Aⱼ)) return False;
12.     GlobalIters++;
13.     if (GlobalIters ≥ K²) return True;
14.     elseif newEdges empty:
15.       LocalIters++;
16.       if (LocalIters ≥ K) return True;
17.     else LocalIters := 0;
18.     run sinkPotSinkDijk(Aj,A'), where A' is act'n. time-point for next contingent link.
19.   end for j = 1, K.
20. end for i = 1, K.
21. return True.
```

counter ever reaches K^2, the algorithm terminates, returning True (Line 13)[6]. LocalIters, initially 0 (Line 4), counts the number of *consecutive* iterations of the inner loop since the last time a new edge was generated. If this counter ever reaches K, then the algorithm terminates, returning True (Line 16)[7].

In the *Speedy* algorithm, the inner loop (Lines 6–19) cycles through the contingent links in the order given by the heuristic until a termination condition is reached. Since potential functions are re-computed after each *inner* iteration (Line 18), the outer loop is provided only for counting purposes. One iteration of the inner loop spans Lines 7–18. (A_j, x_j, y_j, C_j) is the contingent link to be processed, as determined by Order. For the very first iteration of the inner loop, the values, $\mathcal{D}_x(X, A_j)$, that constitute a sink-based potential function (Line 9), are provided by Johnson's algorithm (Line 1); for every other iteration of the inner loop, these values are provided by the *sinkPot/sinkDijk* computation from Line 18 of the previous iteration. This potential function is then used in Line 10 to support a srcDijk traversal of shortest allowable paths emanating from C_j. Because the values, $\mathcal{D}_x(X, A_j)$, are available for all time-points X, any new edge from A_j to some X can be immediately checked for consistency (Line 11). If all new edges are judged to be consistent, then the algorithm increments the

---

[6] This termination condition is analogous to the Morris-N^4 algorithm terminating after K iterations of the *outer* loop.

[7] This termination condition is analogous to the Morris-N^4 algorithm terminating whenever any iteration of the outer loop fails to generate a new edge.

global counter and checks the two termination conditions (Lines 13 and 16). If no termination condition is reached yet, then a sinkPot/sinkDijk computation is run (Line 18), to compute all entries of the form, $\mathcal{D}_x(X, A')$, where A' is the activation time-point for the contingent link to be processed during the next iteration of the inner loop. These values will form the potential function in Line 9 during the next iteration.

The *Speedy* algorithm was evaluated empirically against the Morris-N^4 algorithm [6], which was, at the time, the fastest DC-checking algorithm in the literature. In the special case of so-called *magic loops,* which represent one kind of worst-case scenario involving maximum nesting of lower-case edges [5], the *Speedy* algorithm exhibited an order-of-magnitude improvement over the Morris-N^4 algorithm, which suggests it would be competitive with the more recent $O(N^3)$-time algorithm. For other networks, the degree of improvement appeared to be proportional to the degree of maximal nesting, indicating that the heuristic was working effectively. Furthermore, using the *reverse* of the order suggested by the heuristic lead to significantly worse performance, again indicating that the heuristic was working effectively.

4 *Inky:* Speeding up Incremental DC Checking

This section presents a new incremental DC-checking algorithm, called *Inky,* that applies similar kinds of insights and techniques seen in the previous section to the problem of incremental DC checking. Toward that end, let S be an STNU that is known to be dynamically controllable; let $\mathcal{G}$ be its graph; and let $X \overset{\delta}{\longrightarrow} Y$ be a new edge. The incremental DC-checking problem is to determine whether inserting the new edge into the network will preserve its dynamic controllability.

To avoid redundant computations across multiple invocations, and to bound the number of rounds of edge generation, the *Inky* algorithm maintains an auxiliary K-by-N matrix, called $\mathcal{D}^+$, where for each contingent time-point C and each time-point T, $\mathcal{D}^+(C,T)$ equals the length of the shortest *quasi-allowable* path from C to T (cf. Sect. 2.3). Given a new edge from X to Y, the *Inky* algorithm begins by sorting the lower-case edges such that the values of $\mathcal{D}^+(C_i, X)$ are *non-increasing.* The reason, as shown by the following theorem, is that processing the lower-case edges in this order will require only *one* pass of the outer loop of the DC-checking algorithm, thereby ensuring that the algorithm runs in $O(N^3)$ time.

Theorem 1. *Let $S = (\mathcal{T}, \mathcal{C}, \mathcal{L})$ be a dynamically controllable STNU having K contingent links; and let $\mathcal{G}$ be the graph for S. Let E be a new edge of length δ from X to Y, where $X, Y \in \mathcal{T}$. Let $\mathcal{G}^\dagger$ be the graph obtained by inserting the new edge E into $\mathcal{G}$. As illustrated in Fig. 7, let $\mathcal{P}$ be any semi-reducible path in $\mathcal{G}^\dagger$ whose first edge is a lower-case edge e_i: $A_i \overset{c_i:x_i}{\longrightarrow} C_i$; let U be the final time-point in $\mathcal{P}$; let $\mathcal{P}_i$ be the extension sub-path for e_i in $\mathcal{P}$; and, under the supposition that $\mathcal{P}_i$ contains at least one lower-case edge, let e_j: $A_j \overset{c_j:x_j}{\longrightarrow} C_j$ be the first lower-case edge that occurs in $\mathcal{P}_i$. If $\mathcal{D}^+(C_i, X) \leq \mathcal{D}^+(C_j, X)$, then*

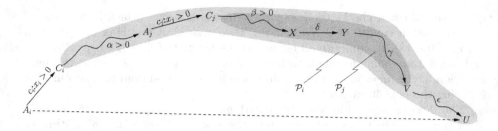

Fig. 7. The scenario addressed by Theorem 1.

the edge from A_i to U generated by using $\mathcal{P}_i$ to reduce away e_i is strictly weaker than an edge (or semi-reducible path) from A_i to U that can be obtained by bypassing that occurrence of e_j in $\mathcal{P}$.

Proof. First, by Morris' Theorem 3, only breach-free extension sub-paths need be considered [8]. Thus, without loss of generality, $\mathcal{P}_i$ is assumed to be breach-free. That is, $\mathcal{P}_i$ does not contain any occurrences of upper-case edges labeled by C_i. Next, since $\mathcal{D}^+(C_i, X) \leq \mathcal{D}^+(C_j, X)$, there must be a quasi-allowable path from C_i to X of some length $\theta = \mathcal{D}^+(C_i, X) \leq \mathcal{D}^+(C_j, X) \leq \beta < \alpha + x_j + \beta$, where α and β are the (positive) lengths shown in Fig. 7. Finally, let $\mathcal{P}_i'$ be the path obtained by replacing the portion of $\mathcal{P}_i$ from C_i to X by the quasi-allowable path of length θ; and let $\mathcal{P}'$ be the concatenation of the edge e_i and the path $\mathcal{P}_i'$. By construction, $\mathcal{P}'$ is a shorter path than $\mathcal{P}$ and it is breach-free. Thus, the semi-reducibility of $\mathcal{P}$ ensures the semi-reducibility of $\mathcal{P}'$. (Morris used a similar argument in the proof of his Theorem 3.) There are two cases to consider. First, if $\mathcal{P}_i'$ is the extension sub-path for e_i in $\mathcal{P}'$, then using $\mathcal{P}_i'$ to reduce away e_i yields a shorter edge from A_i to U than the edge generated by using $\mathcal{P}_i$. On the other hand, if some proper prefix $\mathcal{P}^*$ of $\mathcal{P}_i'$ is the extension sub-path for e_i in $\mathcal{P}'$, then the edge generated by using $\mathcal{P}^*$ to reduce away e_i, followed by the rest of $\mathcal{P}_i'$ is a semi-reducible path from A_i to U that is shorter than the edge generated using $\mathcal{P}_i$.[8]   □

Given Theorem 1, it follows that the only semi-reducible paths in $\mathcal{G}'$ that need to be considered by the incremental algorithm are those in which the order of nesting of LC edges is such that the corresponding $\mathcal{D}^+(C_i, X)$ values are *decreasing*—that is, such that the *innermost* LC edges have smaller $\mathcal{D}^+(C_i, X)$ values. Therefore, to ensure that the innermost LC edges are processed first—in *any* relevant path—the DC-checking algorithm need only process LC edges in the order of *non-decreasing* $\mathcal{D}^+(C_i, X)$ values, where X is the source time-point for the edge being inserted into the network.

The *Inky* algorithm assumes that all $\mathcal{D}^+(C_i, T)$ values are available for the DC network *prior* to the insertion of the new edge from X to Y. Thus, the algorithm must ensure that all such values are *updated* so that they will be available for the next invocation.

---

[8] Any suffix of a breach-free extension sub-path is necessarily semi-reducible [5].

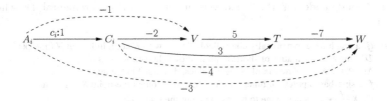

Fig. 8. Alternative ways of reducing away a lower-case edge in an STNU graph.

The APPP Matrix. For each contingent time-point C_i and each time-point T, the *Inky* algorithm also keeps track of whether *all* shortest quasi-allowable paths from C_i to T have the *positive proper prefix* (PPP) property. The reason is that if there is some shortest quasi-allowable path from C_i to T that does *not* have the PPP property, then it is not necessary to generate new edges using allowable paths that contain T, even if such exist. For example, Fig. 8 shows a scenario in which there are two shortest quasi-allowable paths from C_i to T, one having the PPP property and one not. (The single edge from C_i to T has the PPP property; the two-edge path from C_i to V to T does not.) Although a new edge, $A_i \xrightarrow{-3} W$, could be generated by using the shortest allowable path from C_i to T to W to reduce away the lower-case edge, as illustrated on the lower portion of the figure, it is not necessary to do so because the shortest allowable path from C_i to V can be used to reduce away the lower-case edge, generating the new edge, $A_i \xrightarrow{-1} V$, as illustrated on the upper portion of the figure. This edge creates an OU-path from A_i to V to T to W whose length is also -3.

In view of these considerations, the *Inky* algorithm also maintains a K-by-N matrix, called $APPP$, whose values $APPP(C_i, T)$ are all initially *True*. However, if the algorithm ever discovers a shortest quasi-allowable path from C_i to T that does *not* have the PPP property, then it sets $APPP(C_i, T)$ to *False*.

Marking Nodes. While searching through the shortest allowable paths emanating from some contingent time-point, C_i, the *Inky* algorithm marks nodes that are the source time-points for any new edges that have been generated during the current invocation. Initially, the only marked node is X (i.e., the source time-point for the new edge being added to the network). If, during the Dijkstra-like traversal of shortest allowable paths, it happens that all nodes remaining in the priority queue are unmarked and have keys (i.e., shortest quasi-allowable path-lengths) that have not changed, then it is certain that no more shorter extension sub-paths can be found; hence, the processing of that lower-case edge can stop. The algorithm uses simple counter variables to keep track of the numbers of marked nodes in the queue and nodes whose keys have changed.

The Most-basic Form of the Inky Algorithm. Pseudo-code for the most basic form of the *Inky* algorithm is given in Table 4. Its works as follows. First, it sorts the lower-case edges of $\mathcal{G}$ according to their $\mathcal{D}^+(C_i, X)$ values (Line 1). Next, it creates a boolean vector, called mark, that it uses to keep track of the source

Table 4. Pseudo-code for the basic version of the *Inky* incremental DC-checking algorithm.

Inputs: G, a graph for a dynamically controllable STNU with K cont. links and N time-pts.

$\mathcal{D}^+$, the K-by-N matrix of shortest quasi-allowable path-lengths in G

APPP, the K-by-N matrix of boolean values, discussed in the text

f, a sink-based potential function for G that uses some time-point S as its sink

$E\colon X \xrightarrow{\ \delta\ } Y$, a new edge to be inserted into the graph G;

Output: True if inserting E into G preserves its dynamic controllability; False otherwise.

```
 1. order   := lower-case edges of G, sorted into non-decreasing order of D⁺(Cᵢ,X) values.
 2. mark    := boolean vector whose N entries are all initially False, except mark[X] := True.
 3. globalNumMarks := 1;
 4. f := rotatePotentialFunc(G,f,X);   // now X is sink for f
 5. for i = 1 to K:                                                    (Main Loop)
 6.    Let (Aᵢ,xᵢ,yᵢ,Cᵢ) be the iᵗʰ contingent link according to order.
 7.    newEdges := {}.
 8.    Q := empty priority queue of length N.
 9.    forEach node T in G: insert(Q, T, val), where val = f(Cᵢ,D⁺(Cᵢ,T),T);
10.    numMarksInQueue := globalNumMarks;
11.    numChangedInQueue := 0;
12.    while ((numMarksInQueue(Q) > 0) || (numChangedInQueue(Q) > 0)):
13.       T := pop(Q);  // T is next time-point to be processed
14.       if (mark[T]) numMarksInQueue--;
15.       if (T.key < f(Cᵢ,D⁺(Cᵢ,T),T)) numChangedInQueue--;
16.       if ((T ≢ Cᵢ) && APPP(Cᵢ,T) && (f⁻¹(Cᵢ,T.key,T) ≤ 0)):  // ext. sub-path!
17.          newVal := f(Aᵢ,xᵢ,Cᵢ) + T.key;
18.          if (newVal < 0) return False;
19.          if (newVal < currAdjEdgeLen(f,G,Aᵢ,T)):
20.             newEdges += makeEdge(Aᵢ,f⁻¹(Aᵢ,newVal,T),T);  // generate new edge
21.       forEach ordEdge(T,r,W) in successors(G,T):
22.          len := T.key + f(T,r,W);
23.          if (stillInQueue(Q,W) && (len < W.key)):
24.             if (W.key == f(Cᵢ,D⁺(Cᵢ,W),W)) numChangedInQueue++;
25.             decreaseKey(Q,W,len);
26.          if (APPP(Cᵢ,T) && ((f⁻¹(Cᵢ,T.key,T) > 0) || (T ≡ Cᵢ)))
27.             APPP(Cᵢ,W) := True;
28.          if ((len <= W.key) && (T ≢ Cᵢ) && (!APPP(Cᵢ,T)||(f(Cᵢ,T.key,T) ≤ 0)))
29.             APPP(Cᵢ,W) := False;
30.       forEach UC_Edge(T,Cⱼ,r,A_j) in UCsuccessors(G,T):
31.          UC_len := T.key + f(T,r,A_j);
32.          if (UC_len >= f⁻¹(C_i,-x_j,A_j)where xⱼ = lower bound on cont. link for Cⱼ
33.             Process as ordinary path of length UC_len, using Lines 22-29 above
34.          else: UC_val := UC_len + f(Aᵢ,xᵢ,Cᵢ);
35.             if (UC_val < 0) return False;
36.             elseif (UC_val < currAdjUCEdgeLen(f,G,Aᵢ,Cⱼ)):
37.                newEdges += makeUC_Edge(Aᵢ,f⁻¹(Aᵢ,UC_val,A_j),Cⱼ);
38.    if (newEdges):
39.       f := rotatePotentialFunction(G,f,Aᵢ);  // now Ai is sink for f
40.       forEach edge in newEdges:  insertEdge(edge,G);
41.       if (!mark[Aᵢ]):
42.          mark[Aᵢ] := True;
43.          globalNumMarks++;
44. end for i = 1 to K.
45. return True.
```

time-points of any newly created edges (Line 2). Initially, the only new edge is E, the edge to be added to the network. Thus, its source time-point, X, is the only time-point that starts out being marked. Note that once a time-point is marked, it remains marked for the rest of the invocation of the algorithm. The globalNumMarks counter keeps track of how many time-points have been marked. Next, since adding an edge with source time-point X cannot affect a potential function whose sink is X, the initial potential function, f, is *rotated* so that it now uses X as its sink (Line 4). Note that this is done using only the edges from $\mathcal{G}$, ignoring the edge E. This is simply a re-packaging of the sinkPotSinkDijk function seen earlier in Sect. 3.

The main loop of the algorithm, which has exactly K iterations, spans Lines 5 thru 44. The ith iteration processes the contingent link, (A_i, x_i, y_i, C_i), beginning on Line 6. Any edges generated by reducing away the corresponding lower-case edge will be added to the set, newEdges, which is initially empty (Line 7). The Dijkstra-like traversal of shortest quasi-allowable paths emanating from C_i uses a priority queue, Q, which is initially empty (Line 8). However, each time-point T is immediately inserted into the queue using the corresponding $\mathcal{D}^+(C_i, T)$ value (Line 9). Note that the sink-based potential function, f, is used to convert the path length into the appropriate non-negative value. The $\mathcal{D}^+(C_i, T)$ values, of course, do not reflect the presence of the new edge E, but they make good initial values for the queue. The numMarksInQueue and numChangedInQueue counters are initialized in Lines 10–11. The while loop on Line 12 runs as long as it is possible for some shortest allowable path to be discovered that might be used to generate a meaningful new edge by reducing away the ith lower-case edge. Note that if these counters are ever both zero, it would imply that no path to any time-point remaining in the queue could possibly generate a useful new edge.

At Line 13, the next node T is popped from the queue. Lines 14–15 ensure that the two counters numMarksInQueue and numChangedInQueue are updated if necessary. Note that the potential function, f, is used to convert the $\mathcal{D}^+(C_i, T)$ value to its corresponding non-negative value, as seen earlier. Also, T.key denotes length of the shortest quasi-allowable path from C_i to T that has just been discovered, which is the *key* associated with the time-point T in the queue. Line 16 determines whether the path from C_i to T is an extension sub-path (cf. the definition of extension sub-path in Sect. 2.3). The length of the new edge from A_i to T that would be generated by using that extension sub-path to reduce away the lower-case edge is computed in Line 17, and stored in the variable newVal. Note that because the potential function is based solely on shortest paths in the OU-graph, the adjusted length of the *lower-case* edge (i.e., $f(A_i, x_i, C_i)$) will typically be negative. As a result, newVal itself could be negative, which would indicate that a negative semi-reducible loop has been found (i.e., that inserting the edge E into the graph $\mathcal{G}$ made the network *not* dynamically controllable); hence, Line 18 returns False if newVal is negative. Line 19 then checks whether an edge of length newVal would be shorter than any pre-existing edge from A_i to T in the graph. Note that the function, currAdjEdgeLen, uses the potential function f to convert the length of any pre-existing edge to its non-negative counterpart. If the new edge would be shorter, it is then added to the set newEdges (Line 20).

Lines 21–29 then check each *ordinary* edge emanating from T to determine whether any shortest-path values in the queue can be updated. For each successor edge, $T \xrightarrow{r} W$, the length of the quasi-allowable path from C_i to T to W (using non-negative values) is computed (Line 22) to determine whether the key for W that is currently stored in the queue needs to be decreased (Lines 23, 25). Line 24 ensures that the numChangedInQueue counter is properly updated if the key for W is being changed for the first time. Lines 26–27, still in the case where the key for W was decreased, determine whether the new shortest path from C_i to W has the PPP property and, if so, set APPP(C_i, W) to *True*. In contrast, Lines 28–29, which do not necessarily fall within that case, determine whether a quasi-allowable path from C_i to W has been found that does *not* have the PPP property, in which case, the APPP value is set to *False*.

Lines 30–37 then check each *upper-case* successor edge emanating from T in a similar fashion. Lines 32–33 check whether the resulting path can reduce to an ordinary edge, courtesy of the *Label Removal* rule from Table 1. If so, the path is processed by Lines 22–29, as discussed above. Otherwise, Lines 34–37 determine whether the resulting path can be used to generate a new upper-case edge; if so, the edge is added to newEdges. Note that UC_val being negative would imply that a negative semi-reducible loop had been found (i.e., that inserting E into the network made it non-DC, Line 35).

At the end of the ith iteration, if newEdges is non-empty (Line 38), then several steps are taken to incorporate the newly generated edges. First, in anticipation of adding new edges, each of which has A_i as its source, the potential function is rotated so that it uses A_i as its sink (Line 39). Next, the new edges are added to the graph (Line 40). Since each new edge has A_i as its source, those edges cannot disturb the recently rotated potential function. Finally, Lines 41–43 ensure that if this is the first time that new edges with source time-point A_i are being added to the network, then A_i becomes marked and the globalNumMarks counter is updated (Note that A_i might be the activation time-point for multiple contingent links; so this might not be the first time that A_i is marked.)

Finally, if all K iterations of searching for extension sub-paths to generate new edges fail to find a negative loop (cf. Lines 18, 35), then the algorithm returns *True* at Line 45.

Note that in the process of performing the K iterations, the *Inky* algorithm computes all updates of the $\mathcal{D}^+$ matrix that will be needed by subsequent invocations of the algorithm. It also computes updates to the APPP matrix. Furthermore, since the algorithm runs at most K iterations, each of which runs in $O(N^2)$ time (due to the Dijkstra-like traversals), the algorithm runs in $O(N^3)$ time overall.

Improving the Incremental Algorithm. The *Inky* algorithm can be substantially improved as follows. Suppose that C_1 is the first contingent time-point to be processed by the main loop of the algorithm (Lines 5–44). Consider the Dijkstra-like traversal of shortest quasi-allowable paths emanating from C_1. Given that each time-point T is initially inserted into the priority queue, Q, using the corresponding $\mathcal{D}^+(C_1, T)$ value from the previous invocation of the algorithm (Line 9), any time-point that is popped off the queue before X need not have its successor edges processed (cf. Lines 22–29 and 31–37) because following those edges could not possibly change any path-lengths currently stored in the queue. However, once X has been popped off the queue, normal processing of successor edges must resume. For subsequent iterations of the main loop (i.e., when processing other LC edges), a similar approach can be used. In particular, if $\mathcal{A} = \{A_{i_1}, A_{i_2}, \ldots, A_{i_s}\}$ is the set of source time-points for the edges that have been generated so far—all of which must be activation time-points for already-processed LC edges—then until X or some member of $\mathcal{A}$ has been popped off the queue, any other time-point that is popped off the queue need not have its successor edges processed, because doing so could not possibly change any values stored in the queue. However, once X or some member of $\mathcal{A}$ is popped, then normal processing must be resumed. Given that the worst-case complexity of Dijkstra's algorithm is $O(m + N \log N)$, where m is the number of edges in the graph, reducing the number of successor edges that must be followed during the Dijkstra-like traversals necessarily makes the algorithm more efficient.

A similar technique can be used to make the `rotatePotentialFunc` (Line 4) more efficient. In this case, the algorithm must keep track of the *sink* time-point Y of the new edge E, and the *sink* time-points of any edges that have already been generated by prior iterations of the main loop. As long as none of those time-points have been popped off the queue during the sinkDijkstra traversal used by `rotatePotentialFunc`, the predecessor edges of other time-points that are popped off the queue need not be followed. However, once any of those sink time-points is popped, normal processing must resume. Once again, reducing the number of edges that must be followed during a Dijkstra traversal necessarily makes the *Inky* algorithm more efficient.

Finally, when considering whether to generate a new edge from a given allowable path (Lines 19–20), instead of simply checking whether the new edge would be shorter than any pre-existing edge involving the same time-points, the *Inky* algorithm could use an extra, sourceDijkstra traversal in each iteration of the main loop to compute all $\mathcal{D}^*(A_i, T)$ values, where $\mathcal{D}^*$ is the all-pairs-shortest-semi-reducible-path matrix discussed in Sect. 2. On the positive side, this could reduce the number of generated edges, leading to additional savings; on the negative side, this technique requires additional computations to maintain the $\mathcal{D}^*(A_i, T)$ values across invocations. Thus, the viability of this technique, unlike the previous two, must be tested empirically.

5   Conclusions

This paper introduced new techniques for speeding up both full and incremental DC checking for STNUs. The *Speedy* algorithm is the full DC-checking algorithm the results from applying these techniques to the Morris-N^4 algorithm. It has been shown to out-perform the Morris-N^4 algorithm, in some cases by an order of magnitude. It is not yet known whether *Speedy* will be competitive with the more recent $O(N^3)$ algorithm presented after the initial publication of the *Speedy* algorithm.

The main contribution of the paper is the *Inky* algorithm, which is the incremental algorithm that results from applying similar techniques to the incremental DC-checking problem. Like its fastest competitors [9,14], the *Inky* algorithm is $O(N^3)$. However, the techniques it employs to reduce redundant computations, and the fact that its main loop has K iterations, together suggest that *Inky* may out-perform its incremental competitors, each of which can involve up to N iterations. Thus, it is expected that in scenarios where the number of contingent links is relatively small compared to the total number of time-points, *Inky* may perform especially well.

Clearly, a thorough comparative, empirical evaluation of the top incremental DC-checking algorithms is needed. However, given that the competitors have been so recently introduced, such an evaluation must be left to future work.

References

1. Chien, S., Sherwood, R., Rabideau, G., Zetocha, P., Wainwright, R., Klupar, P., Gaasbeck, J.V., Castano, R., Davies, A., Burl, M., Knight, R., Stough, T., Roden, J.: The techsat-21 autonomous space science agent. In: The First International Joint Conference on Autonomous Agents and Multiagent Systems (AAMAS-2002), pp. 570–577, ACM Press (2002)
2. Cormen, T.H., Leiserson, C.E., Rivest, R.L., Stein, C.: Introduction to Algorithms. MIT Press, Cambridge (2009)
3. Dechter, R., Meiri, I., Pearl, J.: Temporal constraint networks. Artif. Intell. **49**, 61–95 (1991). Elsevier
4. Hunsberger, L.: A faster execution algorithm for dynamically controllable STNUs. In: Proceedings of the 20th Symposium on Temporal Representation and Reasoning (TIME-2013) (2013)
5. Hunsberger, L.: Magic loops in simple temporal networks with uncertainty. In: Proceedings of the Fifth International Conference on Agents and Artificial Intelligence (ICAART-2013) (2013)
6. Hunsberger, L.: A faster algorithm for checking the dynamic controllability of simple temporal networks with uncertainty. In: Proceedings of the 6th International Conference on Agents and Artificial Intelligence (ICAART-2014), SciTePress (2014)
7. Hunsberger, L., Posenato, R., Combi, C.: The dynamic controllability of conditional STNs with uncertainty. In: Proceedings of the PlanEx Workshop at ICAPS-2012, pp. 121–128 (2012)
8. Morris, P.: A structural characterization of temporal dynamic controllability. In: Benhamou, F. (ed.) CP 2006. LNCS, vol. 4204, pp. 375–389. Springer, Heidelberg (2006)
9. Morris, P.: Dynamic controllability and dispatchability relationships. In: Simonis, H. (ed.) CPAIOR 2014. LNCS, vol. 8451, pp. 464–479. Springer, Heidelberg (2014)
10. Morris, P., Muscettola, N., Vidal, T.: Dynamic control of plans with temporal uncertainty. In: Nebel, B. (ed.) 17th International Joint Conference on Artificial Intelligence (IJCAI-01), pp. 494–499, Morgan Kaufmann (2001)
11. Morris, P.H., Muscettola, N.: Temporal dynamic controllability revisited. In: Veloso, M.M., Kambhampati, S. (eds.) The 20th National Conference on Artificial Intelligence (AAAI-2005), pp. 1193–1198, MIT Press (2005)

12. Nilsson, M., Kvarnstrom, J., Doherty, P.: Incremental dynamic controllability revisited. In: Proceedings of the 23rd International Conference on Automated Planning and Scheduling (ICAPS-2013) (2013)

13. Nilsson, M., Kvarnstrom, J., Doherty, P.: EfficientIDC: a faster incremental dynamic controllability algorithm. In: Proceedings of the 24th International Conference on Automated Planning and Scheduling (ICAPS-2014) (2014)

14. Nilsson, M., Kvarnstrom, J., Doherty, P.: Incremental dynamic controllability in cubic worst-case time. In: Proceedings of the 21st International Symposium on Temporal Representation and Reasoning (TIME-2014) (2014)

15. Shah, J., Stedl, J., Robertson, P., Williams, B.C.: A fast incremental algorithm for maintaining dispatchability of partially controllable plans. In: Boddy, M., Fox, M., Thiébaux, S. (eds.) Proceedings of the Seventeenth International Conference on Automated Planning and Scheduling (ICAPS 2007), AAAI Press (2007)

16. Stedl, J., Williams, B.C.: A fast incremental dynamic controllability algorithm. In: Proceedings of the ICAPS Workshop on Plan Execution: A Reality Check, pp. 69–75 (2005)

Adaptive Neural Topology
Based on Vapnik-Chervonenkis Dimension

Beatriz Pérez-Sánchez$^{(\boxtimes)}$, Oscar Fontenla-Romero,
and Bertha Guijarro-Berdiñas

Department of Computer Science, Faculty of Informatics, University of A Coruña,
Campus de Elviña s/n, 15071 A Coruña, Spain
{beatriz.perezs,ofontenla,cibertha}@udc.es

Abstract. In many applications, learning algorithms act in dynamic
environments where data flows continuously. In those situations, the
learning algorithms should be able to work in real time adjusting its controlling parameters, even its structures, when knowledge arrives. In a previous work, the authors proposed an online learning method for two-layer
feedforward neural networks which is able to incorporate new hidden neurons during learning without losing the previously acquired knowledge.
In this paper, we present an extension of this previous learning algorithm
which includes a mechanism, based on Vapnik-Chervonenkis dimension,
to adapt the network topology automatically. The experimental study
confirms that the proposed method is able to check whether a modification of the topology is necessary according to the needs of the learning
process.

Keywords: Vapnik-Chervonenkis dimension · Adaptive neural topology · Incremental learning · Optimal neural network structure

1 Introduction

Many real problems in machine learning are of a dynamic nature, this means that
the information flows continuously and the new knowledge could affect previously
learned model [1,2]. As reference we can mentioned important applications such
as, adaptive sensory-motor control systems [3], economic data analysis, brain
control interfaces, natural speech, robotics and, in general, applications dealing
with biological or physiological signals [4]. In those environments, the model
used for the learning process should work in real time and have the ability to
act and react by itself, adjusting its controlling parameters, even its structures,
depending on the requirements of the process. Classical batch learning are not
suitable for handling these types of situations, since they learn the concept from
the beginning and therefore, whenever new samples are available, they need
to discard the existing model and redesign a new one from scratch. This approach presents several problems, the waste of computational resources being the
most important. Therefore, in these situations an online or incremental learning

© Springer International Publishing Switzerland 2015
B. Duval et al. (Eds.): ICAART 2014; LNAI 8946, pp. 194–210, 2015.
DOI: 10.1007/978-3-319-25210-0_12

technique would be a more appropriate approach since it assumes that the information available at any given moment is incomplete and therefore, any learned model is susceptible to modifications. When dealing with neural networks, the adaptation implies changing not only the weights and biases but also the network architecture. The size of the neural network should fit the number and complexity of the data analyzed. The unawareness about the appropriate size of the network to solve a certain problem presents several drawbacks:

- the decision is based on an trial and error approach which is not only computationally expensive but also it does guarantee that the selected number of hidden nodes are close to optimum
- a too large network will lead to overfitting and poor generalization performance but a too small network will not be able to learn the problem well
- if the available number of training samples is large and the necessary number of hidden units is high, it is necessary to reduce the needs of computational resources

Among reasons for finding optimal structure for a neural network we can mention for example, improve and speed up the prediction, obtain better generalization and reduce computational requirements. Recently, several research works have proposed different approaches to modify the network structure as the learning process evolves. There are two general strategies to achieve it: constructive algorithms [5] and pruning methods [6]. The former starts from a small network which later increases its size until a satisfactory solution is found. The latter trains a network that is larger than necessary until an acceptable solution is found and then removes hidden units and weights that are not needed. Generally it is considered that the pruning technique presents several drawbacks with respect to the constructive one [7].

The appropriate number of hidden units is a value very difficult to estimate theoretically; however, different methods have been proposed to dynamically adapt the network structure during the learning process based on several empirical measures. Among others, Ash [8] developed an algorithm for the dynamic creation of nodes where a new hidden neuron is generated when the training error rate is lower than a critical value. Hirose [9] adopted a similar approach but at the same time it can remove units when small values of the error are achieved. Aylward and Anderson [10] proposed a set of rules based in the error rate, the convergence criterion and the distance to the target error. Other studies researched the application of evolutive algorithms to optimize the number of hidden neurons and the value of the weights [11, 12]. Murata [13] studied the problem of determining the optimum number of parameters from a statistical point of view. In [14] new hidden units and layers are included incrementally only when they are needed based on monitoring the residual error that cannot be reduced any further by the already existing network. An approach referred to as error minimized extreme learning machine that can add random hidden nodes was included in [15]. In spite of numerous studies, the authors are not aware of an efficient method for determining the optimal network architecture and nowadays an appropriate selection remains an open problem [16].

In a previous research [17] the authors presented an online learning algorithm for two-layer feedforward neural networks (OANN) which incorporates a factor that weights the errors committed in each of the samples. As a consequence, OANN is effective in stationary context as well as in dynamic environments. In this previous work we included an in-depth study to check and justify the viability of OANN to work with incremental data structures. However, an important drawback is that the modification of the topology was manually forced every fixed number of iterations. Thus, in this paper we extend the previous method including an automatic mechanism to check whether new hidden units should be added depending on the requirements of the online learning process. As a result, we presented a new learning algorithm (automatic-OANN) which is online and incremental both with respect to its learning ability and its topology.

The paper is structured as follows. In Sect. 2 the algorithm is explained. In Sect. 3, its behavior is illustrated by its application to several time series in order to check its performance in different contexts. In Sect. 4 the results are discussed and some conclusions are given. Finally, in Sect. 5 we raise some future lines of research.

2   Description of the Proposed Method

We have to solved two fundamental problems to develop an automatic incremental topology. First, it is necessary to verify whether a learning algorithm can adapt the network topology by incorporating new hidden neurons while maintaining, as much as possible, the knowledge gained in previous stages of learning. Secondly, we must developed a mechanism to know when it is appropriate to modify the network topology by adding new units. The former problem was solved in a previous paper [17] resulting in OANN algorithm, which is summarized in Sect. 2.1. The second challenge is our new contribution and that will be explained in-depth in Sect. 2.2. As a result, the new algorithm, automatic-OANN, will be obtained.

2.1   How to Modify the Structure of the Network

Consider the two-layer feedforward neural network in Fig. 1 where the inputs are represented as the column vector $\mathbf{x}(s)$, the bias has been included by adding a constant input $x_0 = 1$, and outputs are denoted as $\mathbf{y}(s)$, $s = 1, 2, \ldots, S$ where S is the number of training samples. J and K are the number of outputs and hidden neurons, respectively. Functions $g_1, \ldots, g_K$ and $f_1, \ldots, f_J$ are the nonlinear activation functions of the hidden and output layer, respectively.

This network can be considered as the composition of two one-layer subnetworks. As the desired outputs, $z_k(s)$, for each hidden neuron k at the current learning epoch s are unknown then arbitrary values are employed (obtained in base to a previous initialization of the weights using a standard method, for example Nguyen-Widrow [18]). Afterwards, the desired outputs of hidden nodes are not revised during the learning process and they are not influenced by the

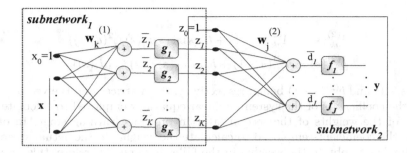

Fig. 1. Two-layer feedforward neural network.

desired output of the whole network. Thus, the training of the first subnetwork is avoided in the learning task. Regarding the second subnetwork, as $d_j(s)$ is the desired output for the j output neuron, that it is always available in a supervised learning, we can use $\bar{d}_j(s) = f_j^{-1}(d_j(s))$ to define the objective function Q for the j output of subnetwork 2 as the sum of squared errors before the nonlinear activation function f_j,

$$Q_j^{(2)}(s) = h_j(s) \left(f_j' \left(\bar{d}_j(s) \right) \left(\mathbf{w}_j^{(2)^T}(s) \mathbf{z}(s) - \bar{d}_j(s) \right) \right)^2 \tag{1}$$

where $j = 1, \ldots, J$, $\mathbf{w}_j^{(2)}(s)$ is the input vector of weights for output neuron j at the instant s and $f_j'(\bar{d}_j(s))$ is a scaling term which weighs the errors. The inclusion of this scaling term is an alternative formulation proposed in [19]. Moreover, the term $h_j(s)$ is included as forgetting function and it determines the importance of the error at the s_{th} sample. Several options can be used as the forgetting function $h_j(s)$; for instance, a linear or an exponential function among others. This function is employed to establish the form and the speed of the adaptation to the new data points in a changing environment. In a stationary context a constant function could be employed to give the same importance to all the samples analyzed during the online learning process. In a non-stationary context this function should be monotonically increasing to take into account the increment in the importance of current information in contrast with the past one. Thanks to the combination of incremental learning property and increasing importance assignment the network forgets quickly in the presence of a change while maintaining a stable behavior when the context is stationary [20].

The objective function presented in Eq. 1 is a convex function, whose global optimum can be easily obtained deriving it with respect to the parameters of the network and setting the derivative to zero [19]. Therefore, we obtain the following system of linear equations,

$$\sum_{k=0}^{K} A_{qk}^{(2)}(s) w_{jk}^{(2)}(s) = b_{qj}^{(2)}(s), \quad q = 0, 1, \ldots, K; \quad j = 1, \ldots, J, \tag{2}$$

where

$$A_{qk}^{(2)}(s) = A_{qk}^{(2)}(s-1) + h_j(s)z_k(s)z_q(s)f_j'^2(\bar{d}_j(s)) \tag{3}$$

$$b_{qj}^{(2)}(s) = b_{qj}^{(2)}(s-1) + h_j(s)\bar{d}_j(s)z_q(s)f_j'^2(\bar{d}_j(s)) \tag{4}$$

$A^{(2)}(s-1)$ and $b^{(2)}(s-1)$ being, respectively, the matrices and the vectors that store the coefficients of the system of linear equations employed to calculate the values of the weights of the second layer in previous learning stage. In other words, the coefficients employed to calculate the weights in the actual stage are used further to obtain the weights in the following one. Therefore, this permits handling the earlier knowledge and using it to incrementally approach the optimum value of the weights. Equation 2 can be rewritten using matrix notation as,

$$\mathbf{A}_j^{(2)}(s)\mathbf{w}_j^{(2)}(s) = \mathbf{b}_j^{(2)}(s), \quad j = 1, \ldots, J, \tag{5}$$

where

$$\mathbf{A}_j^{(2)}(s) = \mathbf{A}_j^{(2)}(s-1) + h_j(s)\mathbf{z}(s)\mathbf{z}^T(s)f_j'^2(\bar{d}_j(s)) \tag{6}$$

$$\mathbf{b}_j^{(2)}(s) = \mathbf{b}_j^{(2)}(s-1) + h_j(s)f_j^{-1}(d_j(s))\mathbf{z}(s)f_j'^2(\bar{d}_j(s)). \tag{7}$$

Finally, from Eq. 5 the optimal weights for the second subnetwork can be obtained as:

$$\mathbf{w}_j^{(2)}(s) = \mathbf{A}_j^{(2)^{-1}}(s)\mathbf{b}_j^{(2)}(s), \forall j. \tag{8}$$

As regards the incremental property of the learning algorithm, the network structure can be adapted depending on the requirements of the learning process. The number of hidden units can be increased by adequately redimensioning the number of weights. Several modifications have to be carried out in order to adapt the current topology to a new one. The fact of increasing the number of hidden neurons implies modifications in both layers of the network. As it can be observed in Fig. 2 the increment of hidden neurons affects not only the first subnetwork (its number of output units increases) but also the second subnetwork (the number of its inputs also grows). As mentioned previously, the training of the first subnetwork is avoided in the training task, therefore we only comment on the modifications corresponding to the second subnetwork. Thus, in Fig. 2 it can be observed that as the number of hidden neurons grows and consequently, all matrices $\mathbf{A}_j^{(2)}$ and the vectors $\mathbf{b}_j^{(2)}$ ($j = 1, \ldots, J$), computed previously modify their size. Therefore in order to adapt them, each matrix $\mathbf{A}_j^{(2)}$ is enlarged by including a new row and a new column of zero values, in order to continue the learning process from this point (zero is the null element for the addition). At the same time, each vector $\mathbf{b}_j^{(2)}$ incorporates a new element of zero value. The rest of elements of the matrices and vectors are maintained without variation, this fact allows us to preserve in some way the knowledge acquired previously with the earlier topology. After these modifications, the latter described matrices and vectors of coefficients allow us to obtain the new set of weights for the current topology of the network.

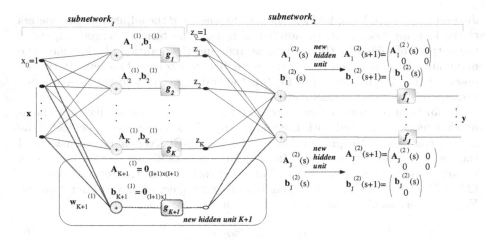

Fig. 2. Incremental Topology.

2.2   When to Change the Structure of the Network

In the previous section we justify the suitability of the learning algorithm OANN to work with adaptive network structures. However, the modification of the topology is manually forced. Therefore, it is necessary to include some mechanism for controlling the growth of the structure automatically. The topology must be changed only if its capabilities are insufficient to satisfy the needs of the learning process.

In statistical learning theory the Vapnik-Chervonenkis dimension [21] is a measure of the capacity of a statistical classification algorithm and it is defined as the cardinality of the largest set of points that the algorithm can shatter. It is said that a classification model g with parameters θ can split a dataset $(x_1, x_2, \ldots, x_n)$ if, for all possible class labels, there is a set θ^* for which the model prevents mistakes in its classification. Because of the Vapnik-Chervonenkis dimension of a model is the maximum cardinality of the set of data that the model is able to divide, this term allows us to predict a probabilistic upper bound on the test error of a classification model based in its complexity. This is an ideal bound that can be calculated according to the training error and the network topology. If the model generalizes properly, the bound indicates the worst test error that the model can obtain. Thus, the bound value establishes the margin to test error and therefore it is possible to use this value to determine whether the current topology is sufficient. In this way, it can be said that an adequate number of hidden units is the one giving the lowest expected risk. Taking into account all these considerations it is established that, with a probability of $1 - \eta$ the smallest test error $(R$, expected risk) is delimited by the following inequality [21],

$$R(d_{VC}, \theta) \leq \frac{\frac{1}{n} \sum_{i=1}^{n} (z_i - \hat{z}_i (d_{VC}, \theta))^2}{\left[1 - \sqrt{\frac{d_{VC} \left(ln \left(\frac{n}{d_{VC}} \right) + 1 \right) - ln \left(\frac{\eta}{4} \right)}{n}} \right]_+} \tag{9}$$

d_{VC} represents the Vapnik-Chervonenkis dimension, θ the adjustable parameters of the learning system, n the number of training patterns, z target outputs, $\hat{z}$ the outputs obtained by the learning system and the notation $[.]_+$ indicates the maximum value between 0 and the term between square brackets. It can be observed that the numerator of Eq. 9 is the training error and the denominator corresponds to a correction function.

In an informal way, we can say that the generalization ability of a classification model is related with the complexity of the structure. Specifically the greater complexity, the greater value of the Vapnik-Chervonenkis dimension and the lesser generalization ability. In [22], authors established certain bounds to a particular network architecture. For a network having the total of M units ($M \geq 2$) and W weights (including biases), they gave an upper bound on the Vapnik-Chervonenkis dimension in the form

$$d_{VC} \leq 2\,W log_2(eM), \tag{10}$$

e being the base of natural logarithms. Although Eq. 10 established an upper bound for the value of the Vapnik-Chervonenkis dimension, this is a valid approach as it allows calculating the error bound in the worst case.

Taking into account all these considerations, we develop an automatic technique to control the neural network growth in the online learning algorithm. The estimation method is based in the ideal bound for the test error calculated thanks to Vapnik-Chervonenkis dimension of the neural network. Then, we obtained a constructive algorithm which adds a new unit to the hidden layer of the network when the obtained test error is higher than the limit established by the ideal bound. The proposed learning algorithm starts from a small dimension network (initially, with only two units in the hidden layer) and adapts its topology (adding only one unit at each time) in base of the needs of learning process. Finally the new proposed learning algorithm, called automatic-OANN, online and incremental with respect to its learning ability and structure is detailed in Algorithm 1.

3   Experimental Results

In a previous work [17], the OANN algorithm was compared to other online algorithms checking and justifying the suitability of OANN to work with adaptive structures without significantly degrading its performance. However as we commented previously, the modification of the topology is manually forced every fixed number of iterations. In this paper, we extend the learning method including an automatic mechanism to check whether new hidden units should be added depending on the requirements of the online learning process. In this experimental study we want to demonstrate the viability of the automatic adjustment mechanism of the structure. Therefore, taking these previous results into account, automatic-OANN will only be compared to the previous OANN version (fixed topology) with the following aims:

- To prove that the performance of the automatic-OANN at the end of the learning process, (when it reaches a final network topology), is similar to the performance that would be obtained by employing this final topology from the beginning to the end of learning process.
- To check if automatic-OANN is able to incorporate hidden units without significant performance degradation with respect to the version without adaptation of the topology.

For this experimental study, we employed distinct system identification problems. Moreover, the behavior of the learning algorithm will be checked when it operates in both stationary and dynamic environments. All experiments shared the following conditions:

- The input data set was normalized, with mean and standard deviation values equal 0 and 1, respectively.
- The neural functions. We employed the logistic sigmoid function for hidden neurons, while for output units a linear function was applied as recommended for regression problems [5].
- In order to obtain significant results, five simulations were carried out. Therefore, mean results will be presented in this section. Moreover, in the case of stationary context 10-fold cross validation was applied.
- In all cases an exponential forgetting function, defined by the following equation,

$$h(s) = e^{\mu s}, s = 1, \ldots, S, \tag{11}$$

was used, where μ is a positive real parameter that controls the growth of the function and thus the response of the network to changes in the environment. When $\mu = 0$ we obtain a constant function, and therefore all errors have the same weight and the forgetting function has no effect. The value of the μ factor for the forgetting ability was set to 0.01.

3.1 Stationary Contexts

In this section, we consider several stationary time series, K.U. Leuven, Kobe and Hénon. The goal is to predict the actual sample based in eight previous patterns, in the case of K.U. Leuven and Kobe series and according to the seven previous ones for Hénon. Next, a brief explanation of each selected time series is given.

- K.U. Leuven Competition Data. The K.U. Leuven time series prediction competition was generated from a computer simulated 5-scroll attractor, resulting from a generalized Chua's circuit which is a paradigm of chaos [23]. We employed the total of 1,800 data points for training.
- Hénon. The Hénon map is a dynamic system that presents a chaotic behavior [24]. The map takes a point (x_n, y_n) in the plane and transforms it according to,

$$x(t + 1) = y(t + 1) - \alpha x(t)^2$$
$$y(t + 1) = \gamma x(t)$$

Algorithm 1. Automatic-OANN algorithm.

Inputs: $\mathbf{x}_s = (x_{1s}, x_{2s}, \ldots x_{Is}); \mathbf{d}_s = (d_{1s}, d_{2s}, \ldots d_{Js}); \; s = 1, \ldots, S.$

Initialization Phase

- The number of hidden units (K) is established as 2
- $\mathbf{A}_j^{(2)}(0) = \mathbf{0}_{(K+1) \times (K+1)}, \quad \mathbf{b}_j^{(2)}(0) = \mathbf{0}_{(K+1)}, \quad \forall j = 1, \ldots, J.$
- The initial weights, $\mathbf{w}_k^{(1)}(0)$, are calculated by means of an initialization method.

For every new sample s ($s = 1, 2, \ldots, S$) and $\forall k = 1, \ldots, K$

- $z_k(s) = g(\mathbf{w}_k^{(1)}(0), \mathbf{x}(s))$

- *For* each output j of the subnetwork 2 ($j = 1, \ldots, J$),
  - The outputs of the hidden layer are calculated in base to the weights obtained in the initialization phase,
  - $\mathbf{A}_j^{(2)}(s) = \mathbf{A}_j^{(2)}(s-1) + h_j(s)\mathbf{z}(s)\mathbf{z}^T(s)f_j'^2(\bar{d}_j(s)),$
  - $\mathbf{b}_j^{(2)}(s) = \mathbf{b}_j^{(2)}(s-1) + h_j(s)f_j^{-1}(d_j(s))\mathbf{z}(s)f_j'^2(\bar{d}_j(s)),$
  - The weights $\mathbf{w}_j^{(2)}(s)$ are obtained by solving the system of linear equations

    $\mathbf{w}_j^{(2)}(s) = \mathbf{A}_j^{(2)^{-1}}(s)\mathbf{b}_j^{(2)}(s)$
  end of For

- Calculate Vapnik-Chervonenkis dimension d_{VC} (see Eq. 10)

- Obtain the test error, MSE_{Test} and the test error *bound* (see Eq. 9)

- *If* $MSE_{Test} > bound$ then a new hidden unit $K+1$ is added
  - The coefficient matrices and vectors are modified as follow

$$\mathbf{A}_j^{(2)}(s) = \begin{pmatrix} & & & (K+1) & \\ \mathbf{A}_j^{(2)}(s-1) & 0 & \cdots & 0 \\ 0 & 0 & \cdots & 0 \\ (K+1) & \vdots & \vdots & \vdots\vdots & \vdots \\ 0 & 0 & \cdots & 0 \end{pmatrix}, \quad \mathbf{b}_j^{(2)}(s) =$$

$$\begin{pmatrix} & \mathbf{b}_j^{(2)}(s-1) \\ & 0 \\ (K+1) & \vdots \\ & 0 \end{pmatrix},$$

  - *For* each new connection,
       calculate the weights, $\mathbf{w}_{K+1}^{(1)}(0)$, by means of some initialization method
    end of For
  - $K = K + 1$
  end of If,

end of For

$x(t)$ being the series value at instant t. In this case the parameters values are established as $\alpha = 1.4$ and $\gamma = 0.3$. The total number of 3,600 patterns are employed for training.

– Kobe[1]. This time series corresponds to the signal of vertical acceleration recorded by a seismograph. The seismograph recorded the Kobe earthquake from the University of Tasmania, Hobart, Australia in 1995 for 51 min at 1 second intervals. The total number of 3,048 observations are available.

Fig. 3 shows the test MSE curves for the time series. In the case of fixed topologies, curves are include for the initial and final ones used for adaptive case. In the case of incremental topology, several peaks are observed in the curves as a result of the topology changes. This fact is due to to the incorporation of a new unit generated perturbations (new elements of zero value in the matrices). However, these degradations are corrected speedily. Apart from that, we can check that the incremental topology achieves results close to those obtained when the final fixed topology is used from the initial epoch. The results are not exactly the same but it should be considered that the proposed method constructs the topology in an online learning scenario and then only a few samples are available to train the last topology. The points over the curves of incremental topology allow us to know when the error obtained by the network exceeds the bound established for the test error. At that moment, the structure is modified in order to response to needs of the learning process.

3.2   Dynamic Environments

In dynamic environments, the distribution of the data could change over time. This situation often occurs when dealing with signals in real-life situations [25]. Moreover, in these types of environments, changes between contexts can be *gradual* if there is a smooth transition between distributions or *abrupt* when the distribution changes quickly [26]. We have considered examples of several changing environments with the aim of checking if the proposed method is able to obtain appropriate behavior when it works in dynamic environments.

Artificial Data Set. The first data set is formed by 4 input random variables that contain values drawn from a normal distribution with zero mean and standard deviation equal to 0.1. The desired output is obtained by a linear mixture of nonlinear functions. Specifically, hyperbolic tangent sigmoid, exponential, sine and logarithmic sigmoid functions were applied, respectively, to the first, second, third and fourth output. Finally, the desired output is obtained by a linear mixture of the transformed inputs. The final set contains a training set of 2,400 samples and 4 test sets, one for each of the changes in the linear mixture of the training set.

---

[1] Available at http://www-personal.buseco.monash.edu.au/~hyndman/TSDL.

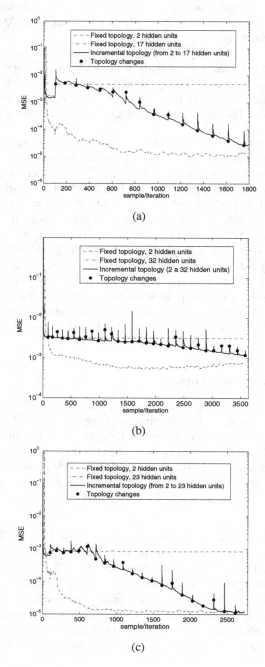

Fig. 3. Test error curves for the time series: (a) K.U. Leuven, (b) Hénon and (c) Kobe.

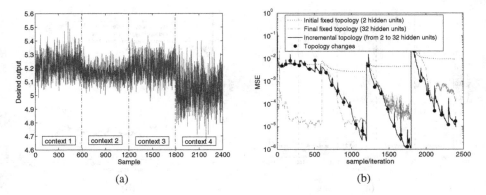

(a)   (b)

Fig. 4. Artificial Data Set: (a) Example of the desired output for the training set, (b) Test error curves.

Figure 4(a) contains the signal employed as desired target during the training process. As can be seen, the signal evolves over time and 4 context changes are generated. Also we created different test sets for each context, so every training sample has associated the test set that represents the context to which it belongs. Thus, for this case we obtain 4 test sets, one for each of the changes in the linear mixture of the process.

Figure 4(b) shows the test error curves obtained by the method with fixed and automatic incremental topologies. The error shown for each point of the signal is the mean value obtained over test set associated to the current training sample. It worth mentioning that the changes between contexts, each 600 samples, cause the large peaks. It can be seen that the initial fixed topology (two hidden neurons) commits a high error because the structure is not sufficient to make a suitable learning. Regarding the incremental topology, although the results at the beginning of the process are not appropriate (the topology is still small), it can be observed as it performs better than the final fixed topology when there are changes of context.

Lorenz Time Series. The proposed approach has been tested on real Lorenz time series [29] which corresponds to a system with chaotic dynamics. Edward Lorenz developed a simplified mathematical model for atmospheric convection. The model is a system described by the solution of three simultaneous differential equations which are shown below,

$$\begin{aligned}
\frac{dx}{dt} &= \sigma(y - x) \\
\frac{dy}{dt} &= x(\rho - z) - y \\
\frac{dz}{dt} &= xy - \beta z
\end{aligned} \tag{12}$$

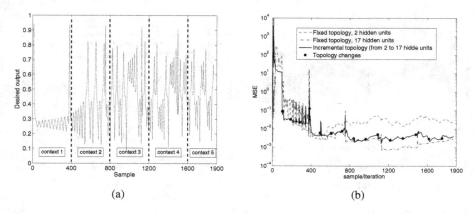

(a)   (b)

Fig. 5. Lorenz time series: (a) Example of the desired output for the training set, (b) Test error curves.

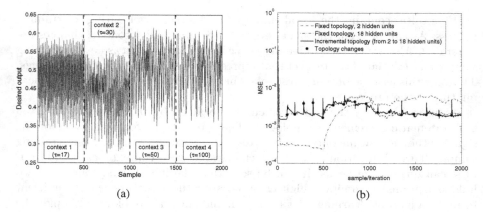

(a)   (b)

Fig. 6. Mackey-Glass time series: (a) Example of the desired output for the training set, (b) Test error curves.

where σ, ρ, β are the system parameters, x, y and z represent the system state and t is time. The system parameters σ, ρ, and β must be positive values, and usually are established as $\sigma = 10$, $\beta = \frac{8}{3}$ and $\rho = 28$. For these values the system exhibits a chaotic behavior. Considering this equation we generated the total of 2,000 observations where 5 different contexts of 400 samples each are included. Every context correspond to different value for the β parameter. The aim of the network is to predict the current sample based on the eight previous ones. Figure 5(a) contains the signal employed as the desired target during the training process.

Figure 5(b) shows the test error curves obtained by the method with fixed and automatic incremental topologies. The spikes observed in the curves are due to two different causes. The changes between contexts, each 400 samples, cause the large peaks while in the case of incremental topology, the small spikes are a result of the variation in the topology. Again the automatic technique allows the initial topology to grow along time adjusting its response of the requirements of the process. Then, the incremental topology show a stable behavior and obtains similar results to the fixed topology even when 17 hidden units are employed from the beginning.

Mackey-Glass Chaotic Time Series. Finally, in this third experiment we employed the real chaotic Mackey-Glass time series, which was originally proposed as a physiological control model for the production of white blood cells [27]. The time series is based on the following differential delay equation,

$$\frac{dy(t)}{dt} = \frac{0.2y(t-\tau)}{1 + y^{10}(t-\tau)} - \rho y(t). \tag{13}$$

Considering this equation and using $\rho = 0.1$, $2,000$ observations are generated by means of the fourth order Runga-Kutta method [28]. It is well-know fact that the chaotic behavior of the Mackey-Glass system grows as the delay coefficient τ increases. Therefore, taking this idea we generated 4 different contexts of 500 samples each, which correspond to different values to the parameter τ (17, 30, 50 and 100). The goal of the network is to predict the current sample based on the eight previous ones. Figure 6(a) contains the signal employed as the desired target during the training process.

Figure 6(b) shows the test error curves obtained by the method with fixed and incremental topologies. It can be seen that the proposed method can obtain similar results to the fixed topology even when 18 hidden units are employed from the beginning. It worth mentioning that the incremental approach presents a more stable behavior than the fixed final topology when a context change appears, specifically for the first two ones. Again, we can check as the automatic technique allows the initial network structure (two hidden units) to evolve in time adapting its answer in function of the needs of the process.

4   Discussion

In this paper we complete our learning algorithm including a mechanism to control the adaptation of the network topology in a automatic way. This mechanism is based of the Vapnkik-Chervonenkis dimension, a good measure of the generalization capacity of a learning model that is related to its complexity. The employment of this measure as reference allows us to predict an upper bound for the test error. In view of the experiments made and the results presented in Sect. 3 for stationary time series as well as for dynamic sets, we can say that the automatic technique based in the Vapnik-Chervonenkis dimension of a neural network allows obtaining an estimation of the appropriate size of the network. Taking as a reference the results reached when a fixed topology is used during the whole learning process, we can check how our developed incremental approach obtains a similar performance without the need to estimate previously the suitable network topology to solve the problem. The proposed method ensures an appropriate size for the network during the learning process maximizing the available computational resources.

5   Conclusions and Future Work

In this work we have presented a modification of the OANN online learning algorithm (proposed in [17]) to control and adjust the network growth in an automatic way according to the needs of the learning process. The main conclusions that can be obtained from the study reported in this paper are,

- The automatic-OANN achieves results close to those obtained when the final fixed topology is used from the beginning of the learning process.
- The network structure begins with the minimum number of hidden neurons and a new unit is added whenever the current topology was not appropriate to satisfy the needs of the process.
- The method allows saving both temporal and spatial resources, an important characteristic when it is necessary to handle a large number of data for training or when the problem is complex and requires a network with a high number of nodes for its resolution.

In spite of these favorable characteristics, there are some aspects that need an in-depth study and will be addressed as future work. Among them, it worth mentioning the following lines of research:

- Improving the addition of hidden units employing different measures, as for example, the increasing tendency of the errors committed.
- Including some pruning technique in order to allow, no only the addition, but also the removal of unnecessary hidden units according to the complexity of learning process.

Acknowledgements. The authors would like to acknowledge support for this work from the Secretaría de Estado de Investigación of the Spanish Government (Grant code TIN2012-37954), partially supported by the European Union FEDER funds.

References

1. Elwell, R., Polikar, R.: Incremental learning in nonstationary environments with controlled forgetting. In: 2009 International Joint Conference on Neural Networks, pp. 1388–1395. IEEE Press, Piscataway (2009)
2. Wang, H., Fan, W., Han, J.: Mining concept-drifting data streams using ensemble classifiers. In: 9th ACM SIGKDD International Conference on Knowledge Discovery and Data Mining, pp. 226–235. ACM, New York (2003)
3. Franklin, D., Wolpert, D.: Computational mechanisms of sensorimotor control. Neuron **72**(3), 425–442 (2011)

4. Mukhopadhyay, S., Sircar, P.: Parametric modelling of non-stationary signals: a unified approach. Sig. Process. **60**(2), 135–152 (1997)
5. Bishop, C.M.: Neural Networks for Pattern Recognition. Oxford University Press, New York (1995)
6. Reed, R.: Pruning algorithms: a survey. IEEE Trans. Neural Netw. **4**, 740–747 (1993)
7. Parekh, R., Yang, J., Honavar, V.: Constructive neural-network learning algorithms for pattern classification. IEEE Trans. Neural Netw. **11**(2), 436–451 (2000)
8. Ash, T.: Dynamic node creation in backpropagation networks. Connection Sci. **1**(4), 365–375 (1989)
9. Hirose, Y., Yamashita, K., Hijiya, S.: Backpropagation algorithm which varies the number of hidden units. Neural Netw. **4**(1), 61–66 (1991)
10. Aylward, S., Anderson R.: An algorithm for neural network architecture generation. In: AIAA Computing in Aerospace Conference VIII (1991)
11. Yao, X.: Evolving artificial neural networks. Proc. IEEE **87**(9), 1423–1447 (1999)
12. Fiesler, E.: Comparative bibliography of ontogenic neural networks. In: Marinaro, M., Morasso, P.G. (eds.) International Conference on Artificial Neural Networks, pp. 793–796. Springer Verlag London Limited, London (1994)
13. Murata, N.: Network information criterion-determining the number of hidden units for an artificial neural network model. IEEE Trans. Neural Netw. **5**(6), 865–872 (1994)
14. Ma, L., Khorasani, K.: A new strategy for adaptively constructing multilayer feed-forward neural networks. Neurocomputing **51**, 361–385 (2003)
15. Islam, M., Sattar, A., Amin, F., Yao, X., Murase, K.: A new adaptive merging and growing algorithm for designing artificial neural networks. IEEE Trans. Neural Netw. **20**(8), 1352–1357 (2009)
16. Sharma, S.K., Chandra, P.: Constructive neural networks: a review. Int. J. Eng. Sci. Technol. **2**(12), 7847–7855 (2010)
17. Pérez-Sánchez, B., Fontenla-Romero, O., Guijarro-Berdiñas, B., Martínez-Rego, D.: An online learning algorithm for adaptable topologies of neural networks. Expert Syst. Appl. **40**(18), 7294–7304 (2013)
18. Nguyen, D., Widrow, B.: Improving the learning speed of 2-layer neural networks choosing initial values of the adaptive weights. In: International Joint Conference on Neural Networks 3, San Diego, USA, pp. 21–26 (1990)
19. Fontenla-Romero, O., Guijarro-Berdiñas, B., Pérez-Sánchez, B., Alonso-Betanzos, A.: A new convex objective function for the supervised learning of single-layer neural networks. Pattern Recogn. **43**(5), 1984–1992 (2010)
20. Martínez-Rego, D., Pérez-Sánchez, B., Fontenla-Romero, O., Alonso-Betanzos, A.: A robust incremental learning method for non-stationary environments. Neurocomputing **74**(11), 1800–1808 (2011)
21. Vapnik, V.: Statistical Learning Theory. John Wiley & Sons, Inc., New York (1998)
22. Baum, E.B., Haussler, D.: What size net gives valid generalization? Neural Comput. **1**(1), 151–160 (1989)
23. Suykens, J.A.K., Vandewalle, J. (eds.): Nonlinear Modeling: Advanced Black-Box Techniques. Kluwer Academic Publishers, Boston (1998)
24. Hénon, M.: A two-dimensional mapping with a strange attractor. Commun. Math. Phys. **50**(1), 69–77 (1976)
25. Wegman, E., Smith, J.: Statistical Signal Processing. Marcel Dekker Inc., New York (1984)
26. Gama, J., Medas, P., Castillo, G., Rodrigues, P.: Learning with drift detection. Intell. Data Anal. **8**, 213–237 (2004)

27. Mackey, M.C., Glass, L.: Oscillation and chaos in physiological control sytems. Science **197**(4300), 287–289 (1977)
28. Butcher, J.: Numerical Methods for Ordinary Differential Equations. Willey, Chichester (2008)
29. Lorenz, E.N.: Deterministic nonperiodic flow. J. Atmos. Sci. **20**, 130–141 (1963)

Towards the Identification of Outliers in Satellite Telemetry Data by Using Fourier Coefficients

Fabien Bouleau[1][(⊠)] and Christoph Schommer[2]

[1] SES Engineering, Chateau de Betzdorf, L-6815 Betzdorf, Luxembourg
fabien.bouleau@gmail.com
http://www.ses.com
[2] Department of Computer Science and Communication,
University of Luxembourg, Luxembourg, Luxembourg
http://www.uni.lu

Abstract. Spacecrafts provide a large set of on-board components information such as their temperature, power and pressure. This information is constantly monitored by engineers, who capture the outliers and determine whether the situation is abnormal or not. However, due to the large quantity of information, only a small part of the data is being processed or used to perform anomaly early detection. A common accepted research concept for anomaly prediction as described in literature yields on using projections, based on probabilities, estimated on learned patterns from the past [6] and data mining methods to enhance the conventional diagnosis approach [14]. Most of them conclude on the need to build a pattern identity chart. We propose an algorithm for efficient outlier detection that builds an identity chart of the patterns using the past data based on their curve fitting information. It detects the functional units of the patterns without apriori knowledge with the intent to learn its structure and to reconstruct the sequence of events described by the signal. On top of statistical elements, each pattern is allotted a characteristics chart. This pattern identity enables fast pattern matching across the data. The extracted features allow classification with regular clustering methods like support vector machines (SVM). The algorithm has been tested and evaluated using real satellite telemetry data. The outcome and performance show promising results for faster anomaly prediction.

Keywords: Data mining · Time series · Machine learning · Pattern identification

1 Introduction

The major concerns for satellite operations are the safety, reliability and durability of the spacecraft fleet. The spacecrafts are being constantly exposed to the space weather: radiations, solar flares, peaks of temperature, etc. Besides, due to the distance, there is no direct visibility on the spacecraft and no way to examine or fix it. The only health information available is the sensors information it sends to earth. It is an instant reading of all the on-board sensors (like a snapshot) sent at regular intervals of one or two seconds. Once rebuilt, each

© Springer International Publishing Switzerland 2015
B. Duval et al. (Eds.): ICAART 2014; LNAI 8946, pp. 211–224, 2015.
DOI: 10.1007/978-3-319-25210-0_13

sequence of data associated to its sensor is a continuous time series expanding over several years.

Anomaly detection and prediction techniques are being constantly developed, in order to perform early detection and avoid the failures, since they have a cost. They may impact the spacecraft lifetime, its capacity, or in the worst case end up with a total loss of control of the satellite. For the most part, expert systems have been built using satellite engineers' knowledge. These systems will trigger an alarm before the anomaly happens. They are thus limited by the satellite engineers knowledge and experience, since they know only a limited part of the model and spacecraft history. A study run by ESOC[1][16] shows that only 10 % of the on-board sensors data is actually being watched. On top of that, the amount of data to process reaches terabytes. Processing the whole set of data to perform detection and classification is nowadays too much time consuming. There is consequently no systematic classification and analysis.

The most common way to tackle anomalies consists in looking at data from the past for similar behavior in order to identify the root cause and to search for indicators to help early detection. Currently, suspicious satellite data is classified manually by the data experts themselves. In this paper, we propose an algorithm for efficient outlier detection that builds a characteristics chart for each patterns using the data from the past using its curve fitting information, in order to enable anomaly detection and eventually prediction. Each detected pattern is thus allotted a characteristics chart with the most relevant statistical elements. This pattern identity chart allows fast pattern matching across the data and pattern classification. In the following section, we will present the state of the art approaches in the space industry. We will then introduce our algorithm for fast pattern matching and the subsequent techniques that can be used for detecting the pattern, fuzzy comparison, to measure the quality of the match and window sliding. The results are presented and discussed in the next part. We will eventually conclude by summarizing the contributions of this paper in the last part.

2   Conventional Approaches to Outlier Detection by Satellite Operator Engineers

Expert systems are built on the knowledge of the satellite engineer, sometimes based on the manufacturer's inputs. They apply to one part of the system only and usually focus on a specific anomaly. Though very accurate, the number of these systems grows fast and each of them requires weeks or sometimes months to be created.

Currently, the model-based approach is handled the following ways. The first consists in identifying the signature of the device instead of the anomaly. The model is then implemented to reproduce its behavior. Anomalies can tentatively

---

[1] European Space Operations Centre, responsible for controlling the European Space Agency (ESA) satellites and space probes.

be reproduced and analyzed by satellite engineers depending on the inputs. The second model-based approach is to build a fully fledged model of the spacecraft, commonly designated as simulator, to either test the maneuvers against it or use its data output as predicted behavior.

The model-based approach nevertheless suffers from its lack of flexibility with regards to internal and environmental reconfigurations. The model needs to be updated as soon as the satellite hardware is altered (broken gyro for instance). On top of that, environmental elements such as space weather may alter the measures. The effort to develop and maintain such a model is merely prohibitive and only some parts are considered due to the overall complexity of the satellite.

Systematic analysis methods emerge, relying classification techniques such as support vector machines for pattern recognition. These techniques are nevertheless subject to performance issues as the cloud of point grows. Besides, most of them require complete reprocessing if only a subset needs to be taken into account.

All these traditional approaches rely on apriori knowledge based on a narrow set of data and the data mining methods suffer from performance issues induced by scalability limitation. The synthesized data based outlier detection approach is been increasingly considered. The concept is to use the curve fitting data to perform pattern matching using specific techniques and properties. Each pattern is described by an identity chart in which appear the curve fitting data and other relevant statistical elements. This identity chart is then used to perform fast pattern matching across the entire database. As for the curve fitting, Fourier series propose an interesting set of properties that allow efficient pattern comparison and match quality measurement. Furthermore, using the sliding window technique as described by [2] would enable efficient reclassification of the patterns while saving reprocessing time and therefore keeping the fast pattern matching performance at its best.

The existing outlier detection techniques of the three categories supervised (like support vector machines), semi-supervised (like transductive support vector machines or heuristics) and unsupervised (like k-Means) all rely on cloud of points rather than a reduced dataset. The order of magnitude for a single parameter over the entire lifetime of one spacecraft (roughly 15 years) is around 500 million points to process. Besides, this data is globally non-stationary: some elements are bound to seasonal effects, some to external factors like solar flares, moon attraction, etc., or simply the orbital position of the satellite. Most algorithms scale with the dimensionality of the input data, inducing a problem of computational cost. To address this issue, approaches like Symbolic Aggregate approXimation [15] as well as the ones described in Data mining in time series databases [12] target the reduction of dimensionality. It nevertheless performs a systematic reduction, regardless the semantics of the data. It obviously does not make sense to compare Volts with Ampers, as does trying to make the intensity signature of battery charge and discharge match. Detecting the different phases of a signal, be it power or thermal signature for instance, is henceforth paramount and will be addressed by our algorithm.

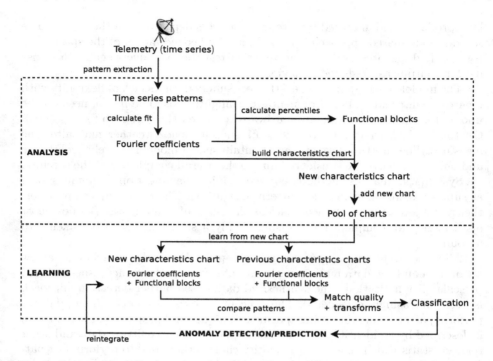

Fig. 1. Pattern identities learning algorithm. The data flows from the telemetry down to anomaly detection/prediction. Our algorithm is represented by the two boxes in-between: the analysis part processes the patterns in order to extract their characteristics charts, composed of the percentiles explained in Sect. 3.3 and other elements listed in Sect. 3.5. The patterns are extracted thanks to the percentiles method and used in the learning part to determine the appropriate class the patterns belong to by comparing them with the pool, as described in the Sects. 3.1 and 3.2. The resulting status vector is then built from the comparisons. Anomaly detection can then be performed within the class elements with cumulative distribution functions as described in Sect. 3.4. The detected outliers can be either notified, ignored or integrated to the training set depending on how it is flagged (respectively anomaly, outlier or normal pattern). It can eventually be reclassified as the training sets evolve, in order to determine if there is a better classification.

Although we are following up the thermal signatures of satellite thrusters dataset only along this paper for the sake of clarity of the explanations, our algorithm has been applied alike over different types of geostationary satellites and different types of sensor measures (battery voltage, tank pressure, etc.).

3   Proposed Outlier Identification System

Our approach to perform outlier identification as summarized on diagram 1 is to extract the features of the time series and enable traditional classification algorithms. Depending on the context, the data analysis may nevertheless differ and require re-classification. Our method provides fast data processing algorithms by using synthesized information.

The first question is which curve fitting technique shall be used in our case in order to preserve efficiency. From our analysis of the different methods, we came to the conclusion that Fourier series is the most suitable method in the case of satellite telemetry. First of all, because of the interesting properties of Fourier with regards to the convolution of two series and how they can be easily factorized that we elaborate below. Besides, due to the oscillating nature of the signals and the background induced by spectrum analysis, most analysis algorithms use this technique. The curve fitting step is therefore already available and normalized in the database.

In this section, we introduce how in our methodology we proceed to compare two patterns using the curve fitting information, along with the interesting properties. We will also show how we measure the quality of our match, the tools we use for horizontal identification and eventually how we define the pattern's characteristics chart.

3.1 Pattern Comparison

Given two Fourier series f_1 and f_2 of the same frequency:

$$f_1(t) = k_1 + \sum_{i=1}^{N} a_{i,1} \cos(i\omega t + \varphi) + b_{i,1} \sin(i\omega t + \varphi) \tag{1}$$

$$f_2(t) = k_2 + \sum_{i=1}^{N} a_{i,2} \cos(i\omega t + \varphi) + b_{i,2} \sin(i\omega t + \varphi) \tag{2}$$

Once factorized, the convolution $R(f_1, f_2, t)$ can then be written the following way:

$$R(f_1, f_2, t) = k_1 - k_2 + \sum_{i=1}^{N} \genfrac{}{}{0pt}{}{(a_{i,1} - a_{i,2}) \cos(i\omega t + \varphi)}{+ (b_{i,2} - b_{i,2}) \sin(i\omega t + \varphi)} \tag{3}$$

The resulting Fourier series represents the distance between the two original Fourier series f_1 and f_2. Let $\widehat{R}$ be the representation of $R(f_1, f_2, t)$ in the frequency domain. We define the quality of the comparison $\rho(\widehat{R})$ by the following equation:

$$\rho(\widehat{R}) = \sum_{i=1}^{N} \frac{\widehat{R}(i)}{i^2} \tag{4}$$

Vertical scaling and translation are the only two purely mathematical transforms we need for the comparison. The horizontal transforms require deeper understanding of the signal itself and will be covered in the next section. Since

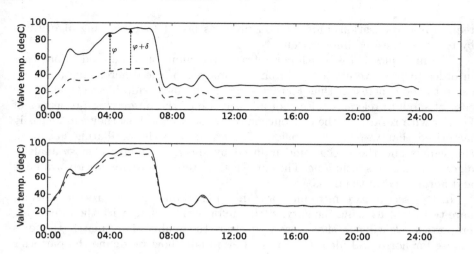

Fig. 2. Fourier series representation of two thermal signatures f_1 (lower curve) and f_2 (upper curve). We apply the affine transforms on the f_1 curve to obtain f_1'. The algorithm evaluated the best fit with $a = 0.502$ and $k = 0.21$, using the points at $180°$ and $190°$. The ranking of this example is bad because of the vertical scaling factor. There is a high probability for this pattern to be either reclassified later on if a fit with better ranking is found or to go in a new class if a correlation with another spacecraft context parameter is found and justifies it.

the nature of the pattern is affected, and henceforth the quality of the comparison, the measures of the transforms will be kept in the characteristics chart of the pattern. The transforms are modeled the following way:

$$a = \frac{f_2(\varphi + \delta) - f_2(\varphi)}{f_1(\varphi + \delta) - f_1(\varphi)}$$
$$k = f_2(\varphi) - a \times f_1(\varphi) \tag{5}$$
$$f_1'(\theta) = a \times f_1(\theta) + k$$

3.2 Pattern Reconsolidation

The second diagram on Fig. 2 shows that even though we have a good performance match after the vertical transforms, the algorithm is still missing it. We are hence introducing the concept of sliding pattern which consists in circularly drifting one of the two series to the right or to the left.

As for the modeling, let θ be the circular drift component defined as $0 \leq \theta < 2\pi$. The f_2 series equation would then be written as follows:

$$f_2(t) = k_2 + \sum_{i=1}^{N} a_{i,2} \cos(i\omega t + \varphi + \theta) + b_{i,2} \sin(i\omega t + \varphi + \theta) \tag{6}$$

The best value of θ is then determined by looking for the minimal $\rho(\widehat{R})$ as per Eq. 4. From there, different approaches are applicable. The most straightforward (and less optimized) is to cycle θ by even steps. Other more accurate techniques can also be applied, such as dichotomy or stochastic research. Stochastic research remains better since it tackles the extrema problem.

3.3   Pattern Functional Units

As previously stated, each parameter of the satellite telemetry comes as a long time series. The individual patterns that will be required for the training set for the classification can be either provided or must be algorithmically determined. The telemetry stream and its curve fit are extracted on a daily basis, regardless the semantics. This is an arbitrary decision based on the satellite engineers as there is one station-keeping maneuver per day. The temperature constantly increases between 12am and 8am that day represents the maneuvers itself, the thruster being idle for the rest of the day.

With regards to classification, the different events need to be isolated in an unsupervised way. The algorithm must hence be capable to learn the pattern structure without apriori knowledge. One intermediate alternative is to use the information databases in which the burn times are scheduled by the engineers. It would nevertheless then relies on user's input and can therefore not adapt to new or unexpected situations. Another approach is to extract the information from the ground control system itself, where the command are actually sent to the spacecraft. If this is more deterministic and accurate, it is still driven by human actions and enters in the semi-supervised category. Some of the actions may furthermore be initiated by the satellite itself and will thus not be captured.

For the reasons aforementioned, if these solutions can be considered as helpers, a proper unsupervised method is still needed. Our approach is to divide the signal horizontally by using the percentiles method. The first element of the percentiles method is the median $\widetilde{m}$. Let $n \in \mathbb{N}$ and p the percentile step ($p = 0.5$ for intervals of 50 %). The horizontal areas are defined by the following thresholds:

$$\left(n - \frac{1}{2}\right) p\widetilde{m} \leq y_n(x) < \left(n + \frac{1}{2}\right) p\widetilde{m} \tag{7}$$

Let $A = ((a_0, b_0), \dots, (a_n, b_n))$ be the Fourier series coefficients. The blocks intervals are delimited the following way:

$$X_n = F(A) \cap \left(n - \frac{1}{2}\right) p\widetilde{m} \tag{8}$$

Curve fitting with Fourier in the context has the drawback of smoothing the data. In order not to miss any outlier, the characteristics chart must therefore enumerate the peaks. In this method, we keep the residuals information per block. With $X_i = \{(x_{i,1}, x'_{i,1}), \dots, (x_{i,n}, x'_{i,n})\}$ the list of block intervals and $f(i)$ the Fourier series, we define S_n as:

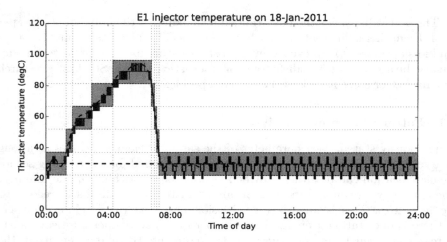

Fig. 3. Original data plot divided by the percentiles method. The thick dashed curve represents the Fourier series and the horizontal thick dashed line the median, used as basepoint to calculate the percentiles. In this example, without apriori knowledge, the blocks captured the idle phase (block around the median) and the "thruster fired" phase. Further analysis show that the fired phase can be split in 3 steps, that the algorithm still needs to learn.

$$S_n = \sum_{j=1}^{|X_n|} \sum_{i=x_{n,j}}^{x'_{n,j}} (y - f(i))^2 \qquad (9)$$

As represented on Fig. 3, the pattern can thus be subdivided into functional blocks, that will figure in the characteristics chart. In our original example, we know by experience that the upper blocks represent the different phases of the thruster burn while the lowest one the idle period. As for the characteristics chart, we will not only keep the quantitative block representation, but the sequence itself. This will help on one hand to split the active from the idle phases and, on the other hand to characterize the remaining steps of the thruster burn, which are "fired", "on-time" and "cooldown".

3.4 Inner Class Comparison for Novelty Detection

Once identified the functional units thanks to percentiles method and a potential class the pattern belongs to using the pattern comparison and reconsolidation methods presented above, our algorithm needs to evaluate whether the introduced pattern presents a novelty amongst the other patterns of the class. To do so, we are using the cumulative distribution functions: equation:

$$F_X(x) = P(X \le x) \qquad (10)$$

In general, we will also consider the tail distribution definition that illustrates better the symmetry of the dispersion.

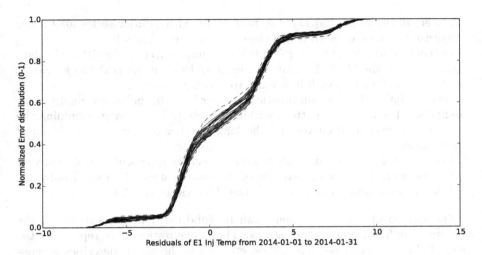

Fig. 4. Cumulative distribution functions (CDF) applied on the residuals of the thruster temperature signature evolution over one month against the calculated Fourier series. The resulting class of CDF elements helps us to determine the upper and lower boundaries of acceptable residuals (dotted lines). Any new pattern for which the CDF exceeds these boundaries is an outlier.

$$\overline{F}_X(x) = 1 - F_X(x)$$
$$= P(X > x) \tag{11}$$

The diagram on Fig. 4 shows how our algorithm determines the upper and lower curve limits. We define a training period of one month during which the CDF are considered without anomaly. From then on, any new CDF outside the upper and lower limit curves raise an outlier detection notification. This one is then flagged as anomaly or not. If not an anomaly, the new upper and lower limits are calculated unless the CDF remains flagged as outlier. It becomes part of the training set otherwise.

3.5 Definition of the Characteristics Chart

The characteristics chart is the element to gather all the features of the studied pattern. The signals however must be put in their original context. As we define it, the chart shall comprise the immutable (or reference) elements:

- Spacecraft context: the configuration of the spacecraft allows our algorithm to segregate the pattern classes depending on the mode the spacecraft is configured, such as eclipse or solstice season, beacon or earth tracking mode, etc.
- Percentiles blocks, as per Eq. (8): our algorithm to subdivides the pattern into functional blocks in order to compare what semantically makes sense (thruster on-time periods shall not be compared with cooldown periods).

- Fourier series, as per Eq. (1): the pattern fit with Fourier series allows our algorithm to determine to which class the pattern belongs to.
- Per-block residuals, as per Eq. (9): this enumerates to our algorithm the data trends within the block. It is later on used to decide if the next block belongs to the same data flow or if it belongs to another phase.
- The cumulative distribution functions, as per Eq. (10): helps our algorithm to define what is a normal pattern within the selected class by determining the upper and lower limit curves and checking whether the new pattern is out of boundaries.
- Timestamp: The timestamp information is usually represented as day of year plus the year. The day of year allows the classification of seasonal patterns, while the year information indicates the elderness of the data.

The spacecraft context elements can be subdivided in two categories: the spacecraft configuration and its status. The configuration part represents the setup of the spacecraft (switch, valves, etc.) while the status describes its condition such as a defective sensor or a broken CPU. This chart remains flexible and additional features can be experimentally added, such as ephemeris data and space weather.

The characteristics charts are then classified and linked with each other in order to preserve the analytical elements:

- Pattern matching quality, as per Eq. (4)
- Transform elements, as per Eqs. (5) and (6)

The training set of each class is dynamic. The main goal of these two elements is then to be able to re-evaluate the classification of a given pattern, in case it later on fits better in another one.

4   Experimental Validation

The validation of our approach is quite difficult for three reasons. First of all, in order to be accurate, the telemetry of the entire lifetime of the satellite should be processed. In this paper we will run it on the most recent subset, that consists in two months of data. Besides, only the propulsion subsystem is analyzed below, the outliers and anomalies of the power subsystem for instance being very difficult for the satellite engineers to detect and explain. The propulsion subsystem henceforth provide a better second sight for validation. Eventually, despite several anomaly detection and curve fitting techniques exist such as [14] and [6] none of them actually plainly address the problematic.

Our experiments dataset is the thrusters temperature telemetry, since related to our original example. We are focusing here on the propulsion subsystem of a single spacecraft. It is composed of 18 parameters, 16 of which representing the thrusters thermal signature, 1 of which is the timestamp (day-of-year) and the last being the thruster identifier. The propulsion subsystem is known to be subject to seasonality, due to the exposition to the sun. The timestamp parameter is

Table 1. Performance of the daily data collection phase for the 16 thermal signatures for a single day. This shows that the data is made available within 2 min.

Collected element	Quantity of processed data	Processing time
Fourier series	~43200 points	8 s
Residuals calculation	~43200 points	75.6 s
Percentile blocks calculation	30 Fourier coefficients	1.38 s
Blocks processing	56 blocks	0.39 s
Cumulated results		85.37 s

therefore relevant in the analysis. The thruster identifier provides its cardinality (north, south, west or east) and correlates with the type of maneuver performed (south-north translation, west-east drift, etc.).

The performance measures depend on how the implementation is performed (programming language, optimizations, etc.) and the hardware it is run on. To obtain our results, we have developed the algorithm using IPython and Matplotlib, since they are accessible for free to everyone. The benchmarks have been

Table 2. Performance of our pattern comparison algorithm for the 16 thermal signatures. The analysis time of the two months of data decreases to 6 min, whereas processing the original cloud of points requires approximately 60 min.

Thruster identifier	Patterns	Vertical scalings	Horizontal drifts	Processing time
E1inj	42	41	54	15.37 s
E2inj	41	24	47	13.92 s
W1inj	45	30	63	22.26 s
W2inj	42	39	54	14.87 s
E1val1	58	59	85	22.87 s
E2val1	58	5	58	18.78 s
N1val1	57	32	67	20.64s
N2val1	58	46	87	24.03 s
N3val1	56	62	101	25.51 s
N4val1	59	91	123	29.96 s
W1val1	57	41	70	20.81 s
W2val1	58	53	67	20.08 s
E1val2	57	55	88	25.26 s
E2val2	59	17	59	20.16 s
W1val2	59	58	91	25.51 s
W2val2	58	50	73	24.23 s
Cumulated results	864	703	1187	344.26 s

performed on a HP Proliant DL580 Intel Xeon E7420 dual CPU quad-core 2 GHz running Debian 7 amd64.

Table 1 shows the data collection performance for one day of telemetry on 16 thrusters. It includes the Fourier fit for the pattern comparison optimization, the percentiles blocks calculation for learning the functional blocks, the Fourier fit residuals per block for compensating the smoothing effect of the Fourier series, and the extraction of statistical elements (median, mode, minimum and maximum) to search for the optimal thresholds for the percentiles.

Table 2 presents the outlier detection performance by only using the data presented above. The original cloud of data at this stage is completely ignored. The performance is the best with minimal transforms and processing time. A higher number of transforms means additional iterations and wasted processing time. The ideal case would be to identify the matching points of f_1 into f_2 and calculate the necessary transforms to achieve it in one iteration only.

As a conclusion, we can observe that the characteristics charts as proposed in our algorithm have been extracted for the last two months of thruster thermal data in less than six minutes and is ready for being processed with regular classification technique. The match rankings are also made available in our database along the charts in order to mitigate the classification and re-classify the case being.

5   Conclusions

In this paper, we have addressed the problem of outlier detection in large data warehouse. For this, we have developed an algorithm using curve fitting information to speed up the patterns comparison and efficiently extracting the patterns features for classification. Processing years of cumulated time series for outlier detection is thus made possible.

We have also addressed the problem of data smoothing induced by Fourier fitting with the percentiles method. The nature of the pattern is then refined using the statistical information of the generated blocks.

In order to keep some flexibility in the analysis, the pattern matching algorithm introduces the concept of match quality (Eq. (4)) on top of pattern transforms. The resulting relevance vector mitigates the results, allows fuzzy classification and provides metrics for re-classification.

The performance of our method is given by experimenting on a relevant subset of data. Measurable efficiency elements are provided in terms of quantity and speed. The results show an acceptable ratio in terms of exploitability and availability of the data: both the data collection and data mining parts are achieved within minutes and the number of iterations is kept minimal.

The horizontal best fit method by sliding the pattern as presented in our algorithm is a topic of on-going and future work. In this respect, the technique can be extended using the sliding window technique described by [2] or by determining the functional units of the blocks definitions as per the percentiles method.

The percentiles method on the other hand is mainly applicable to horizontally shaped time series such as battery charge cycle in the power subsystem.
Improving the semantics detection in differently shaped signals is a topic of on-
going and future work. Clustering techniques on external information such as the
maneuvers schedule and the spacecraft change of state can be used to enhance
the resulting definitions.

As a conclusion to this paper, we will note that our approach provides
accurate characteristics chart for the propulsion subsystem of the spacecraft.
It extracts the essential patterns information to enable systematic processing
in the satellite engineers analysis. Beyond, it preprocesses the pattern matching for classification. This approach provides directions for further fast outlier
identification techniques in time-series data.

Acknowledgements. This work has been made within the research project SPACE,
which is an interdisciplinary research project between the University Luxembourg,
Department of Computer Science and SES Engineering. We thank all the SPACE
members as well as all the SES engineers for their kind support. The views expressed
herein represent the authors' views only and do not in any way bind or commit SES
Engineering itself.

References

1. Azevedo, D.N.R., Ambrósio, A.M.: Dependability in satellite systems: An architecture for satellite telemetry analysis
2. Beringer, J., Hüllermeier, E.: Online clustering of parallel data streams. Data
   Knowl. Eng. **58**(2), 180–204 (2006)
3. Bouleau, F., Schommer, C.: Outlier identification in spacecraft monitoring data
   using curve fitting information. In: European Conference on Data Analysis, p. 158
   (2013)
4. Das, K., Bhaduri, K., Votava, P.: Distributed anomaly detection using 1-class SVM
   for vertically partitioned data. Stat. Anal. Data Min. **4**(4), 393–406 (2011)
5. Deng, K., Moore, A., Nechyba, M.: Learning to recognize time series: combining
   arma models with memory-based learning. In: IEEE International Symposium on
   Computational Intelligence in Robotics and Automation, vol. 1, pp. 246–250 (1997)
6. Fujimaki, R., Yairi, T., Machida, K.: An anomaly detection method for spacecraft
   using relevance vector learning. In: Ho, T.-B., Cheung, D., Liu, H. (eds.) PAKDD
   2005. LNCS (LNAI), vol. 3518, pp. 785–790. Springer, Heidelberg (2005)
7. Fukushima, Y.: Telemetry data mining with SVM for satellite monitoring
8. Isaksson, C., Dunham, M.H.: A comparative study of outlier detection algorithms.
   In: Perner, P. (ed.) MLDM 2009. LNCS, vol. 5632, pp. 440–453. Springer, Heidelberg (2009)
9. Keogh, E.: T5: data mining and machine learning in time series databases (2004)
10. Keogh, E., Chakrabarti, K., Pazzani, M., Mehrotra, S.: Locally adaptive dimen-
    sionality reduction for indexing large time series databases. In: Proceedings of
    the 2001 ACM SIGMOD International Conference on Management of Data, SIG-
    MOD 2001, pp. 151–162. ACM, New York, NY, USA (2001). http://doi.acm.org/
    10.1145/375663.375680

11. Keogh, E., Lonardi, S., chi' Chiu, B.Y.: Finding surprising patterns in a time series database in linear time and space. In: proceedings of the 8th ACM SIGKDD International Conference on Knowledge Discovery and Data Mining, pp. 550–556. ACM Press (2002)
12. Last, M., Kandel, A., Bunke, H.: Data Mining in Time Series Databases. World scientific, Singapore (2004)
13. Létourneau, S., Famili, F., Matwin, S.: Data mining for prediction of aircraft component replacement. Special Issue on Data Mining (1999)
14. Li, Q., Zhou, X., Lin, P., Li, S.: Anomaly detection and fault diagnosis technology of spacecraft based on telemetry-mining. In: 2010 3rd International Symposium on Systems and Control in Aeronautics and Astronautics (ISSCAA), pp. 233–236 (2010)
15. Lin, J., Keogh, E., Lonardi, S., Chiu, B.: A symbolic representation of time series, with implications for streaming algorithms. In: Proceedings of the 8th ACM SIGMOD Workshop on Research Issues in Data Mining and Knowledge Discovery, DMKD 2003, pp. 2–11. ACM, New York, NY, USA (2003). http://doi.acm.org/10.1145/882082.882086
16. Martínez-Heras, J.A., Donati, A., Sousa, B., Fischer, J.: Drmust-a data mining approach for anomaly investigation (2012)
17. Rebbapragada, U., Protopapas, P., Brodley, C.E., Alcock, C.: Finding anomalous periodic time series. Mach. Learn. **74**(3), 281–313 (2009)
18. Saleh, J., Castet, J.: Spacecraft Reliability and Multi-State Failures: A Statistical Approach. Wiley, Chichester (2011)
19. Yairi, T., Kawahara, Y., Fujimaki, R., Sato, Y., Machida, K.: Telemetry-mining: a machine learning approach to anomaly detection and fault diagnosis for space systems. In: Proceedings of the 2nd IEEE International Conference on Space Mission Challenges for Information Technology, SMC-IT 2006, pp. 466–476. IEEE Computer Society, Washington, DC, USA (2006). doi: 10.1109/SMC-IT.2006.79

A Probabilistic Approach to Represent Emotions Intensity into BDI Agents

João Carlos Gluz and Patricia Augustin Jaques[✉]

PIPCA - UNISINOS, Av. Unisinos, 950 - Bairro Cristo Rei, São Leopoldo
CEP 93.022-000, Brazil
{jcgluz,pjaques}@unisinos.br

Abstract. The BDI (Belief-Desire-Intention) model is a well known reasoning architecture for intelligent agents. According to the original BDI approach, an agent is able to deliberate about what action to do next having only three main mental states: belief, desires and intentions. A BDI agent should be able to choose the more rational action to be done with bounded resources and incomplete knowledge in an acceptable time. As humans need emotions to make immediate decisions with incomplete information, some recent works have extending the BDI architecture in order to integrate emotions. However, as they only use logic to represent emotions, they are not able to define the intensity of the emotions. In this paper we present an implementation of the appraisal process of emotions into BDI agents using a BDI language that integrates logic and probabilistic reasoning. Hence, our emotional BDI implementation allows to differentiate between emotions and affective reactions. This is an important aspect because emotions tend to generate stronger response. Besides, the emotion intensity also determines the intensity of an individual reaction. In particular, we implement the event-based emotions with consequences for self based on the OCC cognitive psychological theory of emotions. We also present an illustrative scenario and its implementation.

Keywords: BDI · Emotions · Appraisal · OCC · Bayesian decision networks · BDN

1 Introduction

The role of emotions in cognition has been neglected for many years. However, in the last decades, works of psychologists and neurologists have pointed out the important role of affect in many cognitive activities such as, decision making, memory and learning [7, 11, 13].

These new results about the relation of emotion and cognition have also turning the attention of Artificial Intelligence (AI) researchers to emotions. The emotion synthesis, the simulation of emotions in machines, has been a subject of interest of several AI researches. The main idea is: if humans need emotions to accomplish several cognitive processes that are considered intelligent, so

© Springer International Publishing Switzerland 2015
B. Duval et al. (Eds.): ICAART 2014; LNAI 8946, pp. 225–242, 2015.
DOI: 10.1007/978-3-319-25210-0_14

machines also do. Many works have integrated into agents architectures to infer, respond and show emotions [8,28,32].

A very well known reasoning model in Artificial Intelligence (AI) is the BDI (Belief-Desire-Intention). It views the system as a rational agent having certain mental attitudes of belief, desire, intention, representing, respectively, the informational, motivational and deliberative states of the agent [24,33]. A rational agent has bounded resources, limited understanding and incomplete knowledge on what happens in the environment where it lives in. A BDI agent should be able to choose the more rational action to be done with bounded resources and incomplete knowledge in an acceptable time.

Damasio [7] showed that humans need emotions in order to make immediate decisions with incomplete information. BDI agents also need to decide quickly and with incomplete data from the environment. The BDI is a practical reasoning architecture, which is reasoning directed towards action, employed in the cases when the environment is not fully observable [25]. However, most BDI models do not take into account the agent's emotional mental states in its process of decision making.

Among the several approaches of emotions - for example, basic emotions [9], dimensional models [22], etc., the appraisal theory [18,26,27] appears to be the most appropriate to implement emotions into BDI agents. According to this theory, emotions are elicited by a cognitive process of evaluation called appraisal. The appraisal depends of one's goals and values. These goals and values can be represented as the BDI agents' goals and beliefs. In this way, it is possible to make a direct relation between BDI agents goals and beliefs and appraisals.

When addressing emotional mental states into BDI agents, several research questions should be addressed. An important issue is how to represent and implement the emotional appraisal, the emotions and its properties, such as, intensity. This is a first step before representing how emotions can interfere back in the cognitive processes of the agent, such as decision making. When addressing emotions, one should take into account the emotions intensity, since it defines when an emotion will occur or not. When an affective reaction does not achieve a sufficient intensity threshold, it will not be experienced as an emotion.

The formal logical BDI approach is not appropriate to represent the emotions intensity because it does not allow to represent imprecise data. This is the reason because most emotional BDI models do not take into account the notion of emotion intensity [1,15,32]. In order to represent the emotions intensity, we use AgentSpeak(PL) BDI language [29]. AgentSpeak(PL) is a new agent programming language, which integrates BDI and probabilistic reasoning, i.e., bayesian networks.

This article presents an implementation of the emotional appraisal into BDI agents. We are interested in the implementation of emotions and their intensity using a language that integrates BDI and Bayesian reasoning. Unlike other works, as we use a probabilistic extension of the BDI model that is able to represent the intensity of the affective reactions, our work is able to differentiate between emotions and other affective states with low intensity. Besides, the intensity of the emotion also determines how strong is the response of an individual [26]. Emotions are determined by having a high intensity [26], besides its short duration and its appraisal, which differentiates emotions form other affective states.

In this paper, we focus on the event-based emotions, i.e., emotions that are elicited by the evaluation of the consequences of an event for the accomplishment of a person's goals. We do not formalize emotions in which their appraisals evaluate the consequences for others, such as resentment, pity, gloating and happy-for. We chose to implement the event-based emotions with consequences for self, since these emotions seems to be the most important in the decision making process [2, 12, 23].

This paper is organized as follow. Section 2 presents the OCC model, the psychological emotional model that grounds our work. Section 3 compares the proposed work with related works and highlights its main contribution. In Sect. 4, we describe the AgentSpeak(PL), a language that integrates BDI and Bayesian Decision Network to reason about imprecise data. In Sect. 5, we present how we define the emotion's intensity in the proposed work. In Sect. 6, we cite a scenario and its implementation with AgentSpeak (PL) to illustrate how it can be used to implement emotional probabilistic BDI agents. Finally, in Sect. 6, we present some conclusions.

2   The OCC Model

According to the cognitive view of emotions [27], emotions appear as a result of an evaluation process called appraisal. The central idea of the appraisal theory is that "the emotions are elicited and differentiated on the basis of a person's subjective evaluation (or appraisal) of the personal significance of a situation, event or object on a number of dimensions or criteria" [27].

Ortony, Clore and Collins [19] constructed a cognitive model of emotion, called OCC, which explains the origins of 22 emotions by describing the appraisal of each one. For example, hope appears when a person develops an expectation that some good event will happen in the future. The OCC model assumes that emotions can arise by the evaluation of three aspects of the world: events, agents, or objects. *Events* are the way that people perceive things that happen. *Agents* can be people, biological animals, inanimate objects or abstractions such as institutions. *Objects* are objects viewed *qua* objects. There are three kinds of value structures underlying perceptions of goodness and badness: goals, standards, and attitudes. The events are evaluated in terms of their desirability, if they promote or thwart one's goals and preferences. Standards are used to evaluate actions of an agent according to their obedience to social, moral, or behavioural standards or norms. Finally, the objects are evaluated as appealing depending on the compatibility of their attributes with one's tastes and attitudes. In this paper we refer to the emotions that are generated from the evaluation of an event consequences according to one's goals as *event-based emotions*.

The elicitation of an emotion depends on a person's perception of the world – his *construal*. If an emotion such as distress is a reaction to some undesirable event, the event must be construed as undesirable. For example, when one observes the reactions of players at the outcome of an important game, it is clear that those on the winning team are elated while those on the losing team are devastated. In a real sense, both the winners and losers are reacting to the same objective event. It is their construal of the event that is different. The winners construe it as desirable, while the others construe it as undesirable. It is this construal that drives the emotion system.

A central idea of the model is the type of an emotion. An emotion type is a distinct kind of emotion that can be realized in a variety of recognizably related forms and which are differentiated by their intensity. For example, fear is an emotion type that can be manifested in varying degrees of intensity, such as "concern" (less afraid), "frightened", and "petrified" (more afraid). The use of emotion type has the goal of being language-neutral so that the theory is universal, independent of culture. Instead of defining an emotion by using English words (the author's language), the emotions are characterized by their eliciting conditions.

In the OCC model, the emotions are grouped according to their eliciting conditions. For example, the "attribution group" contains four emotion types, each of which depends on whether the attribution of responsibility to some agent for some action is positive or negative, and on whether the agent is the self or another person. The OCC model is illustrated in Fig. 1. When goals are the source, one may feel pleased if the event is desirable, or displeased if it is not. Which specific emotion arises depends on whether the consequences are for other or for oneself. When concerned for oneself (CONSEQUENCES FOR SELF), the evaluation depends on whether the outcomes are past (PROSPECTS IRRELE-VANT), like joy and distress, or prospective (PROSPECTS RELEVANT), such as hope and fear. If the prospect is confirmed or not, other four emotions may arise, such as satisfaction, disappointment, fear-confirmed and relief. When concern for other (labelled as CONSEQUENCES FOR OTHER), the outcomes are evaluated according to when they are undesirable (UNDESIRABLE FOR OTHER), such as gloating and pity, or desirable for other (DESIRABLE FOR OTHER), such as happy-for and resentment.

When the actions of agents are evaluated according to standards, affective reactions of approval or disapproval arise. The specific emotions depend on whether the action is one's own (SELF AGENT), such as pride and shame; or someone else's (OTHER AGENT), such as admiration and reproach.

The aspects of an object are evaluated according to one's tastes, if one likes or dislikes. In this case, emotions such as love and hate may arise. Finally, emotions like anger and gratitude involve a joint focus on both goals and standards at the same time. For example, one's level of anger depends on how undesirable the outcomes of events are and how blameworthy the related actions are.

The OCC authors believe that this model when implemented in a machine can help to understand what emotions people experience under what conditions.

According to them, it is not the objective of the OCC model to implement machine with emotions, but to be able to predict and explain human cognitions related to emotions. However, Picard [21] disagrees and believes that the OCC model can be used for emotion synthesis in machines. In fact, in computing science research, there already are several works that use OCC in order to implement emotions in machine [8, 10, 14, 28].

3 Related Work

There are other researches that have also been working in the extension of the BDI architecture in order to incorporate emotions. [15] defines an extension of the generic architecture of BDI agents that introduces emotions. However, this work focus on how the emotions influence agents' beliefs and intentions. [32] proposes an extension of the BDI model to integrate emotions, which is based on the OCC model. In this model, the agent's beliefs interfere in its appraisal,

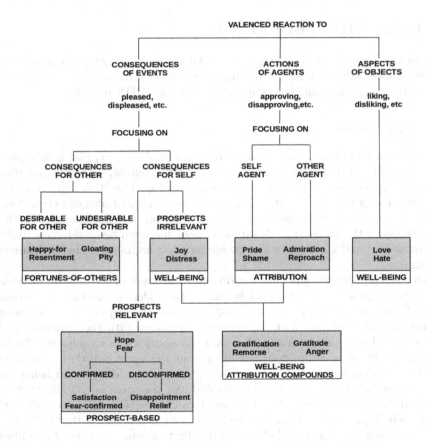

Fig. 1. Global structure of emotion types - OCC model [19].

which will determine its emotions. The emotions, on the other hand, interfere in the choice of intentions. This work focuses on the impact of the emotions in the agent's choice of intentions from desires. Other related works are the ones proposed by [1,30,31]. Both works present a purely logical formalization of the OCC model. They are theoretical works that use a BDI modal logic to describe the OCC's appraisal. They do not show how to create an operational computing model based on the logical formalization.

All the above cited related works do not use a probabilistic model to represent the intensity of the emotions. Their model is not able to represent if the potential of an affective reaction in the appraisal process achieved the necessary threshold to elicit an emotion. In these works, when the appraisal process occurs, an agent always has an emotion. This is not the case, because an emotion only occurs when its intensity achieves a specific threshold [19]. To differentiate emotions from other affective reactions is important because emotions tend to generate stronger response. Besides, the emotion intensity also determines how strong is an individual response [26].

4 Hybrid BDN+BDI Model

This work proposes an extension of the BDI model to also represent emotions and their intensities. We intend to implement the appraisal process of emotions into BDI agents that can represent and reason with BDN models.

4.1 BDI Agents

The BDI model is based on the works of Searle and Dennet, and it was posteriorly generalized by the philosopher Michael Bratman [6], who gave particular attention to the role of intentions in reasoning [34]. The BDI approach views the system as a rational agent having certain mental attitudes of belief, desire, intention, representing, respectively, the information, motivational and deliberative states of the agent [24]. A rational agent has bounded resources, limited understanding and incomplete knowledge on what happens in the environment it lives in.

The *beliefs* represent the information about the state of the environment that is updated appropriately after each sensing action. *Desires* are the motivational state of the system. They have information about the objectives to be accomplished, i.e. what priorities or pay-offs are associated with the various current objectives. They represent a situation that the agent wants to achieve. The fact that the agent has a desire does not mean that the agent will do it. The agent carries out a deliberative process in which it confronts its desires and beliefs and chooses a set of desires that can be satisfied. An *intention* is a desire that was chosen to be executed by a plan, because it can be carried out according to the agents beliefs (it is not rational an agent carries out something that it does not believe). Plans are pre-compiled procedures that depend on a set of conditions for being applicable. The desires can be contradictory to each other, but the intentions cannot [34]. Intentions represent the goals of the agent, defining the chosen course of action. An agent will not give up on its intentions – they will persist, until the agent believes it has successfully achieved them, it believes it cannot achieve them or because the purpose of the intention is no longer present.

In the BDI model, a belief is defined as a two-state logical proposition: or the agent believes that a certain event is true or it believes that the event is false. Nowdays, programming languages and tools available to develop BDI agents do not work with the concept of probabilistic beliefs [4], i.e. they do not allow agents to understand, infer or represent degrees of belief (or degrees of uncertainty) about a given proposition. A degree of belief is defined by the subjective probability assigned to a particular belief.

4.2   Bayesian Decision Networks

The concept of Bayesian Networks (BN) [20] fits in this scenario, allowing to model the probabilistic beliefs of some agent. BN alone are excellent tools to represent probabilistic models of agents, but, with the addition of utility, and decision nodes, it is possible to use the full spectrum of Decision Theory to model agent's behaviour. BN extended with utility and decision nodes are called Bayesian Decision Networks (BDN) [25].

The integration between the current agent programming languages and the concept of belief probabilities can be approached in several ways and at different levels of abstraction. In practical terms, it is possible to make an *ad-hoc* junction of both kind of models in the actual programming code of the agents. Both BDI and BDN agent programming environments rely on libraries and development frameworks, with a standard Application Programming Interface (API). Thus, a hybrid agent can be designed and implemented by combining calls from different sets of APIs, each one from distinct programming environments. The more abstract level to address this issue of integration is usually treated by Probabilistic Logics [16]. Although Probabilistic Logics have the ability to represent both logical beliefs and probabilistic beliefs, there are notorious problems related to the tractability of the resulting models [16]. Another approach is to extend logic programming languages (essentially Prolog) to handle probabilistic concepts. P-Log [3] and PEL [17] fit in this category. They offer interesting ideas, but they lack the full integration with BDI programming languages, like AgentSpeak(L) [5].

4.3   The AgentSpeak(PL) Language

In the present work we consider a programming approach that fully integrates the theoretical and practical aspects of BDN and BDI models. We use AgentSpeak(PL) [29] to implement a probabilistic process, for the appraisal of (some of) the OCC's emotions. AgentSpeak(PL) is a new agent programming language, which is able to support BDN representation and inference in a seamless integrated model of beliefs and plans of an agent. AgentSpeak(PL) is based on the language AgentSpeak(L) [5], inheriting from it all BDI programming concepts.

AgentSpeak(PL) is supported by JasonBayes [29], an extension of the Jason [5] agent development environment. The main changes of AgentSpeak(PL) [29], in respect to AgentSpeak(L) [5], were:

- Inclusion of a probabilistic decision model of the agent, consisting of the specification of a BDN.
- Inclusion of events/triggers based on probabilistic beliefs.
- Inclusion of achievement and test goals, based on probabilistic beliefs.
- Inclusion of actions, which are able to update the probabilistic model.

With these modifications the agent can add to its beliefs a BDN defined by a set of equations $eq_1, ..., Eq_2, ..., eq_n$, which specify:

- The prior probability table for probabilistic nodes (variables) without parents. The table is defined by a set of equations with probability operator '%' applied to node states. For instance: %dirty(true) = 0.5, %dirty(false)=0.5.
- The conditional probability table for probabilistic nodes that have parents. The table is defined through the combined use of the probability operator '%' and the conditional operator '|'. For instance: %cleaning(true) |smash(true) &dirty(true) = 0.0.
- Utility function nodes defined with the utility operator '$'. For instance: $clean_val |cleaning(true) = 80.0.
- Decision variables (or action nodes) defined with the decision operator '?'. For instance: ?%move(yes), ?%move(no) or ?%jump(yes) |barrier(true), ?%jump(yes) |barrier(false).

In AgentSpeak(PL), all BDN variables are represented by unary predicates in the form $p(t)$. The name of the predicate p defines the variable name, while the term t identifies the possible states that the variable can assume. These states are mutually exclusive by definition, i.e. it is not possible to assume two states at same time. Utility functions are defined by the operator '$' to assign utilities to states (events) of the BDN variables. The name of an utility function is a simple literal (atom).

Besides the BDN model, the AgentSpeak(PL) probabilistic inference engine maintains a list of values of model's variables, which contains the current probabilities and utilities assigned, respectively, to probabilistic nodes and utility functions of the BDN model. This list of values is updated when new evidence is aggregate to agent's beliefs base. A new evidence can be a non-probabilistic literal belief $p(t)$ formed by a unary predicate that is exactly the same format (which can be unified) that a BDN variable. In this case, if the non-probabilistic belief is being added to the base $(+p(t))$, the list of values is updated so that the probability assigned to p(t) is 1 (%p(t) = 1), and the probability assigned to other states of the variable p is 0. When the agent is running, it is possible to consult the probability value of any event or state $p(t)$ using the probabilistic operator $\%(p(t))$. The value of some utility function u can be consulted through the operator $\$(u)$.

In terms of the agent's plans programming, probabilistic variables can be used both as trigger events associated with the plans, or in the context conditions of the plans. In the first case, when the probability of some event of state $p(a)$ from the BDN achieves the value 1.0, then a corresponding AgentSpeak action of - $+p(t)$ is generated, adding $p(a)$ to the belief base and deleting any other event of the form $p(x)$ (recall that BDN states or events are mutually exclusive). This action can be used to trigger AgentSpeak plans through the $+p(a)$ expression. In the same form, when the probability of $p(a)$, which was previously 1.0, is reduced, then the corresponding $-p(a)$ action is executed, and $-p(a)$ triggers can be used to activate AgentSpeak plans. The condition can be used as a trigger of AgentSpeak plans, this condition is true when the probability of $p(a)$ is changed. In the case of context condition for plans, it is possible to use the $\%(p(t))$ consult operator to retrieve the probability assigned to some event or state $p(t)$.

Decision variables or action nodes must be related to utility functions in the BDN model, and can be either conditions of other probability nodes or be conditioned by some other node. In any case, after the probabilistic inference engine has updated the list of probabilities and utilities values, decision variables states $p(a)$ that yield the greater utility are added to the belief base of the agent using $-+p(a)$ operations, which can be tracked by AgentSpeak plan triggers.

5   The Appraisal for Joy and Fear Emotions

Emotions in the OCC model [19] depend on several cognitive *variables*, related to the agent's mental states. These variables are instrumental in the process of emotion's arousal, because they determine the intensity of the emotions. The *desirability* is the main variable that determines the arousal and the intensity of several emotions, particularly in the case of joy (and distress) emotion. In fact, there are other global variables that also interfere in all emotions intensity, such as sense of reality, proximity, and unexpectedness, but desirability is a central variable that determines the arousal of the joy and distress emotions, and the other global variables are more concerned to the establishment of the intensity.

The OCC model classifies joy as a prospect-irrelevant event-based emotion, which arouses when someone focuses on the desirability of an event that has happened in respect to this person goals. Joy occurs when a person is pleased about a desirable event that happened. The desirability of an event depends on how a person construes the consequences of an event as being desirable or undesirable. If an event promotes someone goals, it is considered desirable, otherwise it is undesirable. Then, if the consequences of an event are desirable according to a person's goals (it means, they promote someone goals), then one feels joy.

In the OCC model, affective reactions are effectively experienced as emotions only if they achieve a minimum intensity degree. Not all affective reactions (evaluation of an event as un/desirable) are necessarily emotions: "Whether or not these affective reactions are experienced as emotions depends upon how intense they are" [19, p. 20]. Before that, these affective reactions have only a *potential* for the emotion. However, after this potential surpasses the minimum threshold, the emotion starts to be felt with an intensity related to the *potential*. For this purpose, we need some threshold level coefficients to formalize emotions. The coefficients *min_joy* and *min_fear* provide the necessary absolute value threshold limits for joy and fear emotions respectively. They represent the minimum intensity for an affective reaction to be experienced as the emotion of joy or fear.

In [19, p. 182–186] there is a set of rules, which are suggested as computational tools able to determine the potential and the intensity of several emotions, including, joy, fear, and relief. These rules do not have a precise formal semantics, but they can be used as a rough guide to the definition of functions able to estimate the potential and intensity of emotions. In this context, we define the potential of the joy emotion for some perceived event e as the desirability of this event, only if the desirability of e is greater than zero. Otherwise, this potential is zero, meaning that the affective reaction will not be experienced as an emotion. Following the rules that estimate the intensity of joy emotion [19, p. 183], we assumed that joy will be felt with some minimum intensity, only if its potential surpasses the threshold level for joy (*min_joy*). If this is true, then the intensity of joy will be defined as the difference between its potential and this threshold. This entails the following definition of *joy_poten(e)* and *joy_inten(e)* functions that will determinate the potential and intensity of the joy in respect to some event e:

$$joy\_poten(e) = \begin{cases} desirability(e), if\ Perceived(e) \wedge desirability(e) > 0 \\ 0, \quad otherwise \end{cases} \quad (1)$$

$$joy\_poten(e) = \begin{cases} joy\_poten(e), -\textbf{\textit{min\_joy}},\ if\ joy\_poten\ >\ \textbf{\textit{min\_joy}} \\ 0, \quad\quad otherwise \end{cases} \quad (2)$$

Unlike joy, fear is a prospect-relevant event-based emotion. This kind of emotion results from reacting to the prospect of positive or negative events. It means that one is reacting to the possibility or expectation that an event will happen. When someone expects that an event will occur, but it did not happened yet, then one will experiment hope or fear depending upon, respectively, if the event is desirable or undesirable.

In the case of prospective relevant emotions, the *likelihood* of the occurrence of the event becomes an important factor on the determination of the potential of the emotion. The rules defined in [19, p. 185] assume that a (non specified) function will combine undesirability and likelihood. In our appraisal process, we assume that the likelihood variable works like a *filter* of the undesirability, "modulating" the value of undesirability, according to the chance of the occurrence of the event. Note that the likelihood estimation (*likelihood(e)*) is related to, but it is not logically determined by the prospective condition expressed by *Prospect(e)*. Also note that, to simplify the appraisal process, we are working only with positive values for undesirability, the negative valence can be always obtained by $-undesirability(e)$.

$$fear\_poten(e) = \begin{cases} undesirability(e) \times likelihood(e), \\ \quad if\ Prospect(e) \wedge undesirability(e) > 0 \\ 0, \quad otherwise \end{cases} \quad (3)$$

$$fear\_inten(e) = \begin{cases} fear\_poten(e) - \boldsymbol{min\_fear}, \\ \quad fear\_poten(e) > \boldsymbol{min\_fear} \\ 0, \quad otherwise \end{cases} \quad (4)$$

6 A Test Scenario

In this section, we present a scenario to illustrate the process of appraisal that happens with an agent, which has an emotional BDI+BDN architecture. Let us consider a vacuum cleaner robot example, with its environment of a grid with two cells. Besides the desires to work and clean the environment, our agent, whose name is SCR (Super Cleaner Robot), also has the desire of protecting its own existence with a stronger priority.

While SCR is cleaning Cell B, Nick, a clumsy researcher who also works at the same laboratory, goes towards SCR without noticing it. When SCR perceives that Nick is going to stomp on itself, the desire of self-protection becomes an intention and SCR feels fear of being damaged.

In order to alert the awkward scientist, SCR emits an audible alarm. Nick perceives SCR (almost behind his feet) and step on the other cell, avoiding trample over SCR. When SCR perceives that it is not in danger any more, finally, it feels relief. It can continue to do its work; at least while Nick does not decide to come back to his office.

In the next section we present how the process of emotion appraisal, with the corresponding behaviour consequences, can be implemented in AgentSpeak(PL), and how this process will evolve in the first reasoning cycles of the agent.

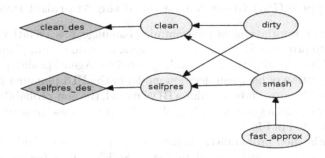

Fig. 2. SCR's BDN model about desirability and likelihood of events.

6.1   A BDN Model for Desirabilitiy and Likelihood

Desirabilities are not utilities. From an ontological perspective they are based on distinct things: the prospective gains of some event in respect to the subjective value of agent goals versus the functional representation of rational preferences for states of the world. However, from a purely formal point of view, it is possible to use utilities to estimate the desirability of some event. If we consider a subjective model for the state of the world, then it is possible to assume that utilities (as a kind of evaluation) could enter in the formation of the desirability. The desirability of some event could be calculated as the difference between the utility of the state of the world previous to the occurrence of the event, and the utility after this event happens, as hypothesized by the agent. Using this interpretation, it is possible to build a BDN model for the states of the world with associated utilities, that can be used to estimate desirability.

Unlike desirability, the likelihood of some event can be best represented by the probability that this event will occur soon, given some evidence that can be observed now. A fully formal model will require some modal or temporal probabilistic logic, however, it is relatively simple to integrate and assess the likelihood of some event in a BDN, by including nodes to represent the evidences that the agent will be smashed. The BDN diagram shown in Fig. 2 integrates desirabilities about SCR's goals to work and to continue to exist, with an estimation of the likelihood to be smashed, given evidence that someone is moving fast toward the agent.

The tables presented in Table 1 define the quantitative model for the BDN presented in Fig. 2. Note that this BDN model does not define the desirability of some event, but the desirability (as utility) of SCR's goals. The model can be instrumental to estimate how much the fact of finding dirt will advance cleaning and self-preservation goals and, thus, can be used to estimate the desirability of these events. Together with the knowledge that any smashing will severely hinder the agent's goals, this model also allows SCR to estimate the undesirability of some expected (prospected) smashing event and if it feels fear about this.

6.2   AgentSpeak(PL) Implementation of the Appraisal Process

Now, using an appropriate agent programming language, it is a relatively straightforward task to program SCR's beliefs, emotions, intentions, and plans. To do so, we will use AgentSpeak(PL) [29], which generalizes AgentSpeak(L), allowing a seamless integration of bayesian decision model in the BDI planning process. An AgentSpeak(PL) source code contains a BDN model, the non-probabilistic beliefs, the agent's goals, and its corresponding plans. Figure 3 shows the source code[1] for the BDN presented in Fig. 2.

SCR's initial non-probabilistic beliefs, its primary goals, and the plans to handle its perceptions are defined in Fig. 4. SCR's initial beliefs include the

---

[1] This code can be programmed by hand, or it can be automatically generated from a graphical model similar to the BDN presented in Fig. 2 (see [29] for details).

Table 1. Probability tables for SCR's BDN model.

clean	U(clean_des)	selfpres	U(selfpres_des)
true	80	true	100
false	0	false	0

fast_approx	P(fast_approx)
true	0.5
false	0.5

smash	fast_approx	P(smash |fast_approx)
true	true	0.9
true	false	0.0
false	true	0.1
false	false	1.0

dirty	P(dirty)
true	0.5
false	0.5

clean	smash	dirty	P(clean | smash & dirty)
true	true	true	0.0
true	true	false	0.0
true	false	true	0.9
true	false	false	0.2
false	true	true	1.0
false	true	false	1.0
false	false	true	0.1
false	false	false	0.8

selfpres	smash	dirty	P(selfpres | smash & dirty)
true	true	true	0.1
true	true	false	0.1
true	false	true	1.0
true	false	false	1.0
false	true	true	0.9
false	true	false	0.9
false	false	true	0.0
false	false	false	0.0

```
// Probabilistic model
// Prior probabilities - standard syntax
%dirty(true) = 0.5.
%dirty(false) = 0.5.
// Prior probabilities - compact syntax
%fast_approx(true,false) = [0.5, 0.5].
// Conditional probabilities - standard syntax
%smash(true) | fast_approx(true) = 0.9 .
%smash(true) | fast_approx(false) = 0.0 .
%smash(false) | fast_approx(true) = 0.1 .
%smash(false) | fast_approx(false) = 1.0.
// Conditional probabilities - compact syntax
%clean(true,false) | smash & dirty =
            [0.0, 0.0, 0.9, 0.2, 1.0, 1.0, 0.1, 0.8].
%selfpres(true,false) | smash & dirty =
            [0.1, 0.1, 1.0, 1.0, 0.9, 0.9, 0.0, 0.0].
// Utility function - standard syntax
$clean_des | cleaning(true) = 80.0 .
$clean_des | cleaning(false) = 0.0 .
// Utility function - compact syntax
$selfpres_des | selfpreserv = [100.0, 0.0].
```

Fig. 3. SCR's BDN model programmed in AgentSpeak(PL).

```
min_joy(50). min_fear(100). located(cell_b). fast_approx(false).
+dirty_in_floor <-         +perceived(dirty); +!cleaning.
+floor_is_clean <-         -perceived(dirty).
+human_very_near <-        +prospect(smash); +!selfpreserv.
+human_fast_approx <-      +prospect(smash); +fast_approx(true); +!selfpreserv.
+human_fast_depart <-      -prospect(smash); -+fast_approx(false).
+!cleaning : perceived(dirty) <-
                           +dirty(true);
                           !joy_inten(dirty,JI); JI > 0;
                           !fear_inten(smash,FI); FI = 0;
                           .cleaning_action.
+!selfpreserv : prospect(smash) <-
                           !fear_inten(smash,FI);FI > 0;
                           .alarm_action.
```

Fig. 4. SCR's initial beliefs, goals, and plans.

threshold levels for joy and fear, the fact it is located in cell B and that nobody is moving fast to this position. Raw perceptions of SCR include seeing if the cell floor is dirty or not, and if somebody is near, moving away or moving towards current position. The perception plans start intentions about what to do if some event is perceived, incorporating the common sense knowledge of SCR about what happens when dirt is detected or when some human is approximating fast.

The planning knowledge of SCR is relatively simple: do some cleaning if dirty is perceived and do some self preservation action if it has fear of something. The AgentSpeak(PL) programming is straightforward. The only small issue is that in AgentSpeak(PL) it is necessary to start goals and plans to calculate the desirability of events. This is due to how the marginal probabilities and the utility values of the BDN model are calculated: whenever some evidence is added to the beliefs of the agent, the bayesian inference engine is activated and all probabilities and utilities are recalculated. An evidence is simply a non-probabilistic belief, like dirty(true) or fast_approx(false), that are identical in name to some variable (node) of the BDN model.

SCR's code is completed by the plans to detect if it feels some joy, or fear, presented in the Fig. 5. These plans implement the functions $joy\_poten(e)$, $joy\_inten(e)$, $fear\_poten(e)$ and $fear\_inten(e)$ as defined in Sect. 5, to estimate the potential and the intensity of these emotions.

6.3 First Reasoning Cycle

Initially, SCR is located in a dirty cell (cell B), it perceives this (dirty_in_floor), register the perception(+perceived(dirty)) and selects the goal to clean the cell (+!cleaning) as an intention. In the first cycle, when dirt is found, but not smashing event expected, then the variable U1 in joy_poten(dirty,JP) plan will give the utility if it is assumed that dirt is found. The value of U1 will be 172, with the utility function returning: $(clean_des) = 72, and $(selfpres_des)=100. The variable U2 will give the utility, if we assume that dirt is not found. Its value is 116, with $(clean_des) = 16, and $(selfpres_des)=100. The difference JP will be 56. With this value for the

```
+!joy_poten(dirty,JP): not dirty(true) <-
    J = 0.
+!joy_poten(dirty,JP): dirty(true) <-
    U1 = $(clean_des) + $(selfpres_des); -+dirty(false);
    U2 = $(clean_des) + $(selfpres_des); -+dirty(true);
    JP = U1 - U2.
+!joy_inten(dirty,JI) <-
    !joy_poten(dirty,JP); ?min_joy(MJ);
    if (JP>MJ) {JI = JP - MJ} else {JI = 0}.

+!fear_poten(smash,J): not prospect(smash) <-
    J = 0.
+!fear_poten(smash,J): prospect(smash) <-
    U1 = $(clean_des) + $(selfpres_des); -+smash(false);
    U2 = $(clean_des) + $(selfpres_des); -smash(false);
    F = U2 - U1.
+!fear_inten(dirty,FI) <-
    !fear_poten(smash,FP); ?min_fear(MF);
    if (FP>MF) {FI = FP - MF} else {FI = 0}.
```

Fig. 5. SCR's plans to estimate its joy, and fear emotions.

potential of joy, then $joy\_inten(dirty,JI)$ plan will return JI=6, indicating that SCR is feeling joy and could start to clean.

6.4   Second Reasoning Cycle

In the second cycle, while it is cleaning, SCR receives from its vision sensor the perception that a human is moving fast toward its cell (human_fast_approx). SCR also knows that human approximation can damage a robot, because the human can trample the robot. It learned that from a previous bad-succeed experience with Nick. So, besides to register the perception (+fast_approx(true)), it will add the belief that it expects to be smashed (+prospect(smashed)).

Now, the variable U1 in the fear_poten(smash,FP) plan, which estimates the state of the world utility when some smashing is expected to occurs, will be reduced to 26.2, with $(clean_des) = 7.2, and $(selfpres_des)=19. Note that the calculus of the likelihood to be smashed is implicit in the BDN. When the evidence (+fast_approx(true)) was added, this changed the probability to be smashed to 90 %. The value of U2 in the same plan, which estimates the utility when no smashing occurs, is 172. This will set the undesirability of being smashed to FP = 145.8, which is the value of the fear potential in this situation. With this value, the fear_inten(dirty,FI) plan will return FI=45.8, indicating that SCR has fear and will start to sound the alarm.

7   Conclusions

In this paper we presented an implementation of the appraisal process of emotions in BDI agents using a BDI language that integrates logic and probabilistic reasoning. Specifically, we implemented the joy and fear event-based emotions with consequences for self based on the OCC cognitive psychological theory of emotions. We also presented an illustrative scenario and its implementation.

One original aspect of this work is that we implemented the emotions intensity using a probabilistic extension of a BDI language, called AgentSpeak(PL). This intensity is defined by the desirability central value, as pointed by the OCC model. In this way, our implementation of emotional BDI allows to differentiate between emotions and affective reactions. This is an important aspect because emotions tend to generate stronger response. Besides, the intensity of the emotion also determines how strong is the response of an individual [26].

We are aware that the implementation of the appraisal of emotions is only a first step. An emotional BDI implementation should address other important dynamic processes between emotions and the mental states of desires, intentions and beliefs in the BDI architecture. As the BDI is a practical reasoning architecture, that is reasoning towards action [34], it is important to discuss how the use of emotions can help the agent to choose the most rational action to be done and how the emotions can improve the way that an agent reasons or decides or acts. These are open questions that we intend to address in a future work. However, we believe that the implementation of the appraisal and the arousal of an emotion depending on the intensity of the affective reaction, presented in this paper, is an important and initial point since the appraisal evaluation explains the origin of an emotion and also differentiates them [27].

Acknowledgements. This work is supported by the following research funding agencies of Brazil: CAPES, CNPq, FAPERGS and RNP/CTIC.

References

1. Adam, C., Herzig, A., Longin, D.: A logical formalization of the OCC theory of emotions. Synthese **168**(2), 201–248 (2009). http://www.springerlink.com/content/8t303657t3110h67
2. Bagozzi, R.P., Dholakia, U.M., Basuroy, S.: How effortful decisions get enacted: the motivating role of decision processes, desires, and anticipated emotions. J. Behav. Deci. Making **16**(4), 273–295 (2003). http://dx.doi.org/10.1002/bdm.446
3. Baral, C., Hunsaker, M.: Using the probabilistic logic programming language p-log for causal and counterfactual reasoning and non-naive conditioning. In: Proceedings of the 20th International Joint Conference on Artifical Intelligence, IJCAI 2007, pp. 243–249. Morgan Kaufmann Publishers Inc., San Francisco (2007). http://dl.acm.org/citation.cfm?id=1625275.1625313
4. Bordini, R.H., Dastani, M., Dix, J., Seghrouchni, A.E.F. (eds.): Multi-Agent Programming: Languages, Platforms and Applications. Multiagent Systems, Artificial Societies, and Simulated Organizations, vol. 15. Springer, New York (2005). http://dro.dur.ac.uk/639/

5. Bordini, R.H., Hübner, J.F., Wooldridge, M.: Programming Multi-Agent Systems in AgentSpeak using Jason. Wiley Series in Agent Technology. Wiley, Chichester (2007)

6. Bratman, M.: What is intention? In: Cohen, P.R., Morgan, J.L., Pollack, M.E. (eds.) Intentions in Communications, pp. 15–31. Bradford books,MIT Press, Cambridge (1990)

7. Damasio, A.R.: Descartes' Error : Emotion, Reason, and the Human Brain. G.P. Putnam, New York (1994)

8. Dias, J., Paiva, A.: I want to be your friend: establishing relations with emotionally intelligent agents. In: Proceedings of the 2013 International Conference on Autonomous Agents and Multi-agent Systems, AAMAS 2013, pp. 777–784. IFMAS, Richland (2013). http://dl.acm.org/citation.cfm?id=2484920.2485041

9. Ekman, P.: An argument for basic emotions. Cogn. Emot. **6**(3–4), 169–200 (1992)

10. Gebhard, P.: Alma: a layered model of affect. In: Proceedings of the Fourth International Joint Conference on Autonomous Agents and Multiagent Systems, AAMAS 2005, pp. 29–36. ACM, New York (2005). http://doi.acm.org/10.1145/1082473.1082478

11. Goleman, D.: Emotional Intelligence. Random House Publishing Group (2012). http://books.google.com.br/books?id=OgXxhmGiRB0C

12. Isen, A.M., Patrick, R.: The effect of positive feelings on risk taking: When the chips are down. Organ. Behav. Hum. Perform. 31(2), 194–202 (1983). http://www.sciencedirect.com/science/article/pii/0030507383901204

13. Izard, C.E.: Emotion-cognition relationships and human development. In: Izard, C.E., Kagan, J., Zajonc, R.B. (eds.) Emotions, Cognition, and Behavior, pp. 17–37. Social Science Research Council, Cambridge University Press (1984). http://books.google.com.br/books?id=IpY5AAAAIAAJ

14. Jaques, P.A., Vicari, R., Pesty, S., Martin, J.-C.: Evaluating a cognitive-based affective student model. In: D'Mello, S., Graesser, A., Schuller, B., Martin, J.-C. (eds.) ACII 2011, Part I. LNCS, vol. 6974, pp. 599–608. Springer, Heidelberg (2011)

15. Jiang, H., Vidal, J., Huhns, M.N.: International Conference On Autonomous Agents. ACM, New York (2007)

16. Korb, K., Nicholson, A.: Bayesian Artificial Intelligence. Chapman & Hall/CRC Computer Science & Data Analysis, Taylor & Francis (2003). http://books.google.com.br/books?id=I5JG767MryAC

17. Milch, B., Koller, D.: Probabilistic models for agents' beliefs and decisions. In: Proceedings of the Sixteenth Conference on Uncertainty in Artificial Intelligence, UAI 2000, pp. 389–396. Morgan Kaufmann Publishers Inc., San Francisco (2000). http://dl.acm.org/citation.cfm?id=2073946.2073992

18. Moors, A., Ellsworth, P.C., Scherer, K.R., Frijda, N.H.: Appraisal Theories of Emotion: State of the Art and Future Development. Emotion Review 5(2), 119–124, May 2013. http://emr.sagepub.com/cgi/doi/10.1177/1754073912468165

19. Ortony, A., Clore, G.L., Collins, A.: The Cognitive Structure of Emotions. Cambridge University Press (1990). http://books.google.com.br/books?id=dA3JEEAp6TsC

20. Pearl, J.: Probabilistic Reasoning in Intelligent Systems: Networks of Plausible Inference. Morgan Kaufmann Publishers Inc., San Francisco (1988)

21. Picard, R.W.: Affective Computing. University Press Group Limited (2000). http://books.google.com.br/books?id=GaVncRTcb1gC

22. Plutchik, R.: A general psychoevolutionary theory of emotion. In: Plutchik, R., Kellerman, H. (eds.) Emotion: Theory, Research, and Experience, vol. 1(3), pp. 3–33. Academic Press, New York (1980)

23. Raghunathan, R., Pham, M.T.: All negative moods are not equal: Motivational influences of anxiety and sadness on decision making. Organizational Behavior and Human Decision Processes **79**(1), 56–77 (1999)
24. Rao, A.S., Georgeff, M.: BDI agents: from theory to practice. Technical report Technical Note 56, Melbourne, Australia (1995)
25. Russell, S.J., Norvig, P.: Artificial Intelligence: A Modern Approach. Prentice Hall Series in Artificial Intelligence, Pearson Education/Prentice Hall (2010). http://books.google.com.br/books?id=8jZBksh-bUMC
26. Scherer, K.R.: Psychological models of emotion. In: Borod, J. (ed.) The Neuropsychology of Emotion, Chap. 6, vol. 137, pp. 137–162. Oxford University Press, Oxford (2000)
27. Scherer, K.R.: Appraisal theory. In: Dalgleish, T., Power, M. (eds.) Handbook of Cognition and Emotion, Chap. 30, vol. 19, pp. 637–663. Wiley (1999). http://psycnet.apa.org/psycinfo/2001-06810-001
28. Signoretti, A., Feitosa, A., Campos, A.M., Canuto, A.M., Xavier-Junior, J.C., Fialho, S.V.: Using an affective attention focus for improving the reasoning process and behavior of intelligent agents. In: Proceedings of the 2011 IEEE/WIC/ACM International Conferences on Web Intelligence and Intelligent Agent Technology - Volume 02, WI-IAT 2011, pp. 97–100. IEEE Computer Society, Washington, DC (2011). http://dx.doi.org/10.1109/WI-IAT.2011.81
29. Silva, D., Gluz, J.: AgentSpeak(PL): a new programming language for BDI agents with integrated bayesian network model. In: 2011 International Conference on Information Science and Applications. IEEE (2011)
30. Steunebrink, B.R., Dastani, M., Meyer, J.J.: A formal model of emotions: Integrating qualitative and quantitative aspects. In: European Conference on Artificial Intelligence, ECAI 2008, vol. 178, pp. 256–260. IOS Press, Patras, Greece (2008). doi:10.3233/978-1-58603-891-5-256
31. Steunebrink, B., Dastani, M., Meyer, J.J.: A formal model of emotion triggers: an approach for bdi agents. Synthese 185(1), 83–129 (2012). http://dx.doi.org/10.1007/s11229-011-0004-8
32. Van Dyke Parunak, H., Bisson, R., Brueckner, S., Matthews, R., Sauter, J.: A model of emotions for situated agents. In: Proceedings of the Fifth International Joint Conference on Autonomous Agents and Multiagent Systems AAMAS 2006, p. 993 (2006). http://portal.acm.org/citation.cfm?doid=1160633.1160810
33. Wooldridge, M.: An Introduction to MultiAgent Systems. Wiley, Chichester (2009)
34. Wooldridge, M.: Intelligent agents. In: Weiss, G. (ed.) Multiagent Systems, pp. 27–77. MIT Press, Cambridge (1999). http://dl.acm.org/citation.cfm?id=305606.305607

Revisiting Classical Dynamic Controllability: A Tighter Complexity Analysis

Mikael Nilsson[(⊠)], Jonas Kvarnström, and Patrick Doherty

Department of Computer and Information Science,
Linköping University, 58183 Linköping, Sweden
{mikni,jonkv,patdo}@ida.liu.se, http://www.ida.liu.se

Abstract. Simple Temporal Networks with Uncertainty (STNUs) allow the representation of temporal problems where some durations are uncontrollable (determined by nature), as is often the case for actions in planning. It is essential to verify that such networks are dynamically controllable (DC) – executable regardless of the outcomes of uncontrollable durations – and to convert them to an executable form. We use insights from incremental DC verification algorithms to re-analyze the original, classical, verification algorithm. This algorithm is the entry level algorithm for DC verification, based on a less complex and more intuitive theory than subsequent algorithms. We show that with a small modification the algorithm is transformed from pseudo-polynomial to $O(n^4)$ which makes it still useful. We also discuss a change reducing the amount of work performed by the algorithm.

Keywords: Temporal networks · Dynamic controllability

1 Background

Time and concurrency are increasingly considered essential in planning and multi-agent environments, but temporal representations vary widely in expressivity. For example, Simple Temporal Problems (STPs, [1]) allow us to efficiently determine whether a set of *timepoints* (events) can be assigned real-valued *times* in a way consistent with a set of *constraints* bounding temporal distances between timepoints. The start and end of an action can be represented as timepoints, but its possible durations can only be represented as an STP constraint if the execution mechanism can *choose* durations arbitrarily within the given bounds. Usually, exact durations are instead chosen by nature and agents must generate plans that work regardless of the eventual outcomes.

STPs with Uncertainty (STPUs, [2]) capture this aspect by introducing *contingent* timepoints corresponding to the end of actions. Associated with these are *contingent* temporal constraints that correspond to possible action durations to be decided by nature. One must then find a way to assign times to ordinary *controlled* timepoints (determine when to start actions) so that for *every* possible outcome for the contingent constraints (action durations), all ordinary *requirement* constraints (corresponding to STP constraints) are satisfied.

© Springer International Publishing Switzerland 2015
B. Duval et al. (Eds.): ICAART 2014; LNAI 8946, pp. 243–261, 2015.
DOI: 10.1007/978-3-319-25210-0_15

If an STPU allows us to schedule controlled timepoints (actions to be started) incrementally given that we receive information when a contingent timepoint occurs (an action ends), it is *dynamically controllable* (*DC*) and can be efficiently executed by a dispatching Algorithm [3]. Conversely, guaranteeing that constraints are satisfied when executing a non-DC plan is impossible; it would require information about future duration outcomes.

We will not describe dispatch algorithms for STPs and STPUs here. They can be found in [3,4]. The idea behind them is to keep track of which events that are allowed to be executed, since their predecessors have executed, and to execute these in a way that satisfies all constraints towards predecessors. When we in the future mention that a network is *dispatchable* we mean that it is in a form which can directly be executed by one of the existing dispatch algorithms.

Although several algorithms for verifying the dynamic controllability of STPUs have been published [4–7] we will focus our attention on the first which is often referred to as classical or MMV [4]. The algorithm is easily implemented, it captures the intuition behind STPUs and has a direct correctness proof. It is also the entry level algorithm for verification. We will show that its run-time is not as thought before, pseudo-polynomial, but $O(n^4)$ through a small modification – the algorithm merely needs to stop earlier. This result shows that the algorithm is quite fast and still useful.

The intuition behind the analysis is that not all of MMV's derivations and tightenings are necessary: Only a certain *core* of derivations actually matters for verifying dynamic controllability, and when the STPU is DC, this core is free of cyclic derivations. This can be exploited through a small change to MMV. Stopping at the right time also preserves another aspect of MMV: the result is dispatchable.

Outline. After providing some fundamental definitions (Sect. 2), we describe the MMV algorithm (Sect. 3). We also present the FastIDC algorithm, which will provide intuitions for our analysis of MMV (Sect. 4). We compare the derivations made by the two algorithms (Sect. 5) and analyze the length of FastIDC derivation chains (Sect. 6), resulting in the new algorithm GlobalDC (Sect. 7) which runs in $O(n^4)$. GlobalDC is in fact identical to a slightly modified MMV algorithm (Sect. 8).

2   Temporal Problems

We start with defining some fundamental concepts.

Definition 1. *A **Simple Temporal Problem** (**STP**, [1]) consists of a number of real variables $x_1, \ldots, x_n$ and constraints $T_{ij} = [a_{ij}, b_{ij}]$, $i \neq j$ limiting the temporal distance $a_{ij} \leq x_j - x_i \leq b_{ij}$ between the variables.*

We will work with STPs in graph form, with timepoints represented as nodes and constraints as labeled edges. They are then referred to as Simple Temporal Networks (STNs). We will also make use of the fact that any STN can be represented as an equivalent *distance graph* [1]. Each constraint $[u, v]$ on an edge

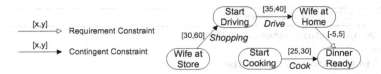

Fig. 1. Example STNU.

AB in an STN is represented as two *corresponding* edges in its distance graph: AB with weight v and BA with weight $-u$. Computing the all-pairs-shortest-path (APSP) distances in the distance graph yields a *minimal representation* containing the tightest distance bounds that are implicit in the original problem [1]. This directly corresponds to the tightest interval constraints $[u', v']$ implicit in the original STN.

If the distance graph has a negative cycle, then no assignment of timepoints to variables satisfies the STN: It is *inconsistent*. Otherwise it is consistent and can be *executed*: Its events can be assigned time-points so that all constraints are satisfied.

Definition 2. *A **Simple Temporal Problem with Uncertainty (STPU)** [2] consists of a number of real variables $x_1, \ldots, x_n$, divided into two disjoint sets of* controlled timepoints R and *contingent timepoints C. An STPU also contains a number of* requirement constraints $R_{ij} = [a_{ij}, b_{ij}]$ *limiting the distance $a_{ij} \leq x_j - x_i \leq b_{ij}$, and a number of* contingent constraints $C_{ij} = [c_{ij}, d_{ij}]$ *limiting the distance $c_{ij} \leq x_j - x_i \leq d_{ij}$. For the constraints C_{ij} we require that $x_j \in C$, $0 < c_{ij} < d_{ij} < \infty$.*

STPUs in graph form are called STNs with Uncertainty (STNUs). An example is shown in Fig. 1. In this example a man wants to cook for his wife. He does not want her to wait too long after she returns home, nor does he want the food to wait too long. These two requirements are captured by a single requirement constraint, whereas the uncontrollable durations of shopping, driving home and cooking are captured by the contingent constraints. The question is whether the requirements can be guaranteed regardless of the outcomes of the uncontrollable durations.

In addition to the types of constraints already existing in an STNU, some algorithms can also generate *wait constraints* that make certain implicit requirements explicit for use in further computations.

Definition 3. *Given a contingent constraint between A and B and a requirement constraint from A to C, the $< B, t >$ annotation on the constraint AC indicates that execution of the timepoint C is not allowed to take place until after **either** B has occurred **or** t units of time have elapsed since A occurred. This constraint is called a **wait constraint** [4], or **wait**, between A and C.*

As there are events whose occurrence we cannot fully control, consistency is not sufficient for an STNU to be executable. However, suppose that for a given STNU

Algorithm 1. The MMV Algorithm.

Boolean **procedure** determineDC()
repeat
   if not pseudo-controllable **then**
     | **return** false
   else
     forall the *triangles ABC* **do**
      | tighten *ABC* using the tightenings in Fig. 2
     end

until *no tightenings were found*
return true

---

there exists a **dynamic execution strategy** [4] that can assign timepoints to controllable events during execution, given that at each time, it is known which contingent events have already occurred. The STNU is then **dynamically controllable** [4] (**DC**) and can be executed. In Fig. 1 a dynamic execution strategy is to start cooking 10 time units after receiving a call that the wife starts driving home. This guarantees that cooking is done within the required time, since she will arrive at home 35 to 40 time units after starting to drive and the dinner will be ready 35 to 40 time units after she started driving.

3   The MMV Algorithm

Algorithm 1 shows a reformulated and clarified [5] version of the classical "MMV" algorithm [4]. Note that these versions share the same worst case complexity.

The algorithm builds on the concept of *pseudo-controllability* [4], a necessary but not sufficient requirement for dynamic controllability. To test for pseudo-controllability the STNU is first converted to an STN by converting all contingent constraints into requirement constraints. The STN then has to be put in its minimal representation (see Sect. 2). If the STN is inconsistent, the corresponding STNU cannot be consistently executed and is not DC. If the STN is consistent but a constraint corresponding to a contingent constraint in the STNU became tighter in the minimal representation, the contingent constraint is *squeezed* [4]. Nature can then place the uncontrollable outcome of the contingent constraint outside what is allowed by the STN representation, causing execution to fail. Therefore the STNU is not DC. Conversely, if the minimal representation is consistent and does not squeeze any corresponding contingent constraint, the STNU is *pseudo-controllable*, but may still fail to be DC.

MMV additionally uses STNU-specific *tightening rules*, also called *derivation rules*, which make constraints that were previously implicit in the STNU explicit (Fig. 2). Each tightening rule can be applied to a "triangle" of nodes if the constraints and requirements of the rule are matched. The result of applying a tightening is a new or tightened constraint, shown as bold edges in the leftmost part of the triangle.

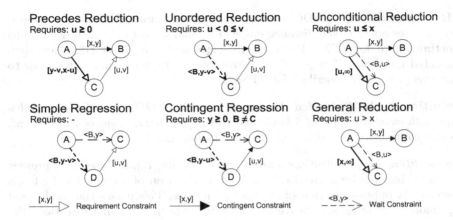

Fig. 2. Tightenings (derivations) of the MMV algorithm.

Algorithm 1 is centered around a loop where it first verifies pseudo-controllability and transfers all tighter constraints found by the associated APSP calculation into the STNU, then applies all possible tightenings. If an STNU is not DC, the tightenings will eventually produce sufficient explicit constraints for the pseudo-controllability test to detect this [4].

The complexity of MMV is said to be $O(Un^3)$ where U is a measure of the size of the domain (the number of constraints and the size of constraint bounds). This comes from a cost of $O(n^3)$ per iteration and the fact that each iteration must tighten at least one constraint leading in the extreme to a negative cycle. Since the complexity bound depends on the size of constraint bounds, it is pseudo-polynomial.

If MMV labels an STNU as DC, the processed STNU can be executed by the dispatcher in [4].

4 The FastIDC Algorithm

The property of dynamic controllability is "monotonic" in the sense that if an STNU is not DC, it can never be made DC by further adding or tightening constraints. Therefore, the *non-incremental* verification performed by MMV is equivalent to starting with an empty STNU (which is trivially DC) and *incrementally* adding one edge at a time, verifying at each step that the STNU remains DC.

We will exploit this fact to compare MMV to the incremental FastIDC Algorithm [8,9], which will allow us to draw certain conclusions about MMV. First, though, we will present and explain FastIDC itself, specifically its tightening/edge-addition aspect (since loosening or removing edges will not be required here). As the original version of this algorithm was incorrect in certain cases, we use the corrected version shown in Algorithm 2 as our starting point [10]. A proof that this version is correct can be found in [11].

FastIDC has three main differences compared to the MMV algorithm.

1: Representation. FastIDC does not work in the standard STNU representation but uses an *extended distance graph* [12], analogous to the distance graphs sometimes used for STNs. Requirement edges and contingent edges are then translated into pairs of edges of the corresponding type in a manner similar to what was previously described for STNs.

Definition 4. *An **Extended Distance Graph (EDG)** is a directed multi-graph with weighted edges of 5 kinds: positive requirement, negative requirement, positive contingent, negative contingent and conditional.*

The *conditional* edges mentioned above, first used by [12], are used to represent the *waits* that can be derived by MMV. The direction of a conditional edge is intentionally opposite to that of the wait it encodes. This makes the conditional edge more similar to a negative requirement edge in the same direction, the difference being the condition.

Definition 5. *A **Conditional Edge** CA annotated $< B, -w >$ encodes a conditional constraint: C must execute after B or at least w time units after A, whichever comes first. The node B is called the **conditioning node** of the constraint/edge.*

2: Derivation Rules. Partly due to the new representation, FastIDC uses different derivation rules. These are shown in EDG form in Fig. 3, where we have numbered two rules (D8–D9) that were unnumbered in the original publication, but shown to be needed [11].

3: Traversal Order. FastIDC uses a significantly different graph traversal order. MMV traverses a graph iteratively, and in each iteration, it considers *all* "triangles" in a graph in arbitrary order. FastIDC, in contrast, uses the concept of *focus edges*. A focus edge is an edge that was tightened and may lead to other constraints being tightened. FastIDC only applies derivation rules to focus edges. If this leads to new tightened edges it will recursively continue to apply the derivation rules until quiescence. Intuitively, this guarantees that all possible consequences of any tightening are covered by the algorithm.

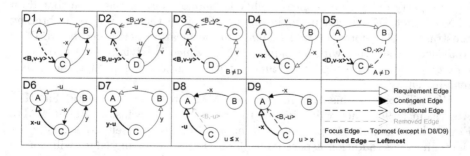

Fig. 3. FastIDC derivation rules D1-D9.

FastIDC Details. Being incremental, FastIDC assumes that at some point a dynamically controllable STNU was already constructed (for example, the empty STNU is trivially DC). Now one or more requirement edges $e_1, \ldots, e_n$ have been added or tightened, together with zero or more contingent edges and zero or more new nodes, resulting in the graph G. FastIDC should then determine whether G is DC.

The algorithm works in the EDG of the STNU. First it adds the newly modified or added requirement edges to a queue, Q (a contingent edge must be added before any other constraint is added to its target node and is then handled implicitly through requirement edges). The queue is sorted in order of decreasing distance to the *temporal reference* (*TR*), a node always executed before all other nodes at time zero. Therefore nodes close to the "end" of the STNU will be dequeued before nodes closer to the "start". This will to some extent prevent duplication of effort by the algorithm, but is not essential for correctness or for understanding the derivation process.

In each iteration an edge e_i is dequeued from Q.

A positive loop (an edge of positive weight from a node to itself) represents a trivially satisfied constraint that can be skipped. A negative loop entails that a node must be executed before itself, which violates DC and is reported.

Algorithm 2. FastIDC – sound version.

---

function FAST-IDC$(G, e_1, \ldots, e_n)$
$Q \leftarrow$ sort $e_1, \ldots, e_n$ by distance to temporal reference
  (order important for efficiency, not correctness)
for *each modified edge e_i in ordered Q* **do**
  if IS-POS-LOOP(e_i) **then** SKIP e_i **if** IS-NEG-LOOP(e_i) **then return** false
  for *each rule (Fig. 3) applicable with e_i as focus* **do**
    if *edge z_i in G is modified or created* **then**
      Update *CCGraph*
      if *Negative cycle created in CCGraph* **then return** false **if** *G is squeezed* **then return** false **if** not FAST-IDC(G, z_i) **then**
        return false
    end
  end
end
return true

If e_i is not a loop, FastIDC determines whether one or more of the derivation rules in Fig. 3 can be applied with e_i as focus. The topmost edge in the figure is the focus in all rules except D8 and D9, where the focus is the conditional edge $< B, -u >$. Note that rule D8 is special: The derived requirement edge represents a stronger constraint than the conditional focus edge, so the conditional edge is removed.

For example, consider rule D1. This rule will be matched if e_i is a positive requirement edge, there is a negative contingent edge from its target B to some other node C, and there is a positive contingent edge from C to B. Then a new constraint (the bold edge) can be derived. This constraint is only added to the EDG if it is strictly tighter than any existing constraint between the same nodes.

More intuitively, D1 represents the situation where an action is started at C and ends at B, with an uncontrollable duration in the interval $[x, y]$. The focus edge AB represents the fact that B, the end of the action, must not occur more than v time units after A. This can be represented more explicitly with a conditional constraint AC labeled $< B, v - y >$: If B has occurred (the action has ended), it is safe to execute A. If at most $v - y$ time units remain until C (equivalently, at least $y - v$ time units have passed *after* C), no more than v time units can remain until B occurs, so it is also safe to execute A.

Whenever a new edge is created, the corrected FastIDC tests whether a cycle containing only negative edges is generated. The test is performed by keeping the nodes in an incrementally updated topological order relative to negative edges. The unlabeled graph which is used for keeping the topological order is called the *CCGraph*. It contains the same nodes as the EDG and has an edge between two nodes iff there is a negative edge between them in the EDG. See [10] for further information.

After this a check is done to see if the new edge *squeezes* a contingent constraint. Suppose FastIDC derives a requirement edge BA of weight w, for example $w = -12$, representing the fact that B must occur at least 12 time units after A. Suppose there is also a contingent edge BA of weight $w' > w$, for example $w' = -10$, representing the fact that an action started at A and ending at B may in fact take as little as 10 time units to execute. Then there are situations where nature may violate the requirement edge constraint, and the STNU is not DC.

If the tests are passed and the edge is tighter than any existing edges in the same position, FastIDC is called recursively to take care of any derivations caused by this new edge. Although perhaps not easy to see at a first glance, all derivations lead to new edges that are closer to the temporal reference. Derivations therefore have a direction and will eventually stop. When no more derivations can be done the algorithm returns true to testify that the STNU is DC. If FastIDC returns true after processing an EDG this EDG can be directly executed by the dispatcher in [4].

5   Comparing FastIDC/MMV

To compare the derivation rules used by MMV to those of FastIDC, we first need a translation into EDG format. This is shown in Fig. 4 where as before the bold edges are derived. *Precedes reduction* is split in two since it adds two edges. *Simple regression* is also split in two, one version regressing over a positive edge and one regressing over a negative edge. All variables used as weights are considered positive, i.e., $-u$ is a negative number (with unconditional reduction as an exception). The additional requirements from Fig. 2 still apply but are omitted for clarity. Most are encoded by the edge types – for instance in unordered reduction, only a positive requirement edge can match the rule, making the $v > 0$ requirement implicit. We now see the following similarities:

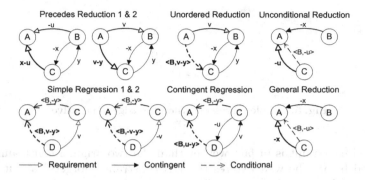

Fig. 4. Classical derivations in EDG format.

- Precedes Reduction 1 (PR1) is identical to D6.
- Unordered reduction is equivalent to D1. However without the extra require-ment ($u \geq 0$) used by MMV to distinguish between applying PR2 and unordered reduction, FastIDC will always apply unordered reduction, even when MMV instead would apply PR2. It can be shown that if the situa-tion calls for an application of PR2, FastIDC derives the same edge as MMV through conversion of the conditional edge resulting from D1 into a require-ment edge (via unconditional reduction, D8). If the application of PR2 directly leads to non-DC detection, FastIDC also detects this directly. So PR and unordered reduction are handled by D1, D6 and D8 together.
- Simple regression 1 is equivalent to D3 and D5. The only difference between D3 and D5 is which edge is regarded as focus.
- Contingent regression is identical to D2.
- Unconditional Reduction is identical to D8.
- General Reduction is identical to D9.

Thus, the only significant differences are:

- FastIDC derivations has no counterpart to Simple Regression 2.
- D4 and D7 have no counterpart rules in MMV. These derive shortest path distances towards earlier nodes in the STNU. This derivation is present and handled by the APSP calculation in MMV.

We see that MMV does everything that the FastIDC derivations do, and also applies SR2 and a complete APSP calculation.

It can in fact be seen that SR2 is not needed, not even by MMV. Figure 5 shows the situation where a conditional edge CA is regressed over an incom-ing negative requirement edge DC. Adding a constraint DA to "bridge" two consecutive negative edges is always redundant both for execution and for DC verification. From an execution perspective this is easily seen since C is always executed before D which ensures that the chain of constraints is respected with-out the addition of DA. From a verification perspective this can be seen since

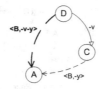

Fig. 5. Simple regression when the edge is negative.

the derived constraint is in fact weaker than the two original constraints. If B is executed before the wait expires the DA constraint "forgets" about the $-v$ part of the constraint which must still be fulfilled. If the wait expires both paths require D to be at least $v + y$ time units after A and the constraint is redundant.

6   Focus Propagation

If we apply rules D1–D9 in Fig. 3, every derived edge has a uniquely defined "parent": The focus edge of the derivation rule. Unless this edge was already present in the original graph, it (recursively) also has a parent. This leads to the following definition.

Definition 6. *Edges that are derived through Fig. 3 derivations are part of a* **derived chain***, where the parent of each edge is the focus edge used to derive it.*

We observe the following:

- A contingent constraint orders the nodes it constrains. In EDG form we see this by the fact that the target of a negative contingent edge is always executed before its source.
- Either D8 or D9 is applicable to any conditional edge. Thus there will always be an order between its nodes set by the negative requirement edge from D8/D9: The target node of a conditional constraint is always executed before its source.

This leads directly to the facts in Table 1. Here, node n_1 is considered *earlier* than n_2 if n_1 must be executed before n_2 in every dynamic execution strategy

Table 1. The derived edges compared to the focus edges.

Rule	Effect
D1	The target of the derived edge is an earlier node
D2, D6	The source of the derived edge is an earlier node
D3, D7	The source of the derived edge is an earlier or unordered node
D4, D5	The target of the derived edge is an earlier node
D8, D9	The derived edge connects the same nodes

and for all duration outcomes. Similarly, node n_1 is considered *unordered* relative to n_2 if their order can differ depending on strategy or outcome.

We now consider the structure of derived chains in DC STNUs. The focus will be on the direction and weight of each derived edge, ignoring whether edges are negative, positive, requirement or conditional (but still keeping track of contingent edges).

Lemma 1. *Suppose all rules in Fig. 3 are applied to the EDG of a dynamically controllable STNU until no more rules are applicable. Then, all derived chains are **acyclic**: No derivation rule has generated an edge having the same source and target as an ancestor of its parent edge along the current chain.*

Proof. Note that by the definition of acyclicity we allow "cycles" of length 1. These can only be created by applications of D8–D9 in a DC STNU.

For D1–D7, each derived edge shares one node with its parent focus edge, but has another source or target. We can then track how the source and target of the focus edge changes through the chain.

Table 1 shows that only derivation rules D1, D4 or D5 result in a different *target* for the derived edge compared to the focus edge. The new target has always "moved" along a negative edge, so it must be executed earlier than the target of the focus edge. Since the STNU is DC, its associated STN cannot have negative cycles. Thus, if the target changes along a chain, it cannot "cycle back" to a previously visited target.

Rules D2, D3, D6 and D7 result in a different *source* for the derived edge. This source may be earlier *or* later than the source of the focus edge, so these rules can be applied in a sequence where the source of the focus edge "leaves" a node n and eventually "returns". Suppose that this happens and the target n' has not changed. This must occur through applications of rules D2, D3, and/or D6–D9. No such derivation step decreases the weight of the focus edge. Therefore, when the source returns to n, the new edge to be derived between n and n' cannot be tighter than the one that already exists. No new edge is actually derived. Thus, if the source changes along a chain, it cannot "cycle back" to a previously visited source. □

This fact together with the previous lemma limits the length of a derived chain to $2n^2$ since we have at most n^2 distinct ordered source/target pairs and can at most have one application of D8/D9 in-between source/target movements. The use of chains to reach an upper bound on iterations is inspired by [5] where an upper bound of $O(n^5)$ is reached for MM.

Note that FastIDC derivations together with local consistency checks and global cycle detection is sufficient to guarantee that all implicit constraints represented by a chain of negative edges are respected, or non-DC is reported. There is no need to add these implicit constraints but the next proof will make use of the fact that they exist.

Some derivations carried out by FastIDC can be proven not to affect the DC verification process, and hence we would like to avoid doing these. These can both be derivations of weaker constraints and constraints that are implicitly

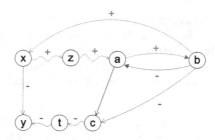

Fig. 6. Situation where D2 or D6 is applied.

checked even if they are not explicitly present in the EDG. In order to single out the needed derivations we define *critical chains*.

Definition 7. *A **critical chain** is a derived chain in which all derivations are needed to correctly classify the STNU. If any derivation in the chain was missing, a non-DC STNU might be misclassified as DC.*

Given a focus edge, one or more derivations may be applicable. Those that would extend the current critical chain into a non-critical one can be skipped without affecting classification. We therefore identify some criteria that are satisfied in all critical chains.

Theorem 1. *Given a DC STNU:*

1. *A D1 derivation for a specific contingent constraint C can only be part of a critical chain once.*
2. *At most one derivation of type D2 and D6 involving a specific contingent constraint C can be part of a critical chain.*

Proof (Proof Sketch). Part 1 is shown as in the proof of Lemma 1: The target cannot come back for another D1 application to the same contingent node.

We use Fig. 6 to illustrate the situation when D2 or D6 is applied over the contingent ab constraint. The rightmost part of this figure is an arbitrary triangle abc where one of the rules is applicable, while the leftmost part is motivated by the proof below.

In the following we do not care if the edges are conditional or requirement: Only the weights of the derived edges are important. We follow a critical chain and see how the source and target change as we continually derive new edges. Applying D2 or D6 gives a new edge ac where the source changes from b to a. We now investigate how derivations can move the source back to b and show that all derivations using the edge which resulted from moving the source back to b are redundant. We already know that the source can only move back to b if the target moves from c. Otherwise there would be a cycle contradicting Lemma 1. So there must be a list, $\langle c, \ldots, y \rangle$ of one or more nodes that the target moves along. Since the source moves only over positive edges (using the weight of the negative in case of contingent) there must be another list $\langle a, \ldots, x \rangle$ that the source moves over before reaching b again. The final edge derived before reaching b is xy, whose edge will be a sum of negative weights along $\langle c, \ldots, y \rangle$ where negative requirement edges and positive contingent edges contribute, and positive weights along $\langle a, \ldots, x \rangle$ where positive requirement edges and negative contingent edges contribute. For the source to return to b, the weight of xy must be negative and there must be a positive edge bx. Then we can apply a rule deriving the edge by. We can determine that this edge is redundant by applying derivations to it. If by is positive it is redundant since there is a tighter implicit constraint along the strictly negative bcy path, as discussed before the theorem. If by is negative we apply derivation to move the source towards x. In this way we continue to apply derivations until we get a positive edge zy or the source reaches x. If this happens the derived edge must have a larger value than the already present xy edge, and be redundant, or we have derived a cycle contradicting Lemma 1. This can also be seen by observing that derivations start with the weight of xy, which can only increase along the derivation chain.

If we instead get a positive edge zy along the derivations we can show that there is a tighter constraint implicit here. We know $z \neq x$. When first deriving xy there was a negative edge from z to some node t in the $\langle c, \ldots, y \rangle$ list. If $t = y$ we arrive with a larger weighted edge (positive) ty this time and it is redundant. If $t \neq y$ there is an implicit tighter negative constraint zty. So again the zt edge is redundant.

So by is already explicitly or implicitly covered and hence redundant for DC-verification. Therefore it is not part of a critical chain.   □

This entails that along a critical chain each contingent constraint can only be part of at most two derivations: One using D1 and one using D2 or D6.

7   GlobalDC

We will apply the theorem above to the new algorithm GlobalDC (Algorithm 3). Given a full STNU this algorithm applies the derivation rules of Fig. 3 globally, i.e., with all edges as focus in all possible *triangles* (giving an iteration $O(n^3)$ run-time). It does this until there are no more changes detected over a global iteration. The structure of GlobalDC is hence directly inspired by the Bellman-Ford algorithm [13]. Non-DC STNUs are detected in the same way as FastIDC, by checking locally that there is no squeeze of contingent constraints and globally that there is no negative cycle.

This full DC algorithm can be compared with how an incremental algorithm (FastIDC) could be used to verify full DC, i.e., by adding edges from the full graph one at a time and doing derivations until done. Note that the order in which the derivation rules are applied to edges does not affect the correctness of FastIDC, only its run-time.

Given a DC STNU, GlobalDC will use the same derivation rules as FastIDC and therefore cannot generate tighter constraints. Since the same mechanism is used for detecting non-DC STNUs, both FastIDC and GlobalDC will indicate that the STNU is DC.

Algorithm 3. The GlobalDC Algorithm.

---

function GLOBAL-DC($G - STNU$)
Interesting ← {All edges of G}
repeat
   for *each edge e in G* **do**
      Interesting ← *Interesting*\\{e}
      for *each rule (Fig. 3) applicable with e as focus* **do**
         Derive new edges z_i
         for *each added edge z_i* **do**
            Interesting ← *Interesting* ∪ {z_i}
            if not *locally consistent* **then return** false **if** *negative cycle*
            created **then return** false
         end
      end
   end
until *Interesting is empty*
return true

Given a non-DC STNU, there exists a sequence of derivations that will let FastIDC decide this. Since GlobalDC performs all possible derivations in each iteration, it will do all derivations that FastIDC does in the same sequence. Again, the same mechanism is used for detecting non-DC STNUs, and both FastIDC and GlobalDC will indicate that the STNU is non-DC.

The key to analyzing the complexity of GlobalDC is the realization that we can stop deriving new constraints as soon as we have derived all critical chains: These are the only derivations that are required for detecting whether the STNU is DC or not.

The target of derived edges must eventually move. It can move at most n times, since it always moves to a node guaranteed to execute earlier. In-between two such moves the source can move between at most n nodes. Between each move of the source there can be one application of D8/D9, resulting in a chain of length $2n$ between each of the target moves. Together this bounds the longest critical chain by $2n^2$.

An example will illustrate how we can shrink the length of critical chains. Figure 7 shows a graph where no more derivations can be made. In Fig. 8 a negative edge *ie* is added to the graph and GlobalDC is used to update the graph with this increment.

Figure 9 shows the critical chain of edge *ac* at this point. Here we see as mentioned before that the source of the derived edge can move many times in sequence without the target moving in-between. In the example chain this is shown by the sequential D7 derivations. For requirement edges in general such a sequence may also include D4 derivations. Conditional edges can also induce sequences of moving sources through derivation rules D3 and D5.

All these derivations leading to sequential movement of the source require it to pass over requirement edges. If we had access to the shortest paths along

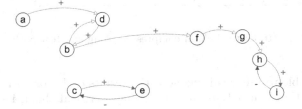

Fig. 7. Example graph in quiescence.

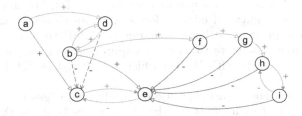

Fig. 8. Derivations resulting from adding the $i \to e$ edge.

requirement edges all these movements could in fact be derived in one global iteration. The source would be moved to all destinations at once and would not be replaced later since it had already followed a shortest path making the derived edge as tight as possible. Of course derivation rules may change the shortest paths, but if we added an APSP calculation to every global iteration we would compress the critical chains so that there would be no repeated application of sources moving along requirement constraints.

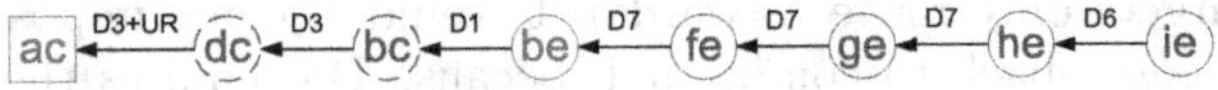

$$ \boxed{ac} \xleftarrow{\text{D3+UR}} (dc) \xleftarrow{\text{D3}} (bc) \xleftarrow{\text{D1}} (be) \xleftarrow{\text{D7}} (fe) \xleftarrow{\text{D7}} (ge) \xleftarrow{\text{D7}} (he) \xleftarrow{\text{D6}} (ie) $$

Fig. 9. The critical chain of edge ac, derived in Fig. 8.

Figure 10 shows how several applications of D7 and two of D3 are compressed by the availability of shortest path edges.

GlobalDC with the addition of APSP calculations in each iteration is still sound and complete since the APSP calculations only make more implicit constraints explicit. The run-time complexity is also preserved since each iteration was already $O(n^3)$ (applying rules to all focus edges). We now give an upper bound of the critical chain length:

Theorem 2. *The length of the longest critical chain in GlobalDC with APSP is $\leq 7n$.*

Fig. 10. Critical chain compressed using shortest paths.

Proof. To be able to prove this we need the results of Theorem 1. We will refer to derivations that can only occur once along a critical chain, i.e. D1, D2 and D6, as *limited derivations*.

What is the longest sequence in a critical chain consisting only of requirement edges such that it does not use any limited derivations? The only non-limited derivation rules that result in a requirement edge are D4, D7 and D8/D9. The last two require a conditional edge as focus, and can therefore only be at the start of such a sequence. We know that due to APSP there can only be one of D4/D7 in a row. Therefore the longest requirement-only sequence not using limited derivations starts with D8/D9 which is followed by D4/D7 for a total length of 2.

The longest sequence consisting of only conditional edges not using limited derivations must start with D5. It can then be continued only by D3. As we have access to shortest paths there can be at most one D3 in any sequence of only conditional edges.

In summary the longest sequences of the same type, requirement or conditional, not using limited derivations, are of length 2.

It is not possible to interleave the length-2 sequences of conditional edges with requirement edges more than once without changing the conditioning node of the conditional edges. To see this suppose we have a requirement edge which derives a conditional edge conditioned on B. This means that the edge is pointing towards A being the start of the contingent duration ending in B. If derivations now takes this edge into a requirement edge this edge must point towards A as well since the only way of going from conditional to requirement is via D8/D9 which preserves the target. If the target of the requirement edge later were to move (such targets only move forwards) it would become impossible to later invoke D5 for going back to conditional, because D5 requires the requirement edge to point towards a node that is after A. So in order for derivations to come back to a conditional edge again by D5 the target must stay at A. But then D5 cannot be applicable, for the same reason: It must point towards a node after A. So it is not possible to interleave these sequences.

This gives us the longest possible sequence without using limited derivations. It starts with a requirement sequence followed by a conditional sequence again followed by a requirement sequence. Such a sequence can have a length of at most 6. An issue here is that if a conditional edge conditioned on for instance B is part of the chain a D1 derivation involving B cannot also occur in the chain since this contingent constraint has already been passed. This means that it does not matter which of derivation D1 or D5 is used to introduce a conditioning node into the chain. The limitation applies to them both.

In conclusion this lets us construct an upper bound on the number of derivations in a critical chain. We have sequences of length 6 and these are interleaved with the n derivations of type D2 and D6 for a total of at most $7n$ derivations. □

Therefore all critical chains will have been generated after at most $7n$ iterations of GlobalDC. If we can iterate $7n$ times without detecting that an STNU is non-DC, it must be DC. With a limitation of $7n$ iterations, GlobalDC verifies DC in $O(n^4)$.

8   A Revised MMV Algorithm

We have described a new algorithm called GlobalDC and seen that it is $O(n^4)$. Compared to MMV, the following similarities and differences exist.

First, GlobalDC and MMV both interleave the application of derivation rules with the calculation of APSP distances and the detection of local inconsistencies and negative cycles. In MMV some of this is hidden in the pseudo-controllability test, but the actual conditions being tested are equivalent.

Algorithm 4. The revised MMV Algorithm.

---

function Revised-MMV($G - STNU$)
  $Interesting \leftarrow \{$All edges of $G\}$
  $iterations \leftarrow 0$
  repeat
    if not pseudo-controllable (G) **then**
      return false
    Compare edges and add all edges which were changed since last iteration to $Interesting$
    for *each edge e in Interesting* **do**
      $Interesting \leftarrow Interesting\backslash\{e\}$
      for *each triangle ABC containing e* **do**
        tighten ABC according to Fig. 4 except SR2
      end
    end
    $iterations \leftarrow iterations + 1$
  until *Interesting is empty* **or** $iterations = 7n$
  return true

Second, GlobalDC works in an EDG whereas MMV works in an STNU extended with wait constraints. These structures represent the same underlying constraints and the difference is not essential.

Third, GlobalDC lacks SR2, which is half of the original Simple Regression (SR) rule. Making this change in MMV will greatly speed it up in practice. Since it runs in an APSP graph it is reasonable to expect, on average, half of the nodes to be after a derived wait. This change will then cut the needed regression in MMV to half of that of the original version.

Fourth, GlobalDC stops after $7n$ iterations. Given the similarities above and the fact that the theorem about critical chain lengths directly carries over, MMV can also stop after $7n$ iterations without affecting correctness. The modified MMV can then decide DC in $O(n^4)$ time. We formulate this as a theorem.

Theorem 3. *The classical MMV algorithm for deciding dynamic controllability of an STNU can, with the small modifications shown in Algorithm 4, decide dynamic controllability in time $O(n^4)$.*

9   Related and Future Work

Recently several papers [14,15] have examined the use of Timed Game Automata (TGA) for both verification and execution of STNUs. These solutions work on a smaller scale and do not exploit the inherent structure of STNUs as distance graphs. Therefore they are more useful in networks that are small in size but involve choice and resources which cannot be handled by pure STNU algorithms.

At the time most of the research presented here was conducted the fastest algorithm for verifying dynamic controllability of an STNU was $O(n^4)$ [6]. Very recent activities however have converged on an algorithm which performs this in $O(n^3)$ [7,16].

Since there are now so many algorithms available for verifying DC it is important to find good benchmarks that can be used both to identify weaknesses but also to establish run-times of the algorithms in relevant use cases. This constitutes a large study that need to be carried out by the community in the near future. Also, execution of the produced networks need to be investigated further in the spirit of [17,18].

10   Conclusion

We have proven that with a small modification the classical "MMV" dynamic controllability algorithm, which in its original form is pseudo-polynomial, finishes in $O(n^4)$ time. The modified algorithm is still a viable option for determining whether an STNU is dynamically controllable. Compared to other algorithms, it offers a simpler and more intuitive theory. It is also an entry level algorithm which many familiarize with before implementing more advanced algorithms. As such it is an excellent choice for regression testing of the more complicated algorithms.

In this paper we also showed that there is no reason for MMV to regress over negative edges, a result that can be used to improve performance further.

Acknowledgements. This work is partially supported by the Swedish Research Council (VR) Linnaeus Center CADICS, the ELLIIT network organization for Information and Communication Technology, the Swedish Foundation for Strategic Research (CUAS Project), the EU FP7 project SHERPA (grant agreement 600958), and Vinnova NFFP6 Project 2013-01206.

References

1. Dechter, R., Meiri, I., Pearl, J.: Temporal constraint networks. Artif. Intell. **49** (1–3), 61–95 (1991)
2. Vidal, T., Ghallab, M.: Dealing with uncertain durations in temporal constraint networks dedicated to planning. In: Proceedings of ECAI (1996)
3. Muscettola, N., Morris, P., Tsamardinos, I.: Reformulating temporal plans for efficient execution. In: Proceedings of the 6th International Conference on Principles of Knowledge Representation and Reasoning (KR) (1998)
4. Morris, P., Muscettola, N., Vidal, T.: Dynamic control of plans with temporal uncertainty. In: Proceedings of IJCAI (2001)
5. Morris, P., Muscettola, N.: Temporal dynamic controllability revisited. In: Proceedings of AAAI (2005)
6. Morris, P.: A structural characterization of temporal dynamic controllability. In: Benhamou, F. (ed.) CP 2006. LNCS, vol. 4204, pp. 375–389. Springer, Heidelberg (2006)
7. Morris, P.: Dynamic controllability and dispatchability relationships. In: Simonis, H. (ed.) CPAIOR 2014. LNCS, vol. 8451, pp. 464–479. Springer, Heidelberg (2014)
8. Stedl, J., Williams, B.: A fast incremental dynamic controllability algorithm. In: Proceedings of ICAPS Workshop on Plan Execution (2005)
9. Shah, J.A., Stedl, J., Williams, B.C., Robertson, P.: A fast incremental algorithm for maintaining dispatchability of partially controllable plans. In: Proceedings of the 17th International Conference on Automated Planning and Scheduling (ICAPS) (2007). http://dblp.uni-trier.de/db/conf/aips/icaps2007.html#ShahSWR07
10. Nilsson, M., Kvarnström, J., Doherty, P.: Incremental dynamic controllability revisited. In: Proceedings of ICAPS (2013)
11. Nilsson, M., Kvarnström, J., Doherty, P.: Classical dynamic controllability revisited: a tighter bound on the classical algorithm. In: Proceedings of ICAART (2014)
12. Stedl, J.L.: Managing temporal uncertainty under limited communication: a formal model of tight and loose team coordination. Master's thesis, Massachusetts Institute of Technology (2004)
13. Cormen, T.H., Stein, C., Rivest, R.L., Leiserson, C.E.: Introduction to Algorithms. McGraw-Hill Higher Education, Boston (2001)
14. Cimatti, A., Hunsberger, L., Micheli, A., Roveri, M.: Using timed game automata to synthesize execution strategies for simple temporal networks with uncertainty. In: Proceedings of AAAI (2014)
15. Cesta, A., Finzi, A., Fratini, S., Orlandini, A., Tronci, E.: Analyzing flexible timeline-based plans. In: Proceedings of ECAI (2010)
16. Nilsson, M., Kvarnström, J., Doherty, P.: EfficientIDC: a faster incremental dynamic controllability algorithm. In: Proceedings of ICAPS (2014)
17. Hunsberger, L.: A fast incremental algorithm for managing the execution of dynamically controllable temporal networks. In: Proceedings of TIME (2010)
18. Hunsberger, L.: A faster execution algorithm for dynamically controllable STNUs. In: Proceedings of TIME, pp. 26–33. IEEE (2013)

ReCon: An Online Task ReConfiguration Approach for Robust Plan Execution

Enrico Scala[✉], Roberto Micalizio, and Pietro Torasso

Dipartimento di Informatica, Universita' di Torino, Turin, Italy
{scala,micalizio,torasso}@di.unito.it

Abstract. The paper presents an approach for the robust plan execution in presence of consumable and continuous resources. Plan execution is a critical activity since a number of unexpected situations could prevent the feasibility of tasks to be accomplished; however, many robotic scenarios (e.g. in space exploration) disallow robotic systems to perform significant deviations from the original plan formulation. In order to both (i) preserve the "stability" of the current plan and (ii) provide the system with a reasonable level of autonomy in handling unexpected situations, an innovative approach based on task reconfiguration is presented. Exploiting an enriched action formulation grounding on the notion of execution modalities, ReCon replaces the replanning mechanism with a novel reconfiguration mechanism, handled by means of a CSP solver. The paper studies the system for a typical planetary rover mission and provides a rich experimental analysis showing that, when the anomalies refer to unexpected resources consumption, the reconfiguration is not only more efficient but also more effective than a plan adaptation mechanism. The experiments are performed by evaluating the recovery performances depending on constraints on computational costs.

Keywords: Replanning · Plan repair · Plan execution · Space exploration · Consumable resources · CSP

1 Introduction

The management of a plan for a robotic agent operating in hazardous and extreme environments is a critical activity that has to take into account several challenges. In particular, in the context of space exploration, a planetary rover operates in an environment which is just partially observable and loosely predictable. As a consequence, the rover must have some form of autonomy in order to guarantee robust plan execution (i.e., reacting to unexpected contingencies). The rover's autonomy, however, is typically bounded both because of limitations of on-board computational power, and because the rover is not in general allowed to change significantly the high level plan synthesized on Earth. Space missions therefore exemplify situations where contingencies occur, but plan repair must be achieved through novel techniques trading-off rover's autonomy and the stability of the mission plan.

© Springer International Publishing Switzerland 2015
B. Duval et al. (Eds.): ICAART 2014; LNAI 8946, pp. 262–279, 2015.
DOI: 10.1007/978-3-319-25210-0_16

Robust plan execution has been tackled in two ways: on-line and off-line. On-line approaches, such as [4,13,14,18,20,27], interleave execution and replanning: whenever unexpected contingencies cause the failure of an action, the plan execution is stopped and a new plan is synthesized as a result of a new planning phase. Off-line approaches, such as [3,8], avoid replanning by anticipating, at planning time, the possible contingencies. The result of such a planning phase is a contingent plan that encodes choices between functionally equivalent subplans[1]. At execution time, the plan executor is able to select a contingent plan according to the current contextual conditions. However, as for instance in the work of [24], the focus is mainly on the temporal dimension and they do not consider consumable and continuous resources.

In this paper we propose a novel on-line methodology to achieve robust plan execution, which is explicitly devised to deal with unexpected deviations in the consumption of rover's resources. First, in line with the action-based approach a-la STRIPS [11] and differently from the constrained based planning [12,22], we model consumable resources as numeric fluents (introduced in PDDL 2.1 [11]). Then, we enrich the model of the rover's actions by introducing a set of *execution modalities*. The basic idea is that the propositional effects of an action can be achieved under different configurations of the rover's devices. These configurations, however, may have a different impact on the consumption of the resources. An *execution modality* explicitly models the resource consumption profile when an action is carried out in a given rover's configuration. The integration of *execution modality* at the PDDL level allows a seamless integration between planning and execution.

Relying on the concept of execution modalities, we propose to handle exceptions arising in planetary rover domains as a reconfiguration of action modalities, rather than as a replanning problem. In particular, the paper proposes a plan execution strategy, denoted as ReCon; once (significant) deviations from the nominal trajectory are detected, ReCon intervenes by reconfiguring the modalities of the actions still to be performed with the purpose of restoring the validity of resource constraints imposed by the rover mission.

To accomplish its task ReCon uses Choco[2] as CSP solver, so that it takes advantage of both the power of the constraint programming and the high level representation of PDDL.

After introducing a motivating example, we describe the employed action model, enriched with the notion of execution modality. Then we introduce the ReCon strategy and an example showing how the system actually works in a exploration rover mission. Finally, an experimental section, which evaluates the competence and the efficiency of the strategy w.r.t. a traditional replanning from scratch and the LPG-ADAPT system reported in [15].

---

[1] The notion of alternative (sub)plans is also presented for (off-line) scheduling; for details see [1].

[2] The software is at disposal at http://www.emn.fr/z-info/choco-solver/, while the work has been presented in [23].

2   Motivating Example

Let us consider a planetary rover in charge of exploring (and analyzing) a number of potentially interesting sites and able to transmit information towards the Earth. In doing so the rover is capable of moving, taking pictures, and starting the data upload once the pieces of information must be transmitted. For simplicity reasons, consider the mission plan of Fig. 1, involving *take picture, drive* and *communications* activities. This mission represents a feasible solution for a planning problem with goal: {in(r1,13), mem>=120, pwr>=0, time<=115}; that is, at the end of plan the rover must be located in 13 (propositional fluent), the free memory must be (at least) 120 memory units, there must be a positive amount of power, and the mission must be completed within 115 s.

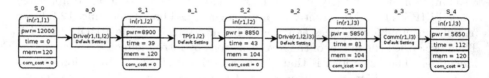

Fig. 1. A simple mission plan.

The figure shows how the four actions (regular boxes) change the status of the rover over the time (rounded-corner boxes)[3]. Note that the status of a rover involves both propositional fluents, (e.g., in(r1, 11) meaning rover r1 is in location 11); and numeric fluents: memory represents the amount of free memory, power is the amount of available power, time is the mission time given in seconds, and com_cost is an overall cost associated with communications.

The estimates about the rover's status are inferred by predicting, deterministically, the effects of the actions. In particular, the numeric fluents have been estimated by using a "default setting" (i.e., a standard modality) associated with each action.

Let us now assume that during the execution of the first drive action the rover has to travel across a rough terrain. Such an unexpected condition affects the drive as the rover is forced to slowdown[4], and as a consequence the drive action takes a longer time to be completed; the effects are propagated till the last snapshot, s_4 where the goal constraint time <= 115 is no longer satisfied.

After detecting this inconsistency, approaches based on a pure replanning step would compute a new plan achieving the goal by changing the original mission. For instance, some actions could be skipped in order to compensate the time lost during the first drive.

---

[3] To simplify the picture, we show in the rover's status just a subset of the whole status variables.

[4] The slowdown command of the rover may be the consequence of a reactive supervisor, which operates as a continuous controller as shown in [21].

However, robotic systems as a planetary rover have typically different configurations of actions to be executed and each configuration can have a different impact on the mission progress. For instance the robotic systems described in [5] and in [21] can perform a drive action in fast or slow modes. Reliable transmission to the earth, for example, can be slow and cheap, or fast and expensive, depending on the devices actually used.

Our proposal is to explicitly represent such different configurations within the action models, and hence try to resolve an impasse via a reconfiguration of the actions still to be performed. Intuitively, our objective is to keep the high level plan structure unchanged, but to adjust the modalities of the actions still to be performed. In Sect. 5 we will see an example of such a reconfiguration.

In the next section we will introduce the rover action model that explicitly expresses the set of *execution modality* at disposal.

3   Modeling Rover's Actions

As shown in the previous section, a planetary rover can perform the same set of actions via different configurations of parameters or devices. To capture this aspect, this section introduces the rover action model adopted in this work. As reported in [25], the model exploits (and extends) the numeric PDDL 2.1 action model [11], i.e. where the numeric fluent notion has been proposed. In particular, we use the numeric fluents to model continuous and consumable resources.

The intuition is that, while actions differ each other in terms of qualitative effects (e.g. a drive action models how the position of the rover changes after the action application), the expected result of an action can actually be obtained in many different ways by appropriately configuring the rover's devices (e.g. the drive action can be performed with several engine configurations). Of course, different configurations have in general different resource profiles and it is therefore possible that the execution of an action in a given configuration would lead to a constraint violation, whereas the same action performed in another configuration would not. We call these alternative configurations *modalities* and we propose to capture the impact of a specific modality by modeling the use of specific configurations in terms of pre/post conditions on the numeric fluents involved; such *modalities* become explicit in the action model definition.

The resulting model expresses the rover actions at two different levels of abstraction. The higher one is the qualitative level indicating "what" the action does. The lower one is the quantitative level expressing "how" the action achieves its effect.

The idea of *alternative behaviors* has also been investigated in (off-line) scheduling, where the notion of Temporal Network with Alternatives has been introduced [1]. It is quite evident however that, as anticipated in the introduction, the concept of execution modality is inspired to an (on-line) action centered approach [4], rather than on a constraints/scheduling based one [6].

By recalling our motivating example, Fig. 2 shows the model of the drive action. The action template drive (?r, ?l1, ?l2) requires a rover ?r to move from a location ?l1 to location ?l2. :modalities introduces the set of modalities associated with a drive; in particular, we express for this action, three alternative modalities:

- safe: the rover moves slowly and far from obstacles; intuitively the action should spend more time but consuming less power
- cruise: the rover moves at its cruise speed and can go closer to obstacles;
- agile: the rover moves faster than cruise, consuming more power but requiring less time.

The :precondition and :effect fields list the applicability conditions and the effects, respectively, and are structured as follows: first a propositional formula encodes the condition under which the action can be applied; the second field (:effect) indicates the positive and the negative effects of the action. For each modality m in :modalities we have the amount of resources required (numeric precondition) or consumed/produced (numeric effect) by the action when performed under that specific modality m.

For instance, the preconditions (reachable ?l1, ?l2) and (in ?r1, ?l1) are two atoms required as preconditions for the application of the action. These two atoms must be satisfied independently of the modality actually used to perform the drive action. While the comparison (safe:(>= (power ?r)(*(safe_cons ?r) (/ (distance ?l1 ?l2) (safe_speed ?r))))) means that the modality safe can be selected when the rover's power is at least larger than a threshold given by evaluating the expression on the right side. Analogously, (safe:(decrease (power ?r)(*(safe_cons ?r) (/(distance ?l1 ?l2) (safe_speed ?r))))) describes in the effects how the rover's power is reduced after the execution of the drive action. More precisely, we have modeled the power consumption as a function depending on the duration of the drive action (computed considering distance and speed) and the average power consumption per time unit given a specific modality. For instance, in safe modality, the amount of power consumed depends on two parameters (safe_cons ?r) and (safe_speed ?r) which are the average consumption and the average speed for the safe modality, respectively, while (distance ?l1 ?l2) is the distance between the two locations ?l1 and ?l2. Finally, note that in the numeric effects of each modality, the model updates also the fluent time according to the selected modality. Also in this case, the duration of the action is estimated by a function associated with each possible action modality.

Analogously to the drive action we model modalities also for the Take Picture (TP) and the Communication (COMM). For TP we have the low (LR) and high (HR) resolution modalities which differ in the quality of the taken picture and the occupied memory. Intuitively, the more the resolution is, the more the memory consumption will be.

Figure 3 reports the model of the communication action; it is worth noticing that execution modalities correspond to two different communication channels: CH1 with low overall comm_cost and low bandwidth, and CH2 with high overall comm_cost but high bandwidth.

```
(:action drive
    :parameters ( ?r - robot ?l1 - site ?l2 - site)
    :modalities (safe,normal,agile)
    :precondition (and (in ?r ?l1) (road ?l1 ?l2)
    (safe: (>= (power ?r) (* (safe_cons ?r)
            (/ (distance ?l1 ?l2) (safe_speed ?r)))))
    (cruise: (>= (power ?r) (* (cruise_cons ?r)
                (/ (distance ?l1 ?l2) (cruise_speed ?r)))))
    (agile: (>= (power ?r) (* (agile_cons ?r)
                (/ (distance ?l1 ?l2) (agile_speed ?r)))))
    )
    :effect
    (and
        (in ?r ?l2) (not (in ?r ?l1))
        (safe: (decrease  (power ?r) (* (safe_cons ?r)
                (/ (distance ?l1 ?l2) (safe_speed ?r))))
                (increase   (time) (/ (distance ?l1 ?l2)) (safe_speed ?r)))
                (increase   (powerC ?r) (* (safe_cons ?r)
                            (/ (distance ?l1 ?l2) (safe_speed ?r))))
        (cruise: (decrease  (power ?r) (* (cruise_cons ?r)
                            (/ (distance ?l1 ?l2) (cruise_speed ?r))))
                (increase   (time) (/ (distance ?l1 ?l2)) (cruise_speed ?r))
                (increase   (powerC ?r) (* (cruise_cons ?r)
                            (/ (distance ?l1 ?l2) (cruise_speed ?r)))))
        (agile: (decrease  (power ?r) (* (agile_cons ?r)
                            (/ (distance ?l1 ?l2) (agile_speed ?r))))
                (increase   (time) (/ (distance ?l1 ?l2)) (agile_speed ?r))
                (increase   (powerC ?r) (* (agile_cons ?r)
                            (/ (distance ?l1 ?l2) (agile_speed ?r)))))
    )
```

Fig. 2. The augmented model of a **drive** action.

The selection of action modalities has to take into account that complex dependencies among resources could exist. For instance, even if a high resolution TP takes the same time as a low resolution TP, the selection has a big impact on the amount of time spent globally, too. As a matter of facts, as long as the amount of stored information increases, the time spent by a (possible) successive COMM grows up accordingly, which means that also the global mission horizon will be revised.

Given the rover's actions defined so far, a rover mission plan is a total ordered set of fully instantiated rover's action templates[5]. Given a particular rover's state S and a given set of goals G to be reached (including both propositional/classical conditions and constraints on the amount of resources), the mission plan is valid iff it achieves G from S.

Executing the Mission Plan. As we have seen in the previous section, the rover's mission can be threatened many times by unexpected contingencies; so the validity of the mission can be easily compromised during its actual execution.

Nevertheless, when the detected unexpected contingency at execution time just invalidates the resource consumption expectations, even if the current modality allocation would not be consistent with the constraints involved in

---

[5] The plan can be also generated automatically by exploiting a numeric planner system, properly modified to handle actions with modalities (e.g., the Metric-FF planning system [17] or LPG [16]).

```
( :action comm
  :parameters ( ?r - robot ?l1 - site )
  :modalities (ch1,ch2)
  :precondition (and(in ?r ?l1)

 (ch1:    (and (> (memoryC ?r) 0) ( >= (power ?r)
                  (/ (memoryC ?r) (bandwith-ch1 ?r)))))
 (ch2:    (and (> (memoryC ?r) 0) ( >= (power ?r)
                  (/ (memoryC ?r) (bandwith-ch2 ?r)))))
  :effect
  (and (infoSent ?r ?l1)
     (ch1:   (assign (memoryC ?r) 0)
         (increase (memory ?r) (memoryC ?r))
            (increase (time) (/ (memoryC ?r)(bandwith-ch1 ?r)))
            (increase (powerC ?r) (* (ch1-cons ?r)
                      (/ (memoryC ?r) (bandwith-ch1 ?r)))
            (decrease (power ?r)   (* (ch1-cons ?r)
                      (/ (memoryC ?r) (bandwith-ch1 ?r)))
            (increase (comm_cost) 1)))
     (ch2:   (assign (memoryC ?r) 0)
         (increase (memory ?r) (memoryC ?r))
         (increase (time) (/ (memoryC ?r)(bandwith-ch2 ?r)))
         (increase (powerC ?r) (* (ch2-cons ?r)
                   (/ (memoryC ?r) (bandwith-ch2 ?r))))
         (decrease (power ?r)   (* (ch2-cons ?r)
              (/ (memoryC ?r) (bandwith-ch2 ?r))))
         (increase (comm_cost) 3))
)
```

Fig. 3. The augmented model of a communication action.

the plan and in the goal, there could be "other" allocations of modalities still feasible. By exploiting this intuition, the next section introduces an adaptive execution technique which, instead of abandoning the mission being executed, tries first to repair the flaws via a reconfiguration of the action modalities. The reconfiguration considers all those actions still to be executed. Given a plan P, to indicate when a plan is just *resource inconsistent*, we will use the predicate *res_incon* over P, i.e. we will say *res_incon(P)*. Otherwise we will say that the plan is valid or structurally invalid. This latter case happens when, given the current plan formulation, at least an action in the plan is not propositional applicable, or there is at least a missing (propositional) goal.

4   ReCon: Adaptive Plan Execution

In this section we describe how the plan adaptation process is actually carried on by exploiting a Constraint Satisfaction Problem representation. The main strategy implemented, namely ReCon, is a continual planning agent [4,9], extended to deal with the rover actions model presented in the previous section. In order to handle the CSP representation, ReCon exploits two further sub-modules: **Update** by means of which new observations are asserted within the CSP representation, and **Adapt** which has the task of making the mission execution adaptive to the incoming situation.

4.1 The Continual Planning Loop

Algorithm 1 shows the main steps required to execute and (just in case) adapt the plan being executed. The algorithm takes in input the initial rover's state S_0, the mission goal $Goal$, and the plan P expressed as discussed in the previous section. Note that each action has to have a particular modality of execution instantiated. The algorithm returns $Success$ when the execution of the whole mission plan achieves the goal; $Failure$ otherwise. In this case, a failure means that there is no way to adapt the current plan in order to reach the goal satisfying mission constraints. To recover from this failure, a replanning step altering the structure of the plan should be invoked, but this step requires the intervention of the ground control station on Earth.

The first step of the algorithm is to build a $CSPModel$ representing the mission plan (line 1). As thoroughly described in [25], our approach inherits the main steps by Lopez et al. in [19] in which the planning problem is addressed as a CSP[6]. As a difference w.r.t. the classical planning, the encoding exploited by our approach needs to store variables for the modalities to be chosen, and variables for the numeric fluents involved in the plan. Numeric fluents variables are replicated as many steps in the plan. The purpose is to capture all the possible evolutions of resources profiles given the modalities that will be selected. The constraints oblige the selection of the modality to be consistent with the resource belonging to the previous and successive time step. Moreover, further constraints allow only reconfigurations consistent with the current observation acquired (which at start-up corresponds to the initial state), and the goals/requirement of the mission.

Once the $CSPModel$ has been built, the algorithm loops over the execution of the plan. Each iteration corresponds to the execution of the i-th action in the plan. At the end of the action execution the process verifies the current observation obs_{i+1} with the rest of the mission to be executed. In case the plan is structurally invalid (some propositional conditions are not satisfied or the goal cannot be reached) ReCon stops the plan execution and returns a failure; i.e., a replanning procedure is required.

Otherwise we can have two other situations. First, there have been no consistent deviations from the nominal predictions therefore the execution can proceed with the remaining part of the plan. Second the plan is just resource inconsistent ($res\_incon(P)$, line 10). In this latter case, ReCon has to adapt the current plan by finding an alternative assignments to action modalities that satisfies the numeric constraints (line 11). If the adaptation has success, a new non-empty

---

[6] Alternative CSP conversions are possible; for instance see [2].

plan $newP$ is returned and substituted to the old one. This new plan is actually the old plan, but with a different allocations of action modalities. Otherwise, the plan cannot be adapted and a failure is returned; in this case, the plan execution is stopped and a new planning phase is needed.

4.2   Update

The **Update** step is sketched in Algorithm 2. The algorithm takes in input the CSP model to update, the last performed action a_i, and the set $NObs$ of observations about numeric fluents. The algorithm starts by asserting within the model that the i-th action has been performed; see lines 1 and 2 in which variable mod_i is constrained to assume the special value $exec$. In particular, a first role of the $exec$ value is to prevent the adaptation process to change the modality of an action that has already been performed, as we will see in the following section. Moreover, $exec$ allows also the acquisition of observations even when the observed values are completely unexpected. In fact, by assigning the modality of action a_i to $exec$, we relax all the constraints over the numeric variables at step $i+1$-th (which encode the action effects). This is done in lines 3–5 in which we iterate over the numeric fluents N^j mentioned in the effects of action a_i, and assign to the corresponding variable at $i+1$-th step the value observed in $NObs$. On the other hand, all the numeric fluents that are not mentioned in the effects of action a_i do not change, so the corresponding variables at step $i+1$ assume the same values as in the previous i-th step (lines 6–8). The idea of the Update is to make the CSP aware of the current new observations and the modalities already executed. In this way, a reconfiguration task does not need to rebuild the structure completely from scratch.

4.3   Adapt

The **Adapt** module, shown in Algorithm 3, takes in input the CSP model, the index i of the last action performed by the rover, the mission goal, and the plan P; the algorithm returns a new adapted plan, if it exists, or an empty plan when no solution exists.

The algorithm starts by removing from $CSPModel$ the constraints on the modalities of actions still to be performed; i.e., each variable mod_k with k greater than i is no longer constrained (a_i is the last performed action and its modality is set to $exec$) (lines 1-2). This step is essential since the current $CSPModel$ is inconsistent; that is, the current assignment of modalities does not satisfies the global constraints. By removing these constraints, we allow the CSP solver to search in the space of possible assignments to modality variables (i.e., the actual decisional variables, since the numeric fluents are just side effects of the modality selection), and find an alternative assignment that satisfies the global constraints (line 3). If the solver returns an empty solution, then there is no way to adapt the current plan and **Adapt** returns no solution. Otherwise (lines 6-10), at least a solution has been found. In this last case, a new assignment of

---

Algorithm 1. ReCon.

    Input: S_0, *Goal*, P
    Output: *Success* or *Failure*
1 $CSPModel = Init(S_0, Goal, P)$;
2 $i = 0$;
3 **while** $\neg P$ *is completed* **do**
4   execute(a_i, $curMod(a_i)$);
5   obs_{i+1} = observe();
6   **if** P *is structurally invalid w.r.t.* obs_{i+1} *and Goal* **then**
7   **return** *Failure*

8   **else**
9   **Update**($CSPModel$,a_i,$num(obs_{i+1})$);
10   **if** $res\_incon(P)$ **then**
11           $newP = $**Adapt**($CSPModel$,$i$,$Goal$,$P$);
12   **if** $newP \neq \emptyset$ **then**
13              $P = newP$

14   **else**
15   **return** *Failure*

16     $i = i + 1$
17 **return** *Success*

Algorithm 2. Update.

    Input: $CSPModel$, a_i,$NObs$
    Output: modified $CSPModel$
1 delConstraint($CSPModel$,mod_i=curMod(a_i));
2 addConstraint($CSPModel$,mod_i=$exec$);
3 **foreach** $N^j \in affected(a_i)$ **do**
4   addConstraint($CSPModel$,
5      $(mod_i$=$exec) \rightarrow N_{i+1}^j$=get($NObs$,$N_{i+1}^j$))
6 **foreach** $N^j \in \neg affected(a_i)$ **do**
7   addConstraint($CSPModel$,
8      $(mod_i$=$exec) \rightarrow N_{i+1}^j$=$N_i^j$)

---

modalities to the variables mod_k $(k : i + 1..|P|)$ is extracted from the solution, and this assignment is returned to the ReCon algorithm as a new plan *newP* such that the actions are the same as in P, but the modality labels associated with the actions $a_{i+1},..,a_{|P|}$ are different.

Note that, in order to keep updated the CSP model for future adaptations, the returned assignment of modalities is also asserted in $CSPModel$; see lines 6 to 10.

Algorithm 3. Adapt.

    Input: $CSPModel$, i,$Goal$,P
    Output: a new plan, if any
1 **for** $k=i+1$ to $|P|$ **do**
2    $\lfloor$ delConstraint($CSPModel\ mod_k$=currentMod(a_k))

3 $Solution$ = solve($CSPModel$);
4 **if** $Solution$ = $null$ **then**
5    $\lfloor$ **return** $\emptyset$

6 **else**
7    $newP$=extractModalitiesVar($Solution$);
8   **for** $k=i+1$ to $|newP|$ **do**
9      $\lfloor$ addConstraint($CSPModel$, mod_i=curMod($newP[i]$))
10    $\lfloor$ **return** $newP$

---

5   Running the Mission Rover Example

Let us consider again the example in Fig. 1, and let us see how ReCon manages its execution. First of all, the plan model must be enriched with the execution modalities as previously explained; Fig. 4 (top) shows the initial configuration of action modalities: the drive actions have `cruise` modalities, the take picture (TP) has `HR` (high resolution) modality, and the communication (`Comm`) uses the low bandwidth channel (`CH1`). This is the enriched plan ReCon receives in input.

Now, let us assume that the actual execution of the first drive action takes a longer time than expected, 47 s instead of 38 s, and consumes more power, 3775 J instead of 3100 J. While the discrepancy on power is not a big issue as it will not cause a failure, the discrepancy on time will cause the violation of the constraint `time <= 115`; in fact, performing the subsequent actions in their initial modalities would require 120 s. In other words, the assignment of modalities to

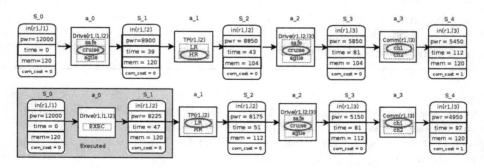

Fig. 4. The initial configuration of modalities (above), and the reconfigured plan (below).

the subsequent actions does not satisfies the mission constraints. This situation is detected by ReCon that intervenes and, by means of the **Adapt** algorithm discussed above, tries to find an alternative configuration of modalities.

Let us assume that communication cost is constrained; that is, the mission goal includes the constraint com_cost = 1; this prevents ReCon from using the fast communication channel. The more intuitive decision is to promote the execution of the *drive* to *agile*. However, this would cause the violation on the constraint concerning the maximum amount of power to be spent. Therefore ReCon has to look for an alternative assignments of modalities.

It is interesting to note that a lower resolution image consumes less memory, meaning that the successive communication, in our case *(COMM R1 L3)*, will need less time (and also less power) for achieving its effects. For this reason ReCon demotes the next activity, i.e. TP, to be execute to modality LR and so the global constraints are now satisfied.

Of course, we assume that mission constraints leave ReCon some room to repair resource inconsistent situations. For instance, if the mission has required an hard constraint on the quality of the taken images, the low resolution would have not been possible, and hence an overall replanning would have been necessary.

In principle, by flattening all the actions and the given modalities as explained in [28], replanning is possible as alternative to the reconfiguration mechanism. In this case, however, the problem to be handled would become much more difficult, since all the possible action sequences applicable starting from the current state could be explored.

To highlight the complexity arising from a replanning formulation, let us assume that in our example there is a connection from location l3 to l4, and from l2 and l4. That is, the rover can move not only from l1 to l2, but also from l2 to l4 and from l4 to l3, for *all* the provided modalities. In addition, for simplicity reasons, assume that from that point (l3), the only possible sequence of actions toward the goal is given by a_2 and a_3.

While the reconfiguration mechanism can focus just on the impact on resources given by the selection of modalities for the next actions (TP, DRIVE, COMM), it is quite evident that a traditional replanner should deal with a larger search space. As matter of fact, it should consider also the (several) possible trajectories of states given by exploring the alternatives ways of reaching location l2 (drive(r1, l2, l4)), for all the possible modalities of execution. That is, it will have to cope with both the propositional and resource constraints of the arising planning problem. For a deeper discussion on this aspect, and on the computational complexity relation between the reconfiguration and the overall numeric planning problem see [25, 28].

As we will see in the next section, this different characterization is crucial for determining the performance of the reconfiguration over replanning from scratch, and even over the state of the art plan repair strategy presented in [10].

6   Experimental Validation

To assess the effectiveness of our proposal, we evaluated two main parameters: (1) the computational cost of reconfiguration, and (2) the competence of ReCon, that is, the ability of completing a mission.

To this aim, we have compared ReCon with three alternative strategies: REPLAN, LPG-ADAPT and NoRep. Whenever the plan becomes resources inconsistent, both REPLAN and LPG-ADAPT stop the execution of the plan and try to recover from the impasse. REPLAN searches a new plan completely from scratch, while LPG-ADAPT uses the old plan as a guidance to speed-up the resolution process[7]. Conversely, NoRep just stops the plan execution as soon as it is no longer valid. We used REPLAN and LPG-ADAPT to better assess the contribution of ReCon w.r.t. the current state of the art in (re)planning dealing with consumable resources.

We have implemented ReCon in Java 1.7 by exploiting the PPMaJaL library[8]; the Choco CSP solver (version 2.1.3)[9] has been used in the **Adapt** algorithm to find an alternative configuration. Concerning the REPLAN strategy, we invoke Metric-FF [17] by converting the rover actions with modalities in PDDL 2.1 actions. In order to study the effectiveness of the strategy in an on-line plan execution context, we allotted each computation with a time deadline; this parameter is critical for the competence of the system being tested. For this reason, we report results obtained with three different time deadlines: 5 s, 30 s and 60 s. Each time deadline corresponds to the maximum time that is given to the reconfiguration/replanning for providing a valid solution, once a plan becomes invalid throughout the whole execution process.

Our tests set consists of 168 plans; each plan involves up to 34 actions (i.e., drives, take pictures, and communications), it is fully instantiated (a modality has been assigned to each action), and feasible since all the goal constraints are satisfied when the plan execution starts.

To simulate unexpected deviations in the consumption of the resources, we have run[10] each test in thirteen different settings. In each of these settings we have noised the amount of resources consumed by the actions. In particular, in setting 1, an action consumes 10 % more than expected at planning time. In setting 2, the noise was increased to 15 %, and so on until in setting 13 where the noise was set to 70 %, i.e. an action consumes 70 % more resources than initially predicted.

On the left of Figs. 5, 6 and 7 we report the competence - measured as the percentage of performed actions in the plan - of the three strategies, in the thirteen settings of noise we have considered. As expected, the competence decreases

---

[7] LPG-ADAPT, [15], is the plan adaptation extension of LPG, [16], one of the more awarded systems throughout the planning competitions of the last decade. LPG-ADAPT can be considered the state of the art in the context of plan adaptation.

[8] http://www.di.unito.ittextasciitildescala.

[9] The Choco Solver implements the state of the art algorithms for constraint programming and has already been used in space applications, see [6]. Choco can be downloaded at http://www.emn.fr/z-info/choco-solver/.

[10] Experiments have run on a 2.53 GHz Intel(R) Core(TM)2 Duo processor with 4 GB.

as long as the amount of noise increases, for all the strategies tested. ReCon resulted more competent than both REPLAN and LPG-ADAPT. Even though REPLAN and LPG-ADAPT can modify all the aspects of the plan structure, and hence they are theoretical more competent than ReCon, the search spaces generated by the overall arising planning problems turned out to be too large from the point of view of REPLAN and LPG-ADAPT. The timeouts are reached by a large number of cases in all cpu-time settings, and this is the reason of a lower competence of REPLAN and LPG-ADAPT. In particular we can observe a large gap between the percentage of plan completed by ReCon and REPLAN in all the cpu-time settings. In our experiments, also with an increased cpu-time at disposal, REPLAN was not able to find solutions for many cases. As refers the comparison with LPG-ADAPT, the gap is more limited for the high level of noise, showing how LPG-ADAPT can effectively takes advantage from the knowledge of the previous plan. It is worth noting that, as expected, this gap decreases as long as the noise increase; this is of course due to the contribution of the flexibility of the search space in which LPG-ADAPT and REPLAN can find a solution.

Observing the differences between the competence of the systems over the various cpu-time setting, it is clear that this parameter is crucial for the competence of LPG-ADAPT, while it does not affect the competence of ReCon, and

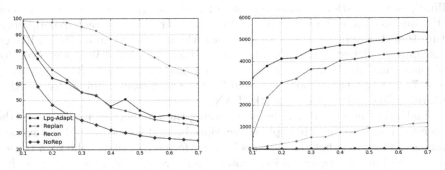

Fig. 5. Competence (left) and cpu-time (right) - 5 s setting.

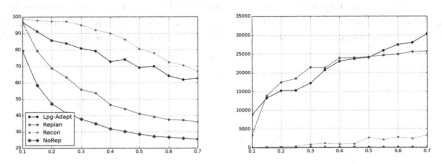

Fig. 6. Competence (left) and cpu-time (right) - 30 s setting.

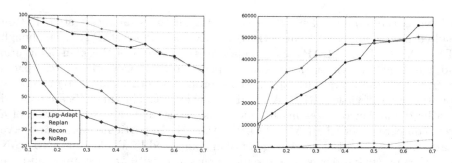

Fig. 7. Competence (left) and cpu-time (right) - 60 s setting.

neither of REPLAN. As expected, the LPG-ADAPT competence is quite competitive with ReCon for the 60 s; in particular in the first 4 settings of noise, ReCon outperforms LPG-ADAPT, while with larger noises, LPG-ADAPT has more or less the same performances of ReCon. In the 5 s setting both LPG-ADAPT and REPLAN are not competitive at all.

The right column of Figs. 5, 6 and 7 reports the computational cost, on average, of the three strategies. Note that for each case considered, the computational cost corresponds to the sum of all the attempts to recovery from the failure (reconfiguration, plan-adaptation or replanning) performed until the end of the mission. Here the advantage of ReCon is very large in each experimental setup. In fact, even for the worst case (when the noise is set to be 70 %), ReCon is extremely efficient, indeed it takes, on average, just 1,2 s. Whereas, even for the cases with small amount of noise, as you can see in Fig. 7, REPLAN takes about 7 s of cpu-time till 50 s employed for the worst cases, while LPG-ADAPT performs a little bit worse than REPLAN.

Finally, in Fig. 8 we conclude by analyzing the number of invocations of the systems throughout the whole plan execution. Basically we collected the average number of attempts that the systems have performed whenever the plan turned out not valid during the execution. Observing the results, it is quite evident that the reconfiguration mechanism is invoked on average more times than the other architectures. This happens because, as long as the plan execution process goes on, the constraints becomes more and more tight, causing the detection mechanism to be invoked more frequently. Differently, each invocation of REPLAN

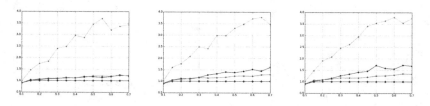

Fig. 8. Average Number of Repairs over all the timeout settings.

generates a completely new plan; therefore the plan execution till the end is not directly related to the previous plan execution problem. This is the reason why REPLAN almost preserves the same amount of invocations throughout the cases we have tested. A similar trend can be found in comparing LPG-ADAPT with ReCon. Also LPG-ADAPT makes on the average less repair than ReCon; the difference of performances between REPLAN and LPG-ADAPT is probably due to the different way the underlying planning systems (LPG and Metric-FF) explores the search space. Of course this should be verified testing other numeric planners.

7   Conclusions

We have proposed in this paper a novel approach to the problem of robust plan execution. Rather than recovering from plan failures via a re-planning step (see e.g., [13,14,18,26]), we have proposed a methodology, called ReCon, based on the re-configuration of the plan actions. ReCon is justified in all those scenarios where a pure replanning approach is unfeasible. This is the case, for instance, of a planetary rover performing a space exploration mission. Albeit a rover must exhibit some form of autonomy, its autonomy is often bounded by two main factors: (1) the on-board computational power is not always sufficient to handle mission recovery problems, and (2) the rover cannot in general deviate from the given mission plan without the approval from the ground control station.

ReCon presents many advantages w.r.t. re-planning. First of all, as the experiments have demonstrated, reconfiguring plan actions is computationally cheaper than synthesizing a new plan from scratch and even trying to adapt it via a classical plan adaptation tool (as the one reported in [15]). Moreover, ReCon leaves the high-level structure of the plan (i.e., the sequence of mission tasks) unchanged, but endows the rover with an appropriate level of autonomy for handling unexpected contingencies. ReCon can be considered as a complementary repair strategy to other works in the context of autonomy for space as those in [7]; as matter of facts, ReCon explores a different dimension of the repair problem, which is based on an action-centered planning representation rather than on a timeline based perspective [12].

The solution described in this paper has been tested on a challenging domain such as a space exploration domain, but its applicability is not restricted to this domain. Many other robotic tasks could benefit of the proposed approach (combined with a generative approach, see [25]), since in many of them the need of adapting the plan execution to the resources constrains is very relevant.

The approach we have presented can be improved in a number of ways. A first important enhancement is the search for an optimal solution. In the current version, in fact, ReCon just finds one possible configuration that satisfies the global constraints. In general, one could be interested in finding the best configuration that optimizes a given objective function. Reasonably, the objective function could take into account the number of changes to action modalities; for instance, in some cases it is desirable to change the configuration as little as possible. Of course, the search for an optimal configuration is justified when the global constraints are not strict, and several alternative solutions are possible.

Acknowledgements. We would like to thank the Choco's team for making freely available the CSP solver, Joerg Hoffman for the Metric-FF planning system as well as Alfonso Gerevini, Alessandro Saetti and Ivan Serina for the LPG-ADAPT system.

References

1. Barták, R., Čepek, O., Hejna, M.: Temporal reasoning in nested temporal networks with alternatives. In: Fages, F., Rossi, F., Soliman, S. (eds.) CSCLP 2007. LNCS (LNAI), vol. 5129, pp. 17–31. Springer, Heidelberg (2008)
2. Barták, R., Toropila, D.: Solving sequential planning problems via constraint satisfaction. Fundamamenta Informaticae **99**(2), 125–145 (2010)
3. Block, S.A., Wehowsky, A.F., Williams, B.C.: Robust execution of contingent, temporally flexible plans. In: Proceedings of National Conference on Artificial Intelligence (AAAI 2006), pp. 802–808 (2006)
4. Brenner, M., Nebel, B.: Continual planning and acting in dynamic multiagent environments. J. Auton. Agent. Multiagent Syst. **19**(3), 297–331 (2009)
5. Calisi, D., Iocchi, L., Nardi, D., Scalzo, C., Ziparo, V.A.: Context-based design of robotic systems. Robot. Auton. Syst. (RAS) **56**(11), 992–1003 (2008)
6. Cesta, A., Fratini, S.: The timeline representation framework as a planning and scheduling software development environment. In: Proceedings of P&S Special Interest Group Workshop (PLANSIG-10) (2009)
7. Chien, S., Johnston, M., Frank, J., Giuliano, M., Kavelaars, A., Lenzen, C., Policella, N.: A generalized timeline representation, services, and interface for automating space mission operations. Technical Report JPL TRS 1992+, Ames Research Center; Jet Propulsion Laboratory, June 2012
8. Conrad, P.R., Williams, B.C.: Drake: an efficient executive for temporal plans with choice. J. Artif. Intell. Res. **42**, 607–659 (2011)
9. desJardins, M., Durfee, E.H., Ortiz Jr., C.L., Wolverton, M.: A survey of research in distributed, continual planning. AI Mag. **20**(4), 13–22 (1999)
10. Fox, M., Gerevini, A., Long, D., Serina, I.: Plan stability: replanning versus plan repair. In: Proceedings of the International Conference on Automated Planning and Scheduling (ICAPS 2006), pp. 212–221 (2006)
11. Fox, M., Long, D.: Pddl2.1: an extension to PDDL for expressing temporal planning domains. J. Artif. Intell. Res. **20**, 61–124 (2003)
12. Fratini, S., Pecora, F., Cesta, A.: Unifying planning and scheduling as timelines in a component-based perspective. Arch. Control Sci. **18**(2), 231–271 (2008)
13. Garrido, A., Guzman, C., Onaindia, E.: Anytime plan-adaptation for continuous planning. In: Proceedings of P&S Special Interest Group Workshop (PLANSIG 2010) (2010)
14. Gerevini, A., Serina, I.: Efficient plan adaptation through replanning windows and heuristic goals. Fundamenta Informaticae **102**(3–4), 287–323 (2010)
15. Gerevini, A., Saetti, A., Serina, I.: Case-based planning for problems with real-valued fluents: kernel functions for effective plan retrieval. In: Proceedings of European Conference on AI (ECAI 2012), pp. 348–353 (2012)
16. Gerevini, A., Saetti, I., Serina, A.: An approach to efficient planning with numerical fluents and multi-criteria plan quality. Artif. Intell. **172**(8–9), 899–944 (2008)

17. Hoffmann, J.: The metric-FF planning system: translating "ignoring delete lists" to numeric state variables. J. Artif. Intell. Res. **20**, 291–341 (2003)
18. van der Krogt, R., de Weerdt, M.: Plan repair as an extension of planning. In: Proceedings of the International Conference on Automated Planning and Scheduling (ICAPS 2005), pp. 161–170 (2005)
19. Lopez, A., Bacchus, F.: Generalizing graphplan by formulating planning as a CSP. In: Proceedings of International Conference on Artificial Intelligence (IJCAI 2003), pp. 954–960 (2003)
20. Micalizio, R.: Action failure recovery via model-based diagnosis and conformant planning. Comput. Intell. **29**(2), 233–280 (2013)
21. Micalizio, R., Scala, E., Torasso, P.: Intelligent supervision for robust plan execution. In: Pirrone, R., Sorbello, F. (eds.) AI*IA 2011. LNCS, vol. 6934, pp. 151–163. Springer, Heidelberg (2011)
22. Muscettola, N.: Hsts: integrating planning and scheduling. Technical Report CMU-RI-TR-93-05, Robotics Institute, Pittsburgh, PA, March 1993
23. Narendra, J., Rochart, G., Lorca, X.: Choco: an open source java constraint programming library. In: CPAIOR 2008 Workshop on Open-Source Software for Integer and Constraint Programming (OSSICP 2008), pp. 1–10 (2008)
24. Policella, N., Cesta, A., Oddi, A., Smith, S.: Solve-and-robustify. J. Sched. **12**, 299–314 (2009)
25. Scala, E., Micalizio, R., Torasso, P.: Robust plan execution via reconfiguration and replanning. AI Communications, to appear (2014)
26. Scala, E.: Numeric kernel for reasoning about plans involving numeric fluents. In: Baldoni, M., Baroglio, C., Boella, G., Micalizio, R. (eds.) AI*IA 2013. LNCS, vol. 8249, pp. 263–275. Springer, Heidelberg (2013)
27. Scala, E.: Numerical kernels for monitoring and repairing plans involving continuous and consumable resources. In: Proceedings of International Conference on Agents and Artificial Intelligence (ICAART 2013), pp. 531–534 (2013)
28. Scala, E.: Reconfiguration and replanning for robust execution of plans involving continuous and consumable resources. Ph.D. thesis, Department of Computer Science - University of Turin (2013)

Combining Semantic Query Disambiguation and Expansion to Improve Intelligent Information Retrieval

Bilel Elayeb[1,3]([⊠]), Ibrahim Bounhas[2], Oussama Ben Khiroun[1],
and Narjès Bellamine Ben Saoud[1,4]

[1] RIADI Research Laboratory, ENSI Manouba University,
2010 Manouba, Tunisia
Bilel.Elayeb@riadi.rnu.tn,
oussama.ben.khiroun@gmail.com,
Narjes.Bellamine@ensi.rnu.tn
[2] LISI Laboratory of Computer Science for Industrial Systems,
ISD Manouba University, 2010 Manouba, Tunisia
Bounhas.Ibrahim@yahoo.fr
[3] Emirates College of Technology, P.O. Box: 41009, Abu Dhabi, UAE
[4] Higher Institute of Informatics (ISI), Tunis El Manar University,
1002 Tunis, Tunisia

Abstract. We show in this paper how Semantic Query Disambiguation (SQD) combined with Semantic Query Expansion (SQE) can improve the effectiveness of intelligent information retrieval. Firstly, we propose and assess a possibilistic-based approach mixing SQD and SQE. This approach is based on corpus analysis using co-occurrence graphs modeled by possibilistic networks. Indeed, our model for relevance judgment uses possibility theory to take advantage of a double measure (possibility and necessity). Secondly, we propose and evaluate a probabilistic circuit-based approach combining SQD and SQE in an intelligent information retrieval context. In this approach, both SQD and SQE tasks are based on a graph data model, in which circuits between its nodes (words) represent the probabilistic scores for their semantic proximities. In order to compare the performance of these two approaches, we perform our experiments using the standard ROMANSEVAL test collection for the SQD task and the CLEF-2003 benchmark for the SQE process in French monolingual information retrieval evaluation. The results show the impact of SQD on SQE based on the recall/precision standard metrics for both the possibilistic and the probabilistic circuit-based approaches. Besides, the results of the possibilistic approach outperform the probabilistic ones, since it takes into account of imprecision cases.

Keywords: Semantic Query Disambiguation · Semantic query expansion · Word sense disambiguation · Information retrieval · Possibility theory · Probability theory · Semantic graph · Semantic proximity

1 Introduction

Information Retrieval Systems (IRS) stay suffer from many challenges especially related to users' queries. In fact, IRS users mainly express their needs within short queries which can also contain ambiguous terms. Consequently, IRS results can

B. Duval et al. (Eds.): ICAART 2014; LNAI 8946, pp. 280–295, 2015.
DOI: 10.1007/978-3-319-25210-0_17

include several irrelevant documents (noise) due to the limit context provided by such queries. This noise decrease search efficiency and open the doors of two problems to be solved: Semantic Query Expansion (SQE) and Semantic Query Disambiguation (SQD) in order to improve search results.

The Semantic Query Disambiguation process [1, 2] is based on Word Sense Disambiguation (WSD) task. Indeed, Word sense disambiguation consists of selecting the suitable sense of a word given its context [3]. In fact, WSD stays as the main problem in natural language processing (NLP) and has a great influence in several related applications such as mono- and cross-language information retrieval, information extraction, machine translation (MT), content analysis, word processing, lexicography and the semantic Web applications.

Recently, WSD field has been mainly improved thanks to SensEval and SemEval competitions. For example, some works confirmed that the efficiency of MT systems has been considerably enhanced, thanks to the incorporation of a WSD task; supporting the translation process [4, 5]. However, in the information retrieval field the WSD task shows also its importance in two ways: (i) query terms can have closely related senses with other words not exist in the query. Consequently, retrieval recall can be enhanced if we take into account of these semantic links between words; and (ii) queries and documents terms can have multiple senses which decrease the retrieval precision [6]. Selecting the correct sense for both queries and documents terms may significantly enhance retrieval precision by decreasing noise in search results.

In general, WSD systems support IR systems (IRS) by identifying the suitable senses of queries and documents terms during search process. On the one side, querying step is improved by identifying the correct sense of each query term given its context. On the other side, correct senses of documents terms should be also identified in order to suitably index them given their context. Both queries and documents terms disambiguation tasks should be done before starting retrieval process. Nevertheless, this conclusion was not approved in some early research works such as [7, 8] in which search effectiveness cannot be improved despite the incorporation of a WSD system in their IRS. On the contrary, other IRS such as in [9–14] justified their efficiency enhancement thanks to the integration of WSD systems.

The Semantic Query expansion is the process of reformulating the set of the original user's query terms adding to them some other terms from their context [15, 16]. This technique aims to enhance search effectiveness in information retrieval task. In case of Web search engine, query expansion includes assessing a user's original query terms and expanding the retrieval query to match further documents. In fact, Query expansion implicates many other methods such as: (i) Re-weighting the original query terms; (ii) Stemming every term in the query in order to identify all the different morphological forms of terms; (iii) Identifying spelling errors and automatically retrieving for the corrected form or proposing it in the results; and (iv) Searching synonyms of original query terms in order to enrich the query context.

However, query expansion task can reformulate the original query by adding some ambiguous terms. This problem cannot be solved without a query disambiguation task. This relationship and dependency between these two tasks prove the need to mix them together for the purpose of improving IR efficiency. [15] and [17, 44] proposed respectively SQE and WSD approaches based on possibilistic networks. However they did not apply their WSD algorithm on query disambiguation. They also used dictionaries as lexical resources.

This paper is a fully revised version of the conference paper [18], in which we briefly presented a combined approach for SQD and SQE tasks using possibilistic networks and applied on an extracted co-occurrence graph. We also tested possibilistic networks for enhancing IR results, by studying many combinations of scenarios of SQD, SQE and relevance feedback. In this paper, we mainly address the following new issues: (i) we explain the theoretical contribution of possibility theory compared to probability theory; (ii) we propose and assess a second probabilistic circuit-based approach mixing SQD and SQE to improve efficiency of intelligent information retrieval. In this approach, both SQD and SQE tasks are based on a dictionary modeled by a graph, in which circuits between its nodes (words) represent the probabilistic scores for their semantic proximities; (iii) we compare the performance of these two approaches by performing our experiments using the standard ROMANSEVAL test collection for the SQD task and the CLEF-2003 test collection for the SQE process in French Monolingual IR evaluation; and (iv) we propose more perspectives for future investigations.

The paper is organized as follows. We review in Sect. 2 previous works using SQD and SQE in intelligent IR. In Sect. 3, we present the co-occurrence graph model used as a resource for both SQE and SQD tasks. Section 4 details the possibilistic and the circuit-based approaches for combining SQE and SQD. Experimental results, comparative study between these two approaches and their discussion are provided in Sect. 5. Finally, Sect. 6 concludes this paper by evaluating our work and proposing some directions for future research.

2   Related Works

In this literature review, we firstly study the most important approaches of WSD improving information retrieval efficiency in Sect. 2.1. Secondly, query expansion techniques and their impact on the performance of IRS are presented in Sect. 2.2. Finally, approaches combining SQD and SQE to improve IR are discussed in Sect. 2.3.

2.1   Semantic Query Disambiguation in IR

Word sense disambiguation (WSD) is a generally known task in natural language processing (NLP) problems and IR [19]. According to the survey presented in [3, 44], WSD seriously depends on knowledge resources which are classified into two groups: structured resources (such as thesauri, electronic dictionaries, etc.) and unstructured resources (such as corpora documents).

Query disambiguation task stays a serious challenge in information retrieval process. That's why several previous works have studied the advantages and the disadvantages of integrating a SQD task in IRS. For example, the authors in [7] matched queries' terms meanings with documents' terms senses in order to take advantage of WSD in IR. However, their results are not very convincing because of the limit sense provided by query terms, which present some disambiguation. In order to confirm the impact of WSD on IR, Sanderson in [20, 21] took advantage of a set of pseudo-words to identify query terms meanings. Nevertheless, he confirmed the important need of high accuracy WSD systems able to improve IR effectiveness.

Schütze and Pedersen in [9] didn't use predefined sense inventories, but they exploited the sense inventory directly from the text retrieval collection. Indeed, and based on the correspondences of their contexts, every word and its occurrences were clustered into senses. Authors proved via their experiments that retrieval effectiveness has been enhanced thanks to the support of WSD task. Besides, IR performance was also increased as a result of using the combination of sense-based ranking and word-based ranking. Nevertheless, the sense inventory is mainly dependent on the used collection. Consequently, it is not easy to enlarge the text collection without re-playing preprocessing task. Further, the clustering process of each word is a hard task and a time consuming step.

On the other hand, the corpus SemCor was manually sense annotated in order to discuss the impact of a wrong WSD on IR [10]. Indeed, authors represented documents and queries with correct meanings as well as synonym sets (synsets) to achieve important enhancements in IR. In fact, thanks to the use of this synset representation, experimental results proved that IR effectiveness still enhanced even they used a WSD with an error rate between 40 % and 50 %. Afterward, the authors in [22] confirmed the discriminative effect of part-of-speech (POS) information in IR tasks.

Besides, senses predefined in hand-crafted sense inventories are also used to disambiguate both queries and documents terms. In fact, identify the correct senses for documents' terms improve indexing task which cannot alone enhance the whole IR performance without a query disambiguation process. For example, and in order to disambiguate the polysemous nouns given their context, Voorhees in [8] took advantage of the hyponymy "IS-A" relation existing in WordNet [23]. All experimental results showed that the stem-based retrieval outperformed the sense-based retrieval. However, these results cannot be improved using a wrong WSD system.

Both documents and queries terms are disambiguated in [11] using a fine-grained sense inventory with an accuracy of 62.1 %. Their experiments using the TREC collections accomplished important enhancements and outperformed a standard term based vector space model. But, the general poor performances of their system and their baseline approach make not easy to objectively evaluate the exact impact of WSD in IR efficiency.

Alternatively, Kim et al. in [12] proposed a coarse-grained sense tagging technique using WordNet to tag words with 25 root senses of nouns. They exploited the stem-based index method and assign a weight to document's term according to its sense matching result with the query. Experimental results, performed using the TREC collections, showed that their coarse-grained sense tagging technique achieved significant improvement since it was flexible and consistent. Moreover, they concluded that drawbacks caused by inaccurate WSD performance can be overcome by the incorporation of senses into the classical stem-based index.

Recently, Zhong and Ng in [14] approved the relevance of WSD task to enhance IR efficiency. Authors presented and tested a technique for senses annotations applied to short queries. In fact, they integrated WSD into the language modeling method to information retrieval [24]. Moreover, they took advantage of sense synonym relationships to more increase the IR effectiveness. Experimental tests using TREC collections proved that supervised WSD performed better results than the two other WSD baselines and considerably enhanced IR performance.

The state-of-the-art IR systems using WSD confirmed that the word sense errors can simply cancel its encouraging effect. Consequently, it is relevant to decrease the destructive effect of wrong disambiguation. One of the possible solutions consists in the incorporation of senses into traditional term index such as stem-based index. Besides, the investigation of semantic relationships between senses considerably improves IR performance. These semantic relations have showed to be useful for query expansion task in IR.

2.2 Semantic Query Expansion in IR

Semantic Query Expansion (SQE) is one of the most popular technique has been used in IR systems to enhance their effectiveness by satisfying their users' needs. Carpineto and Romano in [16] classified SQE into two principal methods: automatic query expansion (ASQE), and interactive query expansion (ISQE), which depend on user assistance. In both cases, SQE can be accomplished by several methods such as utilization of external linguistic resources (thesauri, dictionaries, ontologies, etc.), corpus analysis and relevance feedback techniques [25]. Indeed, Manning et al. in [25] classified SQE methods based on relevance feedback into three principal classes: (i) In the first class called "user relevance feedback", the returned results take into account the user's judgment; (ii) In the second class called "indirect relevance feedback" (or implicit relevance feedback), we took advantage of indirect sources of evidence such as number of hits on web page's links; and (iii) In the third class called "pseudo relevance feedback" (or blind relevance feedback), the IRS exploited the top k most relevant retrieved documents in order to expand the original query. Therefore, a set of candidate terms from these documents is added using often variants of Rocchio algorithm [26]. Even though relevance feedback may decrease noise in IR results, all these methods do not provide a solution to precisely find the suitable sense of the query terms, therefore requiring other techniques for query disambiguation.

Many SQE approaches existing in the literature have used external linguistic resources such as WordNet on English IRS [16, 27, 28]. However, these approaches are based on poor, uncertain and unclear data, while possibility theory is naturally suitable for this type of application; because it permits to express ignorance, imprecision and uncertainty [29]. In fact, it provides two kinds of relevance: (i) plausible relevance quantified by the possibility, trends to remove non-semantically similar terms (irrelevant ones); and (ii) necessity relevance increases our belief in terms not removed by possibility measure. Based on these advantages provided by the possibility theory, Ben Khiroun et al. in [30] proposed and evaluated a possibilistic approach for semantic query expansion. They later extend their approach in [15] by proposing and assessing a new possibilistic IRS which takes advantage, combine and compare the possibilistic and the probabilistic circuit-based approaches for semantic query expansion [31, 32]. Indeed, authors took advantage of the dependencies relationships between the query terms and the articles of a dictionary to model their possibilistic network. Consequently, they investigated possibility and necessity measures to compute the corresponding possibilistic semantic similarity between terms. In fact, the SQE technique consists of injecting into the original query the most possibly and necessarily articles selected from the dictionary. Besides, SQE process was enhanced by incorporating a reweighting model which provides to the original and new query terms some relative importance. The possibilistic and the probabilistic circuit-based approaches for SQE were firstly compared in terms of their impact to IR performance. Secondly, authors mixed these two approaches by assessing two different aggregation methods. They also improved IR efficiency by integration a reweighting query terms technique in the possibilistic matching model existing in [32] to increase the performance of the expansion task. Experimental results using the standard "LeMonde94" test collection and the French dictionary "Le Grand Robert" showed partial enhancement of the results of some test queries. These enhancements, not seen at the global level of analysis, approved that the performance of any semantic query expansion technique depends on the nature of the test queries in the test collection. Moreover, query expansion task can induce noise in the search results because of the injection of polysemous words. To reduce this problem, it is suitable to incorporate a semantic disambiguation mechanism solving the problem of word sense disambiguation before and/or after the expansion task.

2.3   Combining SQD and SQE in IR

Several approaches in the literature studied the impact of SQD with SQE in IR performance using knowledge sources from thesauri. Indeed, some thesauri-based methods accomplished enhancements in IR efficiency by expanding the disambiguated query terms with synonyms and some other information from WordNet [13, 27, 33]. Besides, document expansion also benefited from the investigation of knowledge sources from WordNet which consequently prove enhancements in IRS performance [34, 35].

On the other hand, Pinto and Pérez-sanjulián in [36] exploited WordNet as external linguistic resource for both WSD and SQE. They approved the necessity of incorporating a WSD task in SQE process in order to increase IR performance. Experimental results are achieved using short and long queries from the TREC-8 text collection. These results confirmed that SQE applied on both short and long queries is not sufficient to increase IR efficiency. However, identifying the suitable sense of each ambiguous query term using a set of extracted synonyms from WordNet can mainly contribute to improve IR performance. Consequently, retrieval effectiveness was significantly improved for short queries than long ones.

Moreover, Paskalis and Khodra proposed, tested and evaluated in [2] several scenarios on IR process by using WSD, SQE, stemming and a relevance feedback technique. For WSD task, they investigated an extended implementation of Lesk algorithm [19] in order to identify the correct meaning of each query and document terms. For SQE task, they firstly exploited a co-occurrence based thesaurus built automatically from the documents collection. Secondly, they took advantage of a pseudo relevance feedback technique using a set of top relevant documents in order to extract some representative terms from them. These terms are finally injected in the original query to improve expansion process.

Recently, authors in [17, 44] and [15] proposed and evaluated respectively a possibilistic approach for WSD and a possibilistic approach for semantic query expansion (SQE). Both of them exploited a possibilistic network in order to compute possibilistic scores between French words using the French dictionary "Le Grand Robert" as an external linguistic resource. Indeed, in the possibilistic WSD approach, authors benefited from the double relevance measure (possibility and necessity) between words and their contexts. Experimental results are done using the standard ROMANSEVAL test collection. Experiments proved a promote enhancements in terms of disambiguation rates of French words. This disambiguation performed better on nouns as they are most frequent among the existing words in the context.

In [18], authors studied the impact of Word Sense Disambiguation (WSD) on Query Expansion (SQE) for monolingual intelligent information retrieval. The proposed approaches for WSD and SQE are based on corpus analysis using co-occurrence graphs modeled by possibilistic networks. Indeed, the model for relevance judgment uses possibility theory to take advantages of a double measure (possibility and necessity). Experiments are performed using the standard ROMANSEVAL test collection for the WSD task and the CLEF-2003 benchmark for the SQE process in French monolingual Information Retrieval (IR) evaluation. The results showed the positive impact of WSD on SQE based on the recall/precision standard metrics.

3 Model Architecture and Knowledge Representation

In order to have a generic data representation that can be used for SQE, SQD and relevance feedback, we opted for a graph model that uses co-occurrences between term nodes. These relations are extracted from corpora to model contextual and similarity links. Thus, these relations are useful to compute the similarity between the terms of the queries (in the case of expansion) or between terms and senses (in the case of disambiguation).

To perform co-occurrence graph construction, we consider that two nodes are related if they exist in the same sentence. The edges are bi-oriented and weighted by the normalized co-occurrence frequency of the related terms. On the other hand, ambiguous words are related with their appropriate senses in the dictionary as considered in the following:

- T: the set of terms in the corpus
- S: the set of senses in the dictionary
- A node t_i is related to a node t_j if t_i and t_j co-occur in the same sentence; where $\{t_i, t_j \in T\}$.
- A node t_i is related to a node s_j if t_i is an ambiguous term and s_j represents a sense of t_i; where $\{t_i \in T\}$ and $\{s_j \in S\}$.

The process in Fig. 1 presents the different resources used in the SQD task, SQE and pseudo relevance feedback.

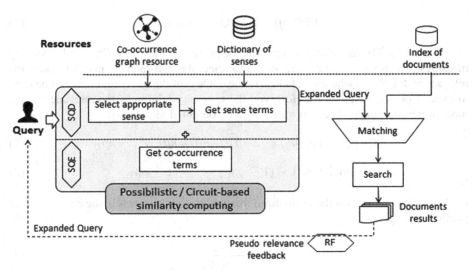

Fig. 1. Sematic query expansion using disambiguation process.

The QE module is executed to generate an expanded query starting from the initial query. In the case of ambiguous terms, the disambiguation module is used before applying QE. Thus, the best sense node having the greater possibilistic or probabilistic score is selected and the terms existing in its definition are used for expanding the original query. For both QE and QD processes, the co-occurrence graph is used to achieve possibilistic and probabilistic circuit-based calculus. Afterwards, the expanded query is matched with documents to achieve results as the classical IR process.

A pseudo relevance feedback is applied at the end of the process by extracting the most significant terms from the top first returned documents. The whole process may be iterated.

4 Possibilistic and Probabilistic Approaches for Combined SQD and SQE

We present in this section two approaches for combined SQD and SQE and we introduce an illustrative example.

4.1 A Possibilistic Approach for SQD and SQE

We based our approach on the possibilistic theory introduced by Zadeh [37] and developed by several authors [38, 39] in order to compute terms similarity in both SQE and SQD tasks. We adapted the possibilistic model architecture of Elayeb in [15] to be applied on co-occurrence graphs. Thus, we define the Degree of Possibilistic Relevance (DPR) for each co-occurrence graph' node n_j given a query $Q = (t_1, t_2, ..., t_T)$ by:

$$DPR(nj) = \Pi(n_j|Q) + N(n_j|Q) \tag{1}$$

Where $\Pi(n_j|Q)$ and $N(n_j|Q)$ represent respectively the possibility and necessity measures. The possibility measure allows to reject the non-relevant nodes identify the relevant nodes (those who are not close to the context of the query and may not be used to expand or disambiguate it). However, the necessity reinforces the relevance of the most important nodes. The two measures are computed as follows:

$$\Pi(n_j|Q) = \Pi(t_1|n_j) * \ldots * \Pi(t_T|n_j) = nft_{1j} * \ldots * nft_{Tj} \tag{2}$$

$$N(n_j|Q) = 1 - (1 - \phi n_{1j}) * \ldots * (1 - \phi n_{Tj}) \tag{3}$$

Where nft_{ij} represents the normalized frequency of query terms in the co-occurrence graph:

$$nft_{ij} = \frac{tf_{ij}}{\max_k(tf_{kj})} \tag{4}$$

In the formula (4), tf_{ij} is the weight of the edge relating the nodes t_i and n_j (i.e. the number of times the two nodes co-occur).
And:

$$\phi n_{ij} = Log_{10}\left(\frac{nCN}{nN_i}\right) * nft_{ij} \tag{5}$$

Where:
nCN = total number of nodes in the co-occurrence graph related to the query terms;
nN_i = number of nodes related to the term t_i.
Using the log function (such as in TF-IDF) allows to compute the discriminative power of the query terms. Thus, we select the graph nodes which are closest to the most discriminative items of the contextual information represented in the query.

4.2 A Probabilistic Approach Using Circuit-Based Measure for SQD and SQE

Elayeb studied in [31, 32] the query expansion problem and its impact on a possibilistic information retrieval system. His method is based on counting circuits in a graph generated from a dictionary. Indeed, in the graph of dictionary words maintain relationships that sometimes make circuits. For a given term t_i of an initial query Q^{old}, using the graph of the dictionary we compute the score of semantic proximity of term t_i with any other term t_j according to following formula [31, 32]:

$$Sem\_Prox(t_i, t_j) = \frac{Number\_of\_Circuits(t_i, t_j)}{Maximum\_Number\_of\_Circuits\_in\_the\_Graph} \tag{6}$$

Where: *Number_of_Circuits(t_i, t_j)* represents the number of circuits starting from the node t_i and passing through the node t_j in the graph of dictionary (i.e. $t_i \rightarrow \ldots \rightarrow t_j \rightarrow \ldots \rightarrow t_i$).

For the SQD task, we consider a sense S_i corresponding to an ambiguous word in the query Q. The semantic proximity of S_i to Q is generalized from the Eq. (6) as follow:

$$Sem\_Pr\,ox(S_i, Q) = \sum_{s_{ij} \in S_i} \sum_{t_k \in Q} Sem\_Pr\,ox(s_{ij}, t_k) \qquad (7)$$

The maximum length of circuit is one of important parameter in this distance. In fact, more the circuit is long more there is chance to mix various groups of meanings. However, taking into account only too short circuits would cause to cluster terms related to the same hyperonym into different groups. More details about the regrouping principle can be found in [31, 32], where author specify that the maximum length of circuit that we can take into account is about 4 edges.

4.3 Illustrative Example

Let us consider the following query admitting that it contains an ambiguous word:

Les règles d'orthographe et de ponctuation pour la langue allemande ont été considérablement simplifiées

Which may be translated as follows:

The rules of spelling and punctuation for the German language has been considerably simplified

The query is tokenized and lemmatized ignoring stop words (like pronouns, articles, etc.) as follow:

règle (**rule**), orthographe (**spelling**), ponctuation (**punctuation**), langue (**language**), allemand (**German**), cosidérable (**considerable**), simple (**simple**)

The output query contains the ambiguous word "simple" (simple). So the WSD is executed and the sense having the best possibilistic score from ROMANSEVAL dictionary is selected (in this example we consider the sense "AII1"):

AI2 Qui n'est formé que [...]
AI3 Qui suffit à soi seul [...]
AII1 Qui est facile à comprendre [...]

Translated as:

AI2 Which is formed only by [...]
AI3 Sufficient to itself alone [...]
AII1 That is easy to understand [...]

So the corresponding terms in the definition "AII1" are injected in query using the possibilistic approach (Fig. 2).

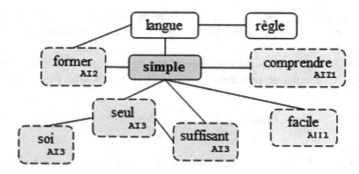

Fig. 2. A sample of the co-occurrence graph.

On the other hand, we consider this sample overview of the graph to compute the semantic proximity by using the circuit-based approach.

When enumerating the number of circuits for the three senses "AI2", "AI3" and "AII1", the sense "AI3" containing the words "seul" (alone), "soi" (itself) and "suffisant" (sufficient) has the highest semantic proximity for circuit-based computation.

Thus, this sense is the best one collating to the query context. So, the terms of the sense "AI3" are selected by the circuit-based approach for SQD task and are added to the query before expansion.

5   Experimental Results

In this section, we evaluate and compare the contribution of the possibilistic and the circuit-based approaches on both SQD and SQE tasks.

5.1   Experimental Settings

We used two test collections to experiment our proposed approach to study the impact of query disambiguation on the expansion process in French language: CLEF2003 and ROMANSEVAL.

On the one hand, the CLEF2003 test collection provides necessary tools for the evaluation of information retrieval systems on large corpora including a set of documents, a set of queries and the list of relevant documents for each query. Each query is represented in the XML format by a title containing its terms, a description and a detailed narrative text. The CLEF2003 collection for French language is composed of Le Monde 94, ATS 94, and ATS95 sub-collections forming 57 test queries and more than 300 MB of data [40].

On the other hand, the ROMANSEVAL test collection is useful for evaluating WSD approaches: it provides the necessary resources for WSD including a set of documents and a list of test sentences containing ambiguous words. A set of 60 ambiguous words distributed on three grammatical categories (20 nouns, 20 adjectives, 20 verbs) were annotated by 6 members in accordance with the senses. Each word occurrence may have one or several labels of sense or none [41].

In all our experiments, we focused only on queries from CLEF2003 test collection which contains ambiguous terms included in ROMANSEVAL test collection.

We used the Terrier experimental platform for IR to evaluate our system [42]. Different common IR measures where used like Recall/Precision, R-precision and Mean Average Precision (MAP) (for more details about state of the art IR measures see [25]). The Okapi (BM25) matching model and the Snowball stemmer (integrated in Terrier) are used for all experimentations.

In order to perform pseudo relevance feedback based on the document collection, we used the Bo1 (Bose-Einstein 1) pseudo relevance feedback method implemented in the Terrier information retrieval platform [42]. The default settings are specified as follows: the number of terms to expand a query is set to 10 and the number of top-ranked documents from which these terms are extracted is limited to 3 documents.

5.2 Evaluating SQD and SQE

This section summarises and discusses the overall performance of the various performed tests. Table 1 reports the main runs and evaluation scores for each one. For both possibilistic and circuit-based approaches, we performed two scenarios: 1- by applying the query expansion alone ("Poss_QE" and "Circuit_QE"); 2- by disambiguating the query before expansion ("Poss_QD&E" and "Circuit_QD&E"). The baseline scenario refers to the initial query without expansion or disambiguation.

Table 1. Overview of the results of the possibilistic and the probabilistic approaches.

Method		MAP	R-precision
Possibilistic	Poss_QE	0,5083	0,4742
	Poss_QD&E	**0,5124**	**0,4760**
Probabilistic	Circuit_QE	0,4920	0,4633
	Circuit_QD&E	0,5071	0,4642
Baseline		0,5487	0,5174

The last two columns present the Mean Average Precision (it is the mean of the average precision scores for each query) and the exact precision (R-Precision is the precision at rank R; where R is the total number of relevant documents) values [25].

The application of query expansion presents a performance decrease for all tests by adding new terms. However, possibilistic expansion method shows slightly better results than the circuit-based expansion method. The application of query disambiguation contributes as well for improving the retrieval results when comparing the 4 tests.

As a preliminary interpretation, this overall negative performance of query expansion (with and without query disambiguation), compared to the baseline test, could be explained by the generation of noise in search results (so lower precision values).

Oviglie et al. noted in [43] that the number of expansion terms for optimal precision varies widely across systems and topic (query) sets. Applying query expansion on long queries (that contain more than 10 words) may produce noisy and non-interpretable results as studied by Pinto and Pérez in [36]. So, we limited the number of expansion terms to the quarter of the query' length in order to reduce the noise phenomenon according to the experimental results in [18].

We conducted a more detailed analysis by examining the Recall/Precision curve in Fig. 3.

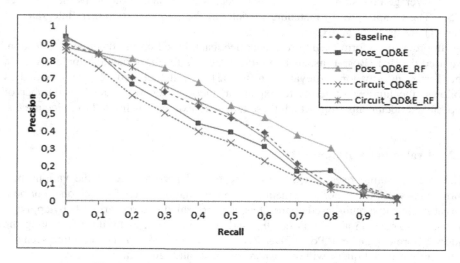

Fig. 3. Recall-Precision curve comparing different tests.

So, focusing on the test scenario "Poss_QD&E", in which we applied both SQD and SQE, we can confirm that the query expansion combined with disambiguation is better than the baseline at high recall levels (i.e. initially better at retrieving the relevant documents).

In these detailed tests, we applied also the pseudo relevance feedback after disambiguating and expanding the queries of the test set. The application of relevance feedback with SQD and SQE improves the information retrieval performance for both possibilistic ("Poss_QD&E_RF") and probabilistic circuit-based ("Circuit_QD&E_RF") approaches. Nevertheless, the possibilistic approach outperforms the circuit-based one. Indeed, the former method refines the search of new terms (respectively senses) for semantic expansion (respectively query disambiguation) by taking into account a double measurement of semantic proximity between the co-occurrence graph nodes.

6 Discussion and Future Works

This work presents and compares possibilistic and probabilistic approaches based on a co-occurrence graph resource. Thus, we compared the impact of word sense disambiguation in IR performance when applying query expansion and relevance feedback. The graph used in the different approaches was prepared from the collection of documents in ROMANSEVAL test collection.

Afterwards, this resource is used to choose the best candidates for disambiguation and expansion tasks. The computed score for semantic similarity depends on the used approach. The results show that the possibilistic one is finer than the probabilistic circuit-based one. This is explained by the fact that possibility and necessity measures increase the relevance of correct senses/terms and penalize the scores of the remaining ones.

Furthermore, we showed the important contribution of pseudo relevance feedback in the presented experiments of this paper. The same positive role of relevance feedback was observed in the works of Paskalis and Khodra [2]. Indeed, we join the fact that it should be better to focus on this technique to improve IR performance in parallel with word sense disambiguation methods.

In order to have a wider comparative study, we aim to compare in future works the impact of changing the knowledge source used for SQD and SQE tasks such as dictionary in place of co-occurrence graphs. However, this may present coverage problem especially for modern terms and proper nouns. As a second perspective, we aim to expand the proposed models from monolingual context to cross-lingual one by using other adapted corpora such as the SemEval corpus.

Acknowledgements. We are grateful to the Evaluations and Language resources Distribution Agency (ELDA) which kindly provided us the Le Monde 94 and ATS 94 document collections of the CLEF 2003 campaign.

References

1. Krovetz, R.: Homonymy and polysemy in information retrieval. In: Proceedings of the 8th Conference on European Chapter of the Association for Computational Linguistics, pp. 72–79. Association for Computational Linguistics, Stroudsburg, PA, USA (1997)
2. Paskalis, F.B.D., Khodra, M.L.: Word sense disambiguation in information retrieval using query expansion. In: International Conference on Electrical Engineering and Informatics (ICEEI), pp. 1–6 (2011)
3. Navigli, R.: Word sense disambiguation: a survey. ACM Comput. Surv. (CSUR) **41**, 1–69 (2009)
4. Chan, Y.S., Ng, H.T.: Word sense disambiguation improves statistical machine translation. In: 45th Annual Meeting of the Association for Computational Linguistics (ACL-2007), pp. 33–40 (2007)
5. Carpuat, M., Wu, D.: Improving statistical machine translation using word sense disambiguation. In: The 2007 Joint Conference on Empirical Methods in Natural Language Processing and Computational Natural Language Learning (EMNLP-CoNLL 2007), pp. 61–72 (2007)
6. Chifu, A.-G., Ionescu, R.-T.: Word sense disambiguation to improve precision for ambiguous queries. Cent. Eur. J. Comput. Sci. **2**, 398–411 (2012)
7. Krovetz, R., Croft, W.B.: Lexical ambiguity and information retrieval. ACM Trans. Inf. Syst. **10**, 115–141 (1992)
8. Voorhees, E.M.: Using WordNet to disambiguate word senses for text retrieval. In: Proceedings of the 16th Annual International ACM SIGIR Conference on Research and Development in Information Retrieval, pp. 171–180. ACM, New York, NY, USA (1993)

9. Schütze, H., Pedersen, J.O.: Information retrieval based on word senses (1995)
10. Gonzalo, J., Verdejo, F., Chugur, I., Cigarrin, J.: Indexing with WordNet synsets can improve text retrieval. In: Proceedings of the COLING-ACL Workshop on Usage of WordNet in Natural Language Processing Systems, pp. 38–44 (1998)
11. Stokoe, C., Oakes, M.P., Tait, J.: Word sense disambiguation in information retrieval revisited. In: Proceedings of the 26th Annual International ACM SIGIR Conference on Research and Development in Information Retrieval, pp. 159–166. ACM, New York, NY, USA (2003)
12. Kim, S., Seo, H., Rim, H.: Information retrieval using word senses: root sense tagging approach. In: Proceedings of the 27th Annual International ACM SIGIR Conference on Research and Development in Information Retrieval, pp. 258–265 (2004)
13. Liu, S., Yu, C., Meng, W.: Word sense disambiguation in queries. In: Proceedings of the 14th ACM International Conference on Information and Knowledge Management, pp. 525–532. ACM, New York, NY, USA (2005)
14. Zhong, Z., Ng, H.T.: Word sense disambiguation improves information retrieval. In: Proceedings of the 50th Annual Meeting of the Association for Computational Linguistics: Long Papers – vol. 1, pp. 273–282. Association for Computational Linguistics, Stroudsburg, PA, USA (2012)
15. Elayeb, B., Bounhas, I., Ben Khiroun, O., Evrard, F., Bellamine-BenSaoud, N.: Towards a possibilistic information retrieval system using semantic query expansion. Int. J. Intell. Inf. Technol. 7, 1–25 (2011)
16. Carpineto, C., Romano, G.: A survey of automatic query expansion in information retrieval. ACM Comput. Surv. (CSUR) 44, 1–50 (2012)
17. Ben Khiroun, O., Elayeb, B., Bounhas, I., Evrard, F., Bellamine-BenSaoud, N.: A possibilistic approach for automatic word sense disambiguation. In: Proceedings of the 24th Conference on Computational Linguistics and Speech Processing (ROCLING), pp. 261–275, Taiwan (2012)
18. Ben Khiroun, O., Elayeb, B., Bounhas, I., Evrard, F., Bellamine-BenSaoud, N.: Improving query expansion by automatic query disambiguation in intelligent information retrieval. In: The 6th International Conference on Agents and Artificial Intelligence (ICAART 2014), pp. 153–160. Angers, Loire Valley, France (2014)
19. Banerjee, S., Pedersen, T.: An adapted lesk algorithm for word sense disambiguation using WordNet. In: Gelbukh, A. (ed.) CICLing 2002. LNCS, vol. 2276, pp. 136–145. Springer, Heidelberg (2002)
20. Sanderson, M.: Word sense disambiguation and information retrieval. In: Croft, B.W., van Rijsbergen, C.J. (eds.) Proceedings of the 17th Annual International ACM SIGIR Conference on Research and Development in Information Retrieval (SIGIR '94), pp. 142–151. Springer, New York (1994)
21. Sanderson, M.: Retrieving with good sense. Inf. Retr. 2, 49–69 (2000)
22. Gonzalo, J., Peñas, A., Verdejo, F.: Lexical ambiguity and information retrieval revisited. In: Proceedings of the Joint SIGDAT Conference on Empirical Methods in NLP and Very Large Corpora (EMNLP/VLC), pp. 195–202 (1999)
23. Miller, G.A., Beckwith, R., Fellbaum, C., Gross, D., Miller, K.J.: Introduction to WordNet: an on-line lexical database. Int. J. Lexicogr. 3, 235–244 (1990)
24. Ponte, J.M., Croft, W.B.: A language modeling approach to information retrieval. In: Proceedings of the 21st Annual International ACM SIGIR Conference on Research and Development in Information Retrieval, pp. 275–281. ACM, New York, NY, USA (1998)
25. Manning, C.D., Raghavan, P., Schütze, H.: Introduction to Information Retrieval. Cambridge University Press, New York (2008)

26. Rocchio, J.: Relevance Feedback in Information Retrieval. The SMART Retrieval System, pp. 313–323. Prentice-Hall, Englewood Cliffs (1971)

27. Voorhees, E.M.: Query expansion using lexical-semantic relations. In: Croft, B.W., van Rijsbergen, C.J. (eds.) Proceedings of the 17th Annual International ACM SIGIR Conference on Research and Development in Information Retrieval (SIGIR '94), pp. 61–69. Springer, New York (1994)

28. Smeaton, A.F.: Using NLP or NLP resources for information retrieval tasks. In: Strzalkowski, T. (ed.) Natural Language Information Retrieval, pp. 99–111. Kluwer Academic Publishers, Dordrecht (1997)

29. Boughanem, M., Brini, A., Dubois, D.: Possibilistic networks for information retrieval. Int. J. Approx. Reason. **50**, 957–968 (2009)

30. Ben Khiroun, O., Elayeb, B., Bounhas, I., Evrard, F., Bellamine Ben Saoud, N.: A possibilistic approach for semantic query expansion. In: The 4th International Conference on Internet Technologies and Applications (ITA 2011), Wrexham Wales (UK), pp. 308–316 (2011)

31. Elayeb, B.: SARIPOD: Système multi-Agent de Recherche Intelligente POssibiliste de Documents Web. Ph.D. thesis, INP Toulouse France (2009)

32. Elayeb, B., Evrard, F., Zaghdoud, M., Ben Ahmed, M.: Towards an intelligent possibilistic web information retrieval using multiagent system. Interact. Technol. Smart Educ. (ITSE), Spec. Issue N. Learn. Support Syst. **6**, 40–59 (2009)

33. Fang, H.: A re-examination of query expansion using lexical resources. In: Proceedings of ACL-08: HLT, pp. 139–147 (2008)

34. Cao, G., Nie, J.-Y., Bai, J.: Integrating word relationships into language models. In: Proceedings of the 28th Annual International ACM SIGIR Conference on Research and Development in Information Retrieval, pp. 298–305. ACM, New York, NY, USA (2005)

35. Agirre, E., Arregi, X., Otegi, A.: Document expansion based on WordNet for robust IR. In: Proceedings of the 23rd International Conference on Computational Linguistics: Posters, pp. 9–17. Association for Computational Linguistics, Stroudsburg, PA, USA (2010)

36. Pinto, F.J., Pérez-sanjulián, C.F.: Automatic query expansion and word sense disambiguation with long and short queries using WordNet under vector model. Actas de los Talleres de las Jornadas de Ingeniería del Software y Bases de Datos. **2**, 17–23 (2008)

37. Zadeh, L.: Fuzzy sets as a basis for a theory of possibility. Fuzzy Sets Syst. **1**, 3–28 (1978)

38. Dubois, D., Prade, H.: Possibility theory and its application: where do we stand. Mathw. Soft Comput. **18**, 18–31 (2011)

39. Dubois, D., Prade, H.: Possibility theory. In: Meyers, R.A. (ed.) Computational Complexity, pp. 2240–2252. Springer, New York (2012)

40. Braschler, M., Peters, C.: CLEF 2003 methodology and metrics. In: Peters, C., Gonzalo, J., Braschler, M., Kluck, M. (eds.) CLEF 2003. LNCS, vol. 3237, pp. 7–20. Springer, Heidelberg (2004)

41. Segond, F.: Framework and results for French. Comput. Humanit. **34**, 49–60 (2000)

42. Ounis, I., Lioma, C., Macdonald, C., Plachouras, V.: Research directions in terrier: a search engine for advanced retrieval on the web. CEPIS Upgrad. J. **8**, 49–56 (2007)

43. Ogilvie, P., Voorhees, E., Callan, J.: On the number of terms used in automatic query expansion. Inf. Retr. **12**, 666–679 (2009)

44. Elayeb, B., Bounhas, I., Ben Khiroun, O., Evrard, F., Bellamine-BenSaoud, N.: A comparative study between possibilistic and probabilistic approaches for monolingual word sense disambiguation. Knowl. Inf. Syst. (2014). doi:10.1007/s10115-014-0753-z

Full-Reference Predictive Modeling of Subjective Image Quality Assessment with ANFIS

El-Sayed M. El-Alfy$^{(\boxtimes)}$ and Mohammed Rehan Riaz

College of Computer Sciences and Engineering,
King Fahd University of Petroleum and Minerals, Dhahran 31261, Saudi Arabia
{alfy,g201106370}@kfupm.edu.sa

Abstract. Digital images often undergo through various processing and distortions which subsequently impacts the perceived image quality. Predicting image quality can be a crucial step to tune certain parameters for designing more effective acquisition, transmission, and storage multimedia systems. With the huge number of images captured and exchanged everyday, automatic prediction of image quality that correlates well with human judgment is steadily gaining increased importance. In this paper, we investigate the performance of three combinations of objective metrics for image quality prediction with an adaptive neuro-fuzzy inference system (ANFIS). Images are processed to extract various attributes which are then used to build a predictive model to estimate a differential mean opinion score for different types of distortions. Using a publicly available and subjectively rated image database, the proposed method is evaluated and compared to individual metrics and an existing technique based on correlation and error measures. The results prove that the proposed method can be a promising approach for predicting subjective quality of images.

Keywords: Image quality assessment · Adaptive neuro-fuzzy inference system · ANFIS · Differential mean opinion score · Human visual system · Subjective assessment · Objective assessment

1 Introduction

Digital images are gaining great importance in the domain of electronic technology in recent years. However, images can be corrupted due to various reasons during acquisition, processing, storage and transmission. With the increasing use of digital imaging systems such as digital cameras, high definition cameras, monitors and printers, Image Quality Assessment (IQA) has attracted great attention in image processing applications [11]. Moreover, a variety of image processing techniques can benefit from image quality assessment for adaptive parameter tuning and prediction of required resources, e.g. [29].

E.M. El-Alfy—On leave from the College of Engineering, Tanta University, Egypt.

B. Duval et al. (Eds.): ICAART 2014; LNAI 8946, pp. 296–311, 2015.
DOI: 10.1007/978-3-319-25210-0_18

Image quality assessment methods can be classified into two broad categories: subjective and objective. The subjective assessment is based on the human perception of image quality and it is preferred when human beings are the ultimate recipients of the image processing applications [27]. To reduce subjectivity, it is typically conducted through a number of human observers who are asked to visually judge the perceived quality of a target image in the presence (full-reference) or absence (no-reference) of its original image. This judgement can be in the form of a rank or score. The results of different observers are averaged and the resulting aggregated metric is called Mean Opinion Score (MOS). This score can be scaled to be in the range from 0 (very low quality) to 1 (very high quality). In the presence of a reference, another evaluation metric is known as DMOS which is the difference between the MOSs assigned to the reference and target images. If we assume the reference image has perfect quality, i.e. its MOS is 1, then the range for DMOS assigned to the target image will be from 0 (very high quality) to 1 (very low quality). Notice that it is the opposite of MOS.

The automation of subjective quality assessment is difficult as it depends on modelling the human visual system (HVS); which is a complex task especially when considering high-level cognitions. In contrast, objective quality assessments use numeric measures to quantify the degree of quality degradation and can benefit from the low-level models of certain features of HVS. Hence, it can be automated to replace the way a human assesses the quality of an image. The majority of objective quality assessment methods are based on pixel difference metrics due to their low computational complexity [2]. However, these methods can suffer from some limitations in dealing with the wide spectrum of image distortion types. Hence, a number of other quality metrics have been proposed in the literature for various situations by different researchers [7].

Whether subjective or objective, image quality assessment techniques can be also classified as no-reference, full-reference or reduced-reference. This classification depends on the availability of information from the original image besides the target or query image. In a no-reference technique, the assessor has only access to the query image; hence it is also termed as blind assessment, e.g. [5,15]. But when the original image is available with the target image, it is termed as full-reference; e.g. [14]. In some applications, only partial information about the original image can made be available, e.g. due restricted bandwidth or storage space, besides the query image and hence it is termed reduced-reference [19].

This paper therefore investigates the ability of the adaptive neuro-fuzzy inference system (ANFIS) approach in predicting the subjective quality of images. It is implemented to estimate an aggregated score from a set of objective metrics for image quality assessment. We consider five types of distortion at different levels including JPEG compression, JPEG 2000 compression, additive pink Gaussian noise (APGN), additive white Gaussian noise (AWGN) and Gaussian blurring (Blur). To evaluate its performance, the predicted value is compared to the actual difference mean opinion score (DMOS). Four performance measures and subsequently computed, namely Pearson's correlation coefficient, Spearman's rank order correlation coefficient, mean absolute error (MAE), and root mean square error (RMSE). This work is a revised and extended version of our earlier work published at the international conference on agents and artificial intelligence [6].

The rest of the paper is structured as follows. Section 2 gives a brief background of the main ANFIS characteristics and how it can be used for function approximation and prediction. The related work is reviewed in Sect. 3. Section 4 provides more details on image quality assessment and defines the quality metrics that are used in this work. Section 5 describes the evaluation dataset and discusses the experimental work and results. Finally, Sect. 6 concludes the paper and highlights future work.

2   ANFIS Background

In the case of fuzzy logic based systems, the mapping of prior human knowledge or experience into the inference process using linguistic variables is an advantage but a cumbersome task. No standard procedure is found to provide an efficient way of this transformation. Usually, a trial and error approach determines the type, size and settings of the input and output membership functions (MFs). Effective tuning methods for the input and output membership functions and reduction of the rule base to the least necessary rules have always been on the list of significant issues to be explored.

Adaptive neuro-fuzzy inference system or ANFIS is emerged to mitigate the above mentioned issues by providing a learning capability to the fuzzy system through its integration with a neural network [8]. Thus, ANFIS combines the advantages of both the fuzzy inference system and the neural network. ANFIS has been widely used to solve several problems in different domains [1,10,17,18].

Typically, the ANFIS system works in two distinct phases. The first phase is a neural-network phase, where a system classifies data and finds patterns. The other phase develops a fuzzy expert system through adaptive tuning of membership functions [10]. Figure 1 shows an illustrative example of a Sugeno-type ANFIS system, with two inputs X and Y, one output F, and two rules. Each input variable is assumed to have two linguistic terms (e.g. small and large). The computations are performed in five layers; where the output from each node in every layer is represented by O_i^l, where l denotes the layer number and i denotes the specific neuron within that layer. The purpose of the first layer is to fuzzify the crisp input values using a set of linguistic terms (e.g., small, medium, and large). Membership functions of these linguistic terms determine the output of this layer as given by:

$$O_{A_i}^1 = \mu_{A_i}(x), O_{B_i}^1 = \mu_{B_i}(y) \tag{1}$$

where $\mu_{A_i}(x)$ and $\mu_{B_i}(y)$ represent the membership functions that establish the degree to which the given input values x and y satisfy the quantifiers A_i and B_i. A variety of membership functions exists such as bell-shaped, trapezoidal, triangular, Gaussian, and sigmoidal. The firing strength for each rule quantifies the extent that any input data belongs to that rule, and is computed in the

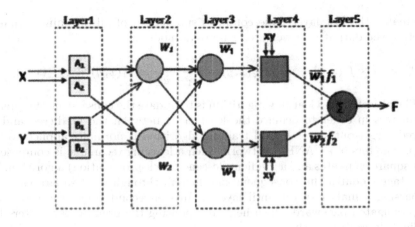

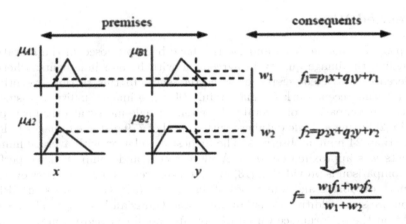

Fig. 1. An illustrative example of ANFIS model architecture and reasoning.

second layer as the multiplication of all the incoming signals at each node as follows:

$$O_i^2 = w_i = \mu_{A_i}(x) \times \mu_{B_i}(y) \tag{2}$$

The nodes in the third layer perform normalization operation by calculating the ratio of the i-th rule's firing strength to the sum of all rules' firing strengths as follows:

$$O_i^3 = \bar{w}_i = \frac{w_i}{w_1 + w_2} \tag{3}$$

In Sugeno-type ANFIS system, the consequent part of each rule is expressed as a linear combination of the inputs. The fourth layer has square-shaped nodes with node functions given as:

$$O_i^4 = \bar{w}_i \times f_i(x, y) = \bar{w}_i \times (p_i x + q_i y + r_i) \tag{4}$$

Finally, the last layer node conducts summation of all incoming signals to generate the output as a weighted sum of all node inputs:

$$O_i^5 = \bar{w}_1 \times f_1(x,y) + \bar{w}_2 \times f_2(x,y) = \frac{1}{w_1 + w_2}[w_1 f_1(x,y) + w_2 f_2(x,y)] \quad (5)$$

The objective of a learning algorithm is to update the consequent and premise parameters in order to achieve the least error between the predicted and the desired target output. A hybrid training algorithm is normally applied to tune the parameters of an ANFIS network. Such a learning technique is composed of least square estimates and a gradient descend (back-propagation) algorithm. The first stage updates the consequent parameters through least-square estimates by passing signals forward until layer 4. In the second stage, the error rates are propagated backward which helps in updating the premise parameters by a gradient descent algorithm.

3   Related Work

In the past, various methods and metrics have been proposed in the literature for assessing the image quality in agreement with human judgments; whether full reference, restricted reference, or no-reference. In this section, we briefly review relevant recent work for mainly full-reference image quality assessment. An extensive comparison of several full-reference single-metric algorithms is presented in [22] using a single database with a total of 779 distorted images judged by more than 24 human observers. The highest correlation with average human judgments was approximately 0.95. A more recent and comprehensive performance comparison is provided in [13] for 18 full-reference quality metrics over six benchmark databases and four types of distortions (JPEG compression, JPEG 2000 compression, additive Gaussian noise, and Gaussian blurring). The results showed that the performance varied widely after nonlinear regression with superior capability for the visual information fidelity (VIF) metric in terms of its correlation to the subjective human ratings.

In [2], the authors proposed a neural network approach for the assessment of image quality. The neural network measured the quality of an image by predicting the mean opinion score (MOS) with the help of six key features extracted from both the reference and target images. These features are the two means, two standard deviations, covariance and mean-square error. The experimental work was carried out using 352 images compressed by JPEG/JPEG2000. The resulting correlation is about 0.9744 between the predicted and actual MOS values. Similar work has been conducted in [9] where a neural network approach is used to predict the subjective image quality score DMOS using statistical features extracted from both the reference and target images. In 2011, Li et al. developed a no-reference image quality assessment using regression neural networks to approximate the functional relationship between a range of distortion types and the human subjective judgment [15].

In [27], the authors developed an image quality assessment method based on structural distortion and image definition. They carried out their experiments on Lena and Barbara original and distorted images. In their work, it was shown that the proposed method is more consistent with human perception. In [12], the authors used characteristics of structural similarity index and artificial neural network for image quality assessment. The experimental results showed that their proposed approach can achieve adaptability for image quality of different types. In [16], the authors conducted a survey on perceptual visual quality metrics, in which they compared 6 image metrics using seven public image databases. In [26], a new full-reference quality assessment metric is proposed to automate the quality assessment of an image in the discrete orthogonal moment domain. This metric was constructed by using image spatial information in terms of low-order moments.

Concerning the recent use of ANFIS approach in the literature, in [1] the authors worked on classifying greenery and non-greenery image classification using an ANFIS technique. They used a hybrid set of parameters which involved texture and color coherence vector (CCV). More recently in [17], ANFIS was used for classification and detection purposes for the brain Magnetic Resonance (MR) images and tumor detection. The decision making was performed in two stages. The first stage involved using feature extraction using principal component analysis (PCA) and in the second stage, ANFIS was trained. The authors mentioned that ANFIS, as a fuzzy logic based paradigm, grasps the learning abilities of neural network to improve the performance of the intelligent system using a priori knowledge. The authors demonstrated that ANFIS can be a promising approach for image classification in the field of medical sciences. In [5], ANFIS is used to assess quality of distorted/decompressed images without reference to the original image using three statistical features as inputs expressed as linguistic variables, namely area, extent and eccentricity.

4   Proposed Approach for Quality Assessment

In this paper, we have developed a full-reference image quality assessment model. The outline of the proposed predictive model is shown in Fig. 2. As a full-reference method, the quality of a query image is compared with a reference image of perfect quality. Image quality is determined through various image quality metrics computed based on features extracted from both the reference and the target images. These features are based on existing studies, e.g. [3,16]. Here, we considered only seven significant full-reference objective quality metrics as follows:

- Peak Signal-to-Noise Ratio (PSNR)
- Universal Quality Index (UQI)
- Mean Structural Similarity Index (MSSIM)
- Weighted Signal to Noise Ratio (WSNR)
- Visual Information Fidelity (VIF)
- Noise Quality Measure (NQM)
- Information Fidelity Criterion(IFC)

These quality metrics are discussed briefly in the following subsections.

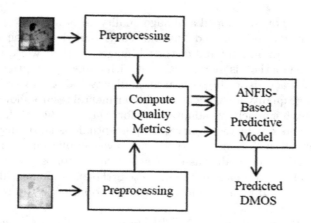

Fig. 2. Outline of the proposed ANFIS-based quality assessment system.

4.1   Peak Signal-to-Noise Ratio (PSNR)

The traditional and most widely-used objective image quality metric for many years is the peak signal to noise ratio (PSNR). PSNR is a pixel-based method, which means that the distorted image is compared to the reference image pixel by pixel. Perhaps, it wide spread use can be attributed to its simplicity and power to assess white noise distortion. However, it has been lately demonstrated that it can be inconsistent to human subjective perception [23]. Moreover, it may not capture the wide spectrum of distortion types.

The peak-signal-to-noise ratio is computed by utilizing the mean square error (MSE) between the reference image and the target image. For a reference image A and a target image B each of size $N \times M$, the mean square error is computed by averaging the squared intensity differences of the pixels of the two images as follows:

$$MSE = \frac{1}{NM} \sum_{i=1}^{N} \sum_{j=1}^{M} (a_{ij} - b_{ij})^2 \qquad (6)$$

where a_{ij} and b_{ij} are the intensities of the pixels at location (i, j) in the reference and target images, respectively. If we assume 8-bit encoding for each pixel, i.e. intensity values represent gray levels in the range from 0 to 255. The maximum gray level of 255 is then used as a scaling factor in computing the $PSNR$ which is defined as:

$$PSNR = 10 \times log(\frac{255^2}{MSE}) \text{ dB} \qquad (7)$$

4.2   Universal Quality Index (UQI)

This metric was suggested by Wang et al. [24,25] by utilizing first and second order statistics of both the reference and the target images. It is based on luminance, contrast, and structural comparisons of both images. The luminance comparison $l(a, b)$ between a reference image A and a target image B is determined in terms of mean values μ_a and μ_b by the relation:

$$l(a, b) = \frac{2\mu_a\mu_b}{\mu_a^2 + \mu_b^2} \tag{8}$$

The contrast comparison $c(a, b)$ is performed utilizing the standard deviations for images A and B as:

$$c(a, b) = \frac{2\sigma_a\sigma_b}{\sigma_a^2 + \sigma_b^2} \tag{9}$$

Utilizing covariance between the images A and B, the structural comparison $s(a, b)$ is given by:

$$s(a, b) = \frac{2\sigma_{ab}}{\sigma_a\sigma_b} \tag{10}$$

Subsequently, the universal quality index is defined as:

$$UQI(a, b) = l(a, b)c(a, b)s(a, b) \tag{11}$$

where the value of $UQI \in [-1, 1]$. It serves as an improved metric when compared to the $PSNR$. However, when the denominator is too small, UQI can become unstable and badly correlate with the subjective evaluations.

4.3 Mean Structural Similarity Index (MSSIM)

An improved metric was later proposed known as structural similarity index ($SSIM$) [25], which is computed in a similar way but adding some constants to overcome the instability of UQI as follows:

$$SSIM(a, b) = \frac{(2\mu_a\mu_b + C_1)(2\sigma_{ab} + C_2)}{(\mu_a^2 + \mu_b^2 + C_1)(\sigma_a^2 + \sigma_b^2 + C_2)} \tag{12}$$

where $C_1 = (K_1L)^2$, $C_2 = (K_2L)^2$, L denotes the dynamic range of pixel values (255 in our case), and K_1 and K_2 are small positive constants. The $SSIM$ index is calculated for the whole image as one block and its value is scaled to the range $[0, 1]$. When features are highly spatially non-stationary, $SSIM$ can be calculated within local windows and the overall image quality is measured by the mean $SSIM$ index as given by:

$$MSSIM = \frac{1}{K} \sum_i \sum_j SSIM(i, j) \tag{13}$$

where K is the total number of local $SSIM$ indices.

4.4 Weighted Signal-to-Noise Ratio (WSNR)

In [4], a different approach to signal-to-noise ratio was used. It is known as weighted signal-to-noise ratio ($WSNR$). This measure is defined as the ratio of average weighted signal power to the average weighted noise power. Here, the contrast sensitivity functions (CSF) are used as weights.

4.5   Visual Information Fidelity (VIF)

VIF metric was proposed in [20]. In this metric the image quality assessment depends upon the amount of information shared between the source (reference) image and the distorted image. A fundamental limit is imposed on how much information can flow from the source image through the channel (i.e., the image distortion process) to the receiver (i.e., human being). VIF is distinctive over traditional image quality assessment methods.

4.6   Noise Quality Measure (NQM)

NQM metric [4] was proposed as a better measure for visual quality than $PSNR$. It considers variation in contrast sensitivity with distance, image dimensions and spatial frequency. It also considers the variation in local luminance, mean and contrast interaction between spatial frequencies, and masking effects. NQM is given by:

$$NQM_{dB} = 10 \times log(\frac{\sum_i \sum_j a_{ij}^2}{\sum_i \sum_j (a_{ij}^2 - b_{ij}^2)^2}) \tag{14}$$

where a_{ij} and b_{ij} denote the (i,j) pixels in the reference and distorted images.

4.7   Information Fidelity Criterion(IFC)

IFC image quality assessment was proposed by [21]. This metric is based on natural scene statistics. The IFC is the mutual information between the source and distorted images. Firstly, the mutual information is derived for one sub-band and then generalized for multiple sub-bands. The IFC quantifies the perceptual quality of the image.

5   Evaluation

The images used in our study are collected from the a recent database released from the Oklahoma State University, Computational Perception and Image Quality Lab. This database is known as Categorical Subjective Image Quality (CSIQ) database [14]. This image database is chosen for our experiments because it has a relatively large number of images distorted with a variety of types. In addition, it was previously used in several image quality assessments in the literature, e.g. [14,28,29].

The adopted dataset has 30 original images and 750 distorted versions of the original images. We chose 5 types of distortions each is taken at five levels (This means there is a total of $5 \times 30 = 150$ images for each distortion type). The considered distortions are JPEG compression, JPEG 2000 compression, additive pink Gaussian noise (APGN), additive white Gaussian noise (AWGN) and Gaussian blurring. Each image in the database is of 512×512 RGB pixels and

Fig. 3. Examples of the images in CSIQ database for two types of distortions JPEG 2000 and Gaussian Blur.

each color has 256 levels (from 0 to 255); a total of 24 bits per pixel. Examples of the images in this database are shown in Fig. 3.

Each distorted image in the database has a subjective rating in the form of DMOS (Difference Mean Opinion Score) ranging from 0 (no distortion or lightly distorted) to 1 (highly distorted). Ratings are conducted by 35 male and female observers with ages from 21 to 35 years. The actual DMOS score for each image pair is also taken from the Oklahoma State University CSIQ image database website. Figure 4 shows distorted images with top ten and bottom ten DMOS ratings including distortion name and index, image name, distortion level, standard deviation of DMOS, and DMOS. It is clear that rating is high when the level of distortion is high and vice versa.

We paired each distorted image with the corresponding original image as a reference. This gave us 750 pairs. Out of the 750 image pairs, we used 600 pairs for training the model, 50 pairs for validating the model and 100 pairs for testing the model. Using MATLAB, we computed the seven image quality measures under consideration (see Sect. 4) using the code developed by their inventors.

dst_idx	dst_type	image	dst_lev	dmos_std	dmos
1	noise	log_seaside	1	0.000	0.000
2	jpeg	boston	1	0.000	0.000
2	jpeg	log_seaside	1	0.000	0.000
3	jpeg 2000	log_seaside	1	0.000	0.000
3	jpeg 2000	boston	1	0.000	0.000
3	jpeg 2000	turtle	1	0.003	0.002
3	jpeg 2000	couple	1	0.000	0.002
2	jpeg	couple	1	0.000	0.002
1	noise	roping	1	0.004	0.003
3	jpeg 2000	shroom	1	0.005	0.004
3	jpeg 2000	aerial_city	5	0.000	0.956
5	blur	sunsetcolor	4	0.023	0.963
5	blur	swarm	5	0.031	0.963
5	blur	lady_liberty	5	0.000	0.966
3	jpeg 2000	swarm	5	0.032	0.967
3	jpeg 2000	fisher	5	0.000	0.972
3	jpeg 2000	family	5	0.000	0.977
3	jpeg 2000	turtle	5	0.001	0.984
5	blur	sunsetcolor	5	0.012	0.999
3	jpeg 2000	sunsetcolor	5	0.021	1.000

Fig. 4. Distorted images with top ten and bottom ten DMOS ratings.

We then built different ANFIS models using subsets of these measures and evaluated their performances. The desired output of the ANFIS network was the crisp DMOS values. The first ANFIS model has only three inputs ($PSNR$, UQI and $MSSIM$) whereas the second ANFIS model has five inputs ($PSNR$, UQI, $MSSIM$, $WSNR$ and VIF). The last ANFIS model has seven inputs ($PSNR$, UQI, $MSSIM$, $WSNR$, VIF, NQM, and IFC). Table 1 shows the ANFIS parameters and their values used for training with 7 input variables. Figure 5 shows a snapshot of the corresponding ANFIS model for the 7 input variables. The other two models use similar parameter types but the values for input and output MFs differs accordingly.

For the purpose of evaluating the performance of each model, we used four measures. The Pearson's linear correlation coefficient ρ is given by:

$$\rho = \frac{Cov(DMOS^a, DMOS^p)}{Var(DMOS^a)Var(DMOS^p)} \qquad (15)$$

where $DMOS^a$ and $DMOS^p$ are vectors containing the actual and predicted values for DMOS. To assess the monotonicity relationship between predicted value and actual value for a particular model, we used Spearman's rank order coefficient ρ_s. This measure is computed suing the same equation for Pearson's coefficient but replacing the raw scores by their ranks. In order to find the error of the model, we used Mean Absolute Error (MAE) and Root Mean Square Error ($RMSE$) which as calculated as follows:

$$MAE = \frac{1}{N} \sum_{i=1}^{N} |DMOS_i^a - DMOS_i^p| \qquad (16)$$

Table 1. ANFIS parameters and their values that are used for training with 7 input variables.

Parameter	Value
Number of training data records	600
Number of validation data records	50
Number of testing data records	100
AND method	Product
OR method	Probabilistic OR (probor)
Implication method	Product
Defuzzification method	Weighted average (wtaver)
Aggregation method	Sum
Output MF function	Linear
Input MFs type	Generalized bell MF (gbellmf)
Number of inputs	7
Number of outputs	1
Number of MFs per variable	2
Number of rules	128

Table 2. Results for a 3-input ANFIS model (using PSNR, UQI and MSSIM).

Distortion	ρ	ρ_s	MAE	$RMSE$
Blur	0.9641	0.9665	5.4136	7.8199
JPEG2000	0.9887	0.9837	3.9443	4.9776
JPEG	0.9477	0.9519	7.7900	10.3173
APGN	0.2413	0.6019	25.7500	60.2763
AWGN	0.9510	0.9562	11.4998	15.3729

$$RMSE = \sqrt{\frac{1}{N} \sum_{i=1}^{N} (DMOS_i^a - DMOS_i^p)^2} \qquad (17)$$

where $DMOS_i^a$ and $DMOS_i^p$ are the actual and predicted values for DMOS for the i-th image.

We started with the three features metrics $PSNR$, UQI and $MSSIM$, selected arbitrarily as inputs to the ANFIS network. The results of our experiment are given in Table 2. In order to study the performance as more features become available, we added two more feature metrics, i.e. $WSNR$ and VIF, and repeated the experiment with a 5-input ANFIS network. The corresponding results are shown in Table 3. We again added two more feature metrics, i.e. NQM and IFC, and repeated the experiment with a 7-input ANFIS network

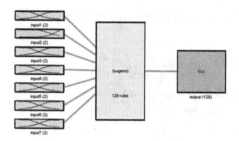

Fig. 5. ANFIS model with 7 inputs.

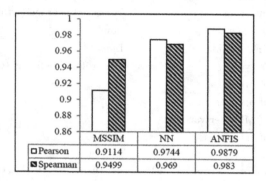

	MSSIM	NN	ANFIS
▢ Pearson	0.9114	0.9744	0.9879
▨ Spearman	0.9499	0.969	0.983

Fig. 6. Average Pearson's and Spearman's correlation results for JPEG/JPEG 2000 distortion types for the proposed ANFIS-based with 7 inputs as compared to neural network (NN) and MSSIM, as reported in [2].

and the yielded results are shown in Table 4. The rationale behind repeating the experiments was to judge the performance of the ANFIS network by increasing the feature metrics incrementally and document the results.

Considering the results in Tables 1, 2 and 3, we can see that the predicted DMOS values are highly correlated with the actual DMOS values for all distortion types except APGN. The correlation improves as more inputs become available. Similar conclusions can be made regarding MAE and $RMSE$.

For the sake of comparison, Fig. 6 shows the average values for the correlation of two types of distortion JPEG/JPEG 2000 for our method and two other methods from the literature: neural network [2] and MSSIM [25]; as reported in [2]. We should mention that the authors for the other works used a different image database and provided the results for only these two types of distortions. For a more relevant comparison on the same dataset (CSIQ), Figure 7 compares the Spearman's correlation resulting from the proposed approach to that individual metrics on the four types of distortions reported in [13]. This figures demonstrates that the proposed approach has consistently higher correlation.

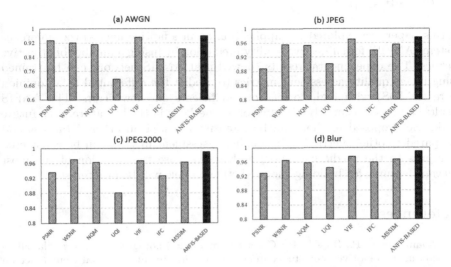

Fig. 7. Comparing the Spearman's correlation results for four distortion types on the same dataset (CSIQ) of individual metrics, as reported in [13] and the proposed ANFIS-Based model with 7 inputs.

Table 3. Results for a 5-input ANFIS model (using PSNR, UQI, MSSIM, WSNR and VIF).

Distortion	ρ	ρ_s	MAE	$RMSE$
Blur	0.9862	0.9811	3.3941	4.8836
JPEG2000	0.9938	0.9881	2.6951	3.5981
JPEG	0.9762	0.9693	6.1521	7.8826
APGN	0.6181	0.7143	94.3118	98.5724
AWGN	0.9646	0.9636	16.7365	22.4688

Table 4. Results for a 7-input ANFIS model (using PSNR, UQI, MSSIM, WSNR, VIF, NQM, and IFC).

Distortion	ρ	ρ_s	MAE	$RMSE$
Blur	0.9937	0.9902	2.2626	3.2177
JPEG2000	0.9944	0.9902	2.4395	3.3292
JPEG	0.9814	0.9758	5.7428	7.1629
APGN	0.7035	0.7338	96.1683	99.8975
AWGN	0.9548	0.9590	16.8990	22.7204

6 Conclusion

In this paper, we explored the application of an adaptive neuro-fuzzy inference system (ANFIS) for subjective quality prediction by fusing a number of objective metrics. The experimental results showed that ANFIS network can be trained using image quality assessment metrics to predict the differential mean opinion score (DMOS) with high correlation coefficients and low errors. The ANFIS results compare favourably with two other methods in the literature. As a future work, the proposed method can be intensively tested on other databases and compared to other metrics. More quality assessment metrics can be considered as inputs to the predictive model and in this case the selection of the most relevant features for building the predictive models will be of interest.

References

1. Balamurugan, P., Rajesh, R.: Greenery image and non-greenery image classification using adaptive neuro-fuzzy inference system. In: International Conference on Computational Intelligence and Multimedia Applications, 2007, vol. 3, pp. 431–435 (2007)
2. Bouzerdoum, A., Havstad, A., Beghdadi, A.: Image quality assessment using a neural network approach. In: Proceedings of the Fourth IEEE International Symposium on Signal Processing and Information Technology, pp. 330–333 (2004)
3. Chetouani, A., Beghdadi, A., Deriche, M.: Image distortion analysis and classification scheme using a neural approach. In: 2nd European Workshop on Visual Information Processing (EUVIP) 2010, pp. 183–186 (2010)
4. Damera-Venkata, N., Kite, T.D., Geisler, W.S., Evans, B.L., Bovik, A.C.: Image quality assessment based on a degradation model. IEEE Trans, Image Process. 9(4), 636–650 (2000)
5. De, I., Sil, J.: No-reference quality prediction of distorted/decompressed images using ANFIS. In: International Conference on Computer Technology and Development, ICCTD 2009, vol. 2, pp. 90–94 (2009)
6. El-Alfy, E.S., Riaz, M.: Image quality assessment using ANFIS approach. In: Proceedings of the 6th International Conference on Agents and Artificial Intelligence (ICAART), vol. 1, pp. 169–177 (2014)
7. He, L., Gao, F., Hou, W., Hao, L.: Objective image quality assessment: a survey. Int. J. Comput. Math. 91(11), 1–15 (2013)
8. Jang, J.S.: ANFIS: adaptive-network-based fuzzy inference system. IEEE Trans. Syst. Man Cybern. 23(3), 665–685 (1993)
9. Kaya, S., Milanova, M., Talburt, J., Tsou, B., Altynova, M.: Subjective image quality prediction based on neural network. In: Proceedings of the 16th International Conference on Information Quality (2011)
10. Khuntia, S.R., Panda, S.: ANFIS approach for SSSC controller design for the improvement of transient stability performance. Math. Comput. Model. 57(12), 289–300 (2013)
11. Kudelka Jr., M.: Image quality assessment. In: Proceedings of Contributed Papers, WDS 2012 Part I. pp. 94–99 (2012)

12. Kung, C.H., Yang, W.S., Huang, C.Y., Kung, C.M.: Investigation of the image quality assessment using neural networks and structure similarity. In: Proceedings of the 3rd International Symposium Computer Science and Computational Technology (2010)
13. Lahoulou, A., Bouridane, A., Viennet, E., Haddadi, M.: Full-reference image quality metrics performance evaluation over image quality databases. Arab. J. Sci. Eng. **38**(9), 2327–2356 (2013)
14. Larson, E.C., Chandler, D.M.: Most apparent distortion: full-reference image quality assessment and the role of strategy. J. Electron. Imaging **19**(1), 011006–011006 (2010)
15. Li, C., Bovik, A.C., Wu, X.: Blind image quality assessment using a general regression neural network. IEEE Trans. Neural Networks **22**(5), 793–799 (2011)
16. Lin, W.: Jay Kuo, C.C.: Perceptual visual quality metrics: a survey. J. Vis. Commun. Image Represent. **22**(4), 297–312 (2011)
17. Sri Meena, R., Revathi, P., Reshma Begum, H.M., Singh, A.B.: Performance analysis of neural network and ANFIS in brain MR image classification. In: Patnaik, S., Yang, Y.-M. (eds.) Soft Computing Techniques in Vision Sci. SCI, vol. 395, pp. 101–113. Springer, Heidelberg (2012)
18. Meharrar, A., Tioursi, M., Hatti, M., Stambouli, A.B.: A variable speed wind generator maximum power tracking based on adaptive neuro-fuzzy inference system. Expert Syst. Appl. **38**(6), 7659–7664 (2011)
19. Rehman, A., Wang, Z.: Reduced-reference image quality assessment by structural similarity estimation. IEEE Trans. Image Process. **21**(8), 3378–3389 (2012)
20. Sheikh, H.R., Bovik, A.C.: Image information and visual quality. IEEE Trans. Image Process. **15**(2), 430–444 (2006)
21. Sheikh, H.R., Bovik, A.C., De Veciana, G.: An information fidelity criterion for image quality assessment using natural scene statistics. IEEE Trans. Image Process. **14**(12), 2117–2128 (2005)
22. Sheikh, H.R., Sabir, M.F., Bovik, A.C.: A statistical evaluation of recent full reference image quality assessment algorithms. IEEE Trans. Image Process. **15**(11), 3440–3451 (2006)
23. Wang, Z., Bovik, A.: Mean squared error: love it or leave it? a new look at signal fidelity measures. IEEE Signal Process. Mag. **26**(1), 98–117 (2009)
24. Wang, Z., Bovik, A.C.: A universal image quality index. IEEE Signal Process. Lett. **9**(3), 81–84 (2002)
25. Wang, Z., Bovik, A.C., Sheikh, H.R., Simoncelli, E.P.: Image quality assessment: from error visibility to structural similarity. IEEE Trans. Image Process. **13**(4), 600–612 (2004)
26. Wee, C.Y., Paramesran, R., Mukundan, R., Jiang, X.: Image quality assessment by discrete orthogonal moments. Pattern Recogn. **43**(12), 4055–4068 (2010)
27. Yi, Y., Yu, X., Wang, L., Yang, Z.: Image quality assessment based on structural distortion and image definition. In: Proceedings of the International Conference on Computer Science and Software Engineering, **6**, 253–256 (2008)
28. Zhang, F., Ma, L., Li, S., Ngan, K.N.: Practical image quality metric applied to image coding. IEEE Trans. Multimedia **13**(4), 615–624 (2011)
29. Zhu, X., Milanfar, P.: Automatic parameter selection for denoising algorithms using a no-reference measure of image content. IEEE Trans. Image Process. **19**(12), 3116–3132 (2010)

A Heuristic Automatic Clustering Method
Based on Hierarchical Clustering

François LaPlante[1]($\boxtimes$), Nabil Belacel[2], and Mustapha Kardouchi[1]

[1] Université de Moncton, Moncton, NB E1A 3E9, Canada
`francois.laplante@nrc-cnrc.gc.ca, mustapha.kardouchi@umoncton.ca`
[2] National Research Council - Information and Communications Technologies,
Moncton, NB E1A 7R9, Canada
`nabil.belacel@nrc-cnrc.gc.ca`

Abstract. We propose a clustering method which produces valid results while automatically determining an optimal number of clusters. The proposed method achieves these results with minimal user input, of which none pertains to a number of clusters. Our method's effectiveness in clustering, including its ability to produce valid results on data sets presenting nested or interlocking shapes, is demonstrated and compared with cluster validity analysis to other methods to which a known optimal number of clusters is provided, and to other automatic clustering methods. Depending on the particularities of the data set used, our method has produced results which are roughly equivalent or better than those of the compared methods.

Keywords: Data-mining · Automatic clustering · Unsupervised learning

1 Introduction

Data clustering, also known as cluster analysis, segmentation analysis, taxonomy analysis [1], is a form of unsupervised classification of data points into groups called clusters. Data points in a same cluster should be as similar to each other as possible and data points in different clusters should be as dissimilar as possible [2].

One common problem across many clustering methods is determining the correct (optimal) number of clusters. One prevalent method to determine an optimal number of clusters involves the use of validity indices. Cluster validity indices are a value computed based on a clustering result and represent a relative quality of this clustering. Often, a clustering method will be applied to the target data set a number of times with a different number of clusters and a validity index will be computed for each resulting clustering. The result which leads to the best index value will be taken as being the most optimal. Given n the number of data points, the number of clusters to try can be a sequence (often from 2 to $\sqrt{n}$), all possible values (1 to n), or a selection of specific values or ranges based on prior knowledge of the data set.

© Springer International Publishing Switzerland 2015
B. Duval et al. (Eds.): ICAART 2014; LNAI 8946, pp. 312–328, 2015.
DOI: 10.1007/978-3-319-25210-0_19

Even with the use of cluster validity indices, it is still required to cluster the data many times and compare the results to determine the optimal clustering. There is a group of clustering algorithms, called automatic clustering algorithms, which determine an optimal number of clusters automatically. These methods, although generally more complex and time consuming, do not need to be run more than once. Some of these algorithms, such as Y-means [3], still require an initial number of clusters from which to start. Others, such as the method proposed by Mok et. al. [4], hereafter referred to as RAC, requires no user input at all regarding the number of clusters. Our goal is to develop an automatic clustering algorithm which requires minimal user input and more specifically does not require to be provided a target number of clusters or an initial number of clusters from which to start.

2 Related Works

2.1 Types of Clustering

Clustering methods can be categorized in many ways such as hard or fuzzy, hierarchical or partitional, and as combinations of these types.

Hard vs. Fuzzy Clustering. Hard clustering, also called crisp clustering, is a type of clustering where every datum belongs to one and only one cluster. In contrast, fuzzy clustering is a form of clustering where data belong to multiple clusters according to a membership function [1]. Hard clustering is generally simpler to implement and has lower time complexity. Hard clustering performs well with linearly separable data but often does not perform very well with non linearly separable data, outliers, or noise. Fuzzy clustering often has a larger memory footprint as it often requires a $c \times n$ matrix to store memberships, where c is the number of clusters and n is the number of data points. Fuzzy clustering is able to handle non-linearly separable data as well as outliers, and noise better than hard clustering.

Hierarchical vs. Partitional Clustering. A hierarchical clustering method yields a dendrogram representing the nested grouping of patterns and similarity levels at which groupings change [2]. A partitional clustering method yields a single partition of the data instead of a clustering structure, such as the dendrogram produced by a hierarchical method [1].

Automatic Clustering. Automatic clustering is a form of clustering where the number of clusters c is unknown and determining its optimal value is left up to the clustering method. Some automatic clustering methods may require an initial number of clusters, from which clusters will be split and merged until a pseudo-optimal number of clusters is achieved. Other methods require no initial value or additional information regarding the number of clusters and will determine a pseudo-optimal value without any user input. Other parameters, such as a fuzzy constant (for fuzzy clustering algorithms) or thresholds, may still be required, but are generally kept to minimum or are optional with good default values.

2.2 Validation Methods

As clustering is by definition an unsupervised method, there is generally no training data with known output values with which to compare results. As such, it requires a different approach to evaluating its results. The quality of clustering is evaluated using a validity index, which is a relative measure of clustering quality based on a number of parameters. There are many clustering validity indices, but the approach to using them generally remains the same and is as follows:

1. Use fixed values for all parameters other than c the number of clusters.
2. Iteratively cluster the data set with the clustering method being evaluated with varying values of c (often from 2 to $\sqrt{2}$).
3. Calculate the validity index for every clustering generated by 2.
4. The clustering for which the validity index presents the best value is considered to be "optimal".

A good index must consider compactness (high intra-cluster density), separation (high inter-cluster distance or dissimilarity) and the geometric structure of data [5].

Xie and Beni Index. Xie and Beni have proposed a validity index which relies on two properties, compactness and separation [6], which was later modified by Pal and Bezdek [7]. This index is defined by

$$V_{XB} = \frac{\sum_{i=1}^{c} \sum_{k=1}^{n} u_{ik}^m \|x_k - v_i\|^2}{n(\min_{i,j \in c, i \neq j} \{v_i - v_j\})} \tag{1}$$

where u is a $n \times c$ matrix such that u_{ik} is the membership of object k to cluster i, m is a fuzzy constant, x_k are data points and v_i are clusters (represented by their centroids).

The numerator of the equation, which is equivalent to the least squared error, is an indicator of compactness of the fuzzy partition, while the denominator is an indicator of the strength of the separation between the clusters. A more optimal partition should produce a smaller value for the compactness and well separated clusters should produce a higher value for the separation. An optimal number of clusters c is generally found by solving $min_{2 \leq c \leq n-1} V_{XB}(c)$.

Fukuyama and Sugeno Index. Fukuyama and Sugeno also proposed a validity index based on compactness and separation [8] defined by:

$$V_{FS} = J_m - K_m$$

$$= \sum_{i=1}^{c} \sum_{k=1}^{n} u_{ik}^m \|x_k - v_i\|^2 - \sum_{i=1}^{c} \sum_{k=1}^{n} u_{ik}^m \|v_i - \bar{v}\|^2 \tag{2}$$

where J_m represents a measure of compactness, K_m represents a measure of separation between clusters and $\bar{v}$ is the mean of all cluster centroids. An optimal number of clusters c is generally found by solving $min_{2 \leq c \leq n-1} V_{FS}(c)$.

Kwon Index. Kwon extends the index of Xie and Beni's validity function to eliminate its tendency to monotonically decrease when the number of clusters approaches the number of data points. To achieve this, a penalty function was introduced to the numerator of Xie and Beni's original validity index. The resulting index was defined as

$$V_K = \frac{\sum_{j=1}^{n} \sum_{i=1}^{c} u_{ij}^m \|x_j - v_i\|^2 + \frac{1}{c} \sum_{i=1}^{c} \|v_i - \bar{v}\|^2}{min_{i,k \in c, i \neq k} \|v_i - v_k\|^2} \tag{3}$$

An optimal number of clusters c is generally found by solving $min_{2 \leq c \leq n-1} V_K(c)$.

PBM Index. Pakhira and Bandyopadhyay [9] proposed the PBM index, which was developed for both hard and fuzzy clustering. The hard clustering version of the PBM index is defined by

$$V_{PBM} = \left(\frac{1}{c} \cdot \frac{E_1}{E_c} \cdot D_c \right)^2 \tag{4}$$

where

$$E_c = \sum_{k=1}^{c} E_k \tag{5}$$

and

$$E_k = \sum_{j=1}^{n} \|x_j - v_k\| \tag{6}$$

with v_k being the centroid of the data set and

$$D_c = \max_{i,j \in c} \|v_i - v_j\| \tag{7}$$

An optimal number of clusters c is generally found by solving $max_{2 < c < n-1} V_{PBM}(c)$.

Compose Within and Between Scattering. The CWB index proposed by Rezaee [10] focusing on both the density of clusters and their separation. Although meant to evaluate fuzzy clustering results, it can be used to evaluate hard clustering by generating a partition matrix u such that memberships have values of 1 or 0 (is a member or is not a member).

Given a fuzzy c-partition of the data set $X = \{x_1, x_2, \ldots, x_n | x_i \in R^p\}$ with c cluster centers v_i, the variance of the pattern set X is called $\sigma(X) \in R^p$ with the value of the pth dimension defined as

$$\sigma_x^p = \frac{1}{n} \sum_{k=1}^{n} (x_k^p - \bar{x}^p)^2 \tag{8}$$

where $\bar{x}^p$ is the pth element of the mean of $\bar{X} = \sum_{k=1}^{n} x_k / n$.

The fuzzy variation of cluster i is called $\sigma(v_i) \in R^p$ with the pth value defined as

$$\sigma_{v_i}^p = \frac{1}{n} \sum_{k=1}^{n} u_{ik} (x_k^p - v_i^p)^n \tag{9}$$

The average scattering for c clusters is defined as

$$Scat(c) = \frac{\frac{1}{n} \sum_{i=1}^{c} \|\sigma(v_i)\|}{\|\sigma(X)\|} \tag{10}$$

where $\|x\| = (x^T \cdot x)^{1/2}$

A dissimilarity function $Dis(c)$ is defined as

$$Dis(c) = \frac{D_{max}}{D_{min}} \sum_{k=1}^{c} \left(\sum_{z=1}^{c} \|v_k - v_z\| \right)^{-1} \tag{11}$$

where $D_{max} = \max_{i,j \in \{2,3,\ldots,c\}} \{\|v_i - v_j\|\}$ is the maximum dissimilarity between the cluster prototypes. The D_{min} has the same definition as D_{max}, but for the minimum dissimilarity between the cluster prototypes.

The compose within and between scattering index is now defined by combining the last two equations:

$$V_{CWB} = \alpha Scat(c) + Dis(c) \tag{12}$$

where α is a weighting factor.

An optimal number of clusters c is generally found by solving $min_{2<c<n-1} V_{CWB}(c)$.

Silhouettes Index. Rousseeuw introduced the concept of silhouettes [11] which represent how well data lie within their clusters. The silhouette value of a datum is defined by

$$S(i) = \begin{cases} 1 - a(i)/b(i), & a(i) < b(i) \\ 0, & a(i) = b(i) \\ b(i)/a(i) - 1, & a(i) > b(i) \end{cases} \tag{13}$$

which can also be written as

$$S(i) = \frac{b(i) - a(i)}{max\{a(i), b(i)\}} \tag{14}$$

where $a(i)$ is the average dissimilarity between a point i and all other points in its cluster and $b(i)$ is the average dissimilarity between a point i and all points of the nearest cluster to which point i is no assigned. The silhouette index for a given cluster is the average silhouette for all points within that cluster and the silhouette index of a clustering is the average of all silhouettes in the data set:

$$V_S = \sum_{i=1}^{n} S(i)/n. \tag{15}$$

An optimal number of clusters c is generally found by solving $max_{2 \leq c \leq n-1} V_S(c)$.

3   Proposed Method

The proposed method, Heuristic Divisive Analysis (HDA), consists of two phases: splitting and merging. The first phase splits the data set into a number of clusters, often leading to more cluster than optimal. The second phase merges (or links) clusters, leading to a more optimal clustering. The reason for this two-step approach is to address one of the larger drawbacks of hard clustering; poor performance when dealing with data which is not linearly separable. Both steps use different approaches to computing the dissimilarity between clusters, which allows for the creation of non-elliptical clusters which may be nested or interlocked.

3.1   Splitting

The splitting algorithm is a divisive hierarchical method based on the DIANA clustering algorithm [12]. However, the proposed method employs a heuristic function to interrupt the hierarchical division of the data set once an "adequate" clustering for this step has been reached.

DIANA. DIANA (DIvisive ANAlysis) is a divisive hierarchical clustering algorithm based on the idea of MacNaughton-Smith et al. [13]. Given $X = x_1, x_2, \ldots, x_n$ a data set consisting of n records and beginning with all points being in one cluster, the algorithm will alternate between separating the cluster in two and selecting the next cluster to split until every point has become its own cluster. To split a cluster in two, the algorithm must first find the point with the greatest average dissimilarity to the rest of the records. The average dissimilarity of a record x_i with regards to X is defined as

$$D_i = \frac{1}{n-1} \sum_{j=1, i=1}^{n} D(x_i, x_j) \tag{16}$$

where $D(x, y)$ is a dissimilarity metric (in this case, we use Euclidean distance). Given $D_{max} = max_{0 \leq i \leq n-1} D_i$, x_{max} is the point with the greatest average dissimilarity which is then split from the cluster. We then have two clusters: $C_1 = \{x_{max}\}$ and $C_2 = X \backslash C_1$. Next, the algorithm checks every point in C_2 to determine whether or not it should be moved to C_1. To accomplish this, the algorithm must compute the dissimilarity between x and C_1 as well as the dissimilarity between x and $C_2 \backslash x$. The dissimilarity between x and C_1 is defined as

$$D_{C_1}(x) = \frac{1}{|C_1|} \sum_{y \in C_1} D(x, y), x \in C_1 \tag{17}$$

where $|C_1|$ denotes the number of records in C_1. The dissimilarity between x and $C_2 \backslash x$ is defined as

$$D_{C_2}(x) = \frac{1}{|C_2 - 1|} \sum_{y \in C_2, y \neq x} D(x, y), x \in C_2 \tag{18}$$

If $D_{C_1} < D_{C_2}$, then x is moved from C_2 to C_1. This process is repeated until there are no more records in C_2 which should be moved to C_1.

To select the next cluster to separate, the algorithm will chose the cluster with the greatest diameter. The diameter of a cluster is defined as

$$Diam(C) = \max_{x, y \in C} D(x, y) \tag{19}$$

Heuristic Stopping Function. The first phase in our method consists of running the DIANA algorithm with a heuristic function in order to stop it once an "adequate" clustering has been reached. This function consists of first calculating the average intra-cluster dissimilarity (again, we use Euclidean distance) of each cluster, defined as

$$AvgIntraCluster\ Distance(C) = \frac{\sum_{x \in C} D(x, \bar{x})}{|C|} \tag{20}$$

where $\bar{x}$ denotes the mean of all points in cluster C. The heuristic index for this clustering is the average of all the average intra-cluster dissimilarities. If the heuristic index for this clustering is lower than that of the previous clustering, the current clustering is considered the most optimal to date. Otherwise, we have reached our "adequate" clustering at the previous step, but we will continue running the DIANA algorithm for a set number of iterations as a preventative measure against falling into a local optimum. We chose this rather simple heuristic instead of one of the many known validity indices because it allowed us to decrease the complexity (as it uses values which our implementation had already calculated) and still produced good results.

3.2   Merging

The splitting phase's result can be non-optimal. This is especially likely when data sets contain clusters which are not linearly separable or have irregular shapes. In these cases, the "adequate" clustering will usually contain instances where what should be one single cluster is divided into many. These many clusters will be very close to each other in relation to the other clusters and it is the goal of this merging phase to collect them into optimal clusters.

For each pair of clusters, we calculate the *average nearest neighbor* dissimilarity, defined as

$$AvgNearestNeighbor(C) = \frac{\sum\limits_{x \in C} \min\limits_{y \in C, y \neq x} D(x,y)}{|C|} \qquad (21)$$

for both clusters and keep the greater of both values as our merging dissimilarity threshold M_T. We then go through each pair of objects with one object from each cluster and if we find a pair where the dissimilarity between the two objects is less than the merging dissimilarity threshold (multiplied by a constant), then the two clusters are merged. We express the test for merging as

$$CanMerge(C_1, C_2) = \begin{cases} true, & \exists x \in C_1, \exists y \in C_2 | D(x,y) < M_T \cdot K \\ false, & otherwise \end{cases} \qquad (22)$$

where K is a merging constant.

As the *average nearest neighbor* dissimilarity of a single-object cluster is zero, a pair of single-object clusters will always have $M_T = 0$ and will never merge. This edge case is handled by defining the merge threshold for this particular merge as follows:

$$M_T = \frac{\sum\limits_{i=1}^{c} \sigma_{C_i}}{c} \qquad (23)$$

where σ_{C_i} is the standard deviation of the dissimilarity between the objects in a cluster C_i to the centroid of that cluster.

Once all merges are completed, we are left with the final clustering. The value of the merging constant can be adjusted depending on the data set and we have found experimentally that a value of 2 generally produces good results.

We have also tested an alternative merging method based on the Y-means approach to merging. Because the Y-means algorithm uses dissimilarities between cluster centroids, merging clusters will relocate the centroids in such a way that is detrimental to our method. To avoid this drawback, we link clusters by attributing them labels instead of merging them until all pairs are linked, after which we merge all linked clusters. We express the test for linking as

$$CanLink(C_1, C_2) = \begin{cases} true & D(C_1, C_2) \leq (\sigma_{C_1} \cdot \sigma_{C_2}) \cdot L \\ false & otherwise \end{cases} \qquad (24)$$

where L is a linking constant. The value of the linking constant can be adjusted depending on the data set and we have found that a value of 0.5 generally produces good results with our method.

4   Results

The proposed method was tested with five data sets. The results were compared to the Y-means, fuzzy c-means [14] and RAC algorithms using the Xie & Beni, Fukuyama & Sugeno, Kwon, CWB, PBM and Silhouette validation indices.

4.1   Data Sets

The first data set was the Iris data set [15], composed of 150 elements in four dimensions belonging to three categories of 50 elements each; however, two of the three categories of the data set are so close as to generally be clustered together.

The second data set, or "nested circles" data set, is composed of 600 elements in two dimensions belonging to two groups. The first group, of 100 elements, is a full circular shape in the center of the plane. The second group, of 500 elements, is a circular shell surrounding the first group. As the centroids of both clusters are approximately identical, it is difficult for clustering methods which use cluster centroids (such as Y-means and fuzzy c-means) to produce an appropriate clustering.

The third data set, or "nested crescents" data set, is composed of 500 elements in two dimensions belonging to two groups of 250 elements each. The two groups form opposing semi-circles which are offset and inset in such a way that one tip of each semi-circle is nested within the other semi-circle.

The fourth data set, or "five groups" data set, is composed of 1500 elements in two dimensions belonging to five groups of 300 elements each. Each group is a roughly circular with an approximately Gaussian distribution. The groups are spread in such a way as to have two pairs of tightly adjacent clusters.

The fifth data set or "Aggregation" data set is a testing data set proposed by Gionis et.al. [16]. This data set presents 7 roughly elliptical groupings, one of which has a concave indentation. Two pairs of these groups a linked by narrow lines of data points.

4.2   Clustering Results

We have compared our method to the Y-means algorithm, another hard automatic clustering method based on the well-known k-means algorithm. Y-means requires an initial number of clusters, as such we provided it with the known optimal number of clusters or the best approximations thereof.

We have also compared our method to the fuzzy c-means algorithm. Although this method belongs to the category of fuzzy clustering, we compared our method to it as our method should be able to correctly treat non-linearly separable data and comparison with a fuzzy method could prove interesting.

As well as the previous two methods, we have compared our method to the RAC method. This method makes use of the fuzzy c-means algorithm as well as graph partitioning concepts to arrive at a hard partition. This automatic clustering method should also be able to correctly treat non-linearly separable data but has a greater time complexity.

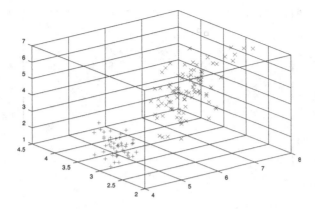

Fig. 1. Clustering result on Iris data set.

Of the validity indices used, Xie & Beni, Fukuyama & Sugeno, Kwon, and CWB should be minimized (lower values indicates a better clustering) while PBM and Silhouette should be maximized (higher value indicates a better clustering).

Iris Data Set. Figure 1 shows the result of clustering the Iris data set with our method. We can discern four clusters, two of which contain one and two members respectively. These two clusters are considered as outliers and the remaining two clusters then approximately correspond to the expected results.

Table 1 shows the validation results of our method and the compared methods for the Iris data set. The results for the proposed method (HDA) were calculated after removing all outliers. We notice that for the XB, Kwon, CWB, and PBM indices, although our method does not produce the best validation result, its results are very near the best. For the XB and Kwon indices, our method outperformed the other hard clustering methods. The small variations in results between our method and the others are partly due to the data points eliminated when removing outliers.

Table 1. Iris data set validation results.

	XB	FS	Kwon	CWB	PBM	Silhouette
KMP (c=2)	0.0654087	592.227	10.0613	0.503325	23.8917	0.690417
KMP (c=3)	1.55946	789.946	239.56	2.7801	14.4217	0.561084
FCM (c=2)	0.0544162	530.501	8.41243	0.503334	17.1528	0.685031
FCM (c=3)	0.137036	509.939	21.966	1.34779	14.8009	0.558518
RAC	0.0654087	592.227	10.0613	0.503325	23.8917	0.690417
HDA	0.061941	568.303	9.35532	0.508643	24.2696	0.697063

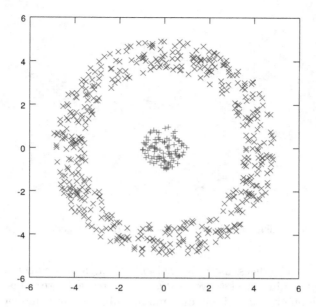

Fig. 2. Clustering result on nested circles data set.

Nested Circles Data Set. Figure 2 shows the result of clustering the nested circles data set with our method and Table 2 shows the validation results of our method and the compared methods for the nested circles data set.

We can observe that the two clusters are correctly identified. However, the validation indices for our method and RAC (which produced the same clustering) are all much worse than those of Y-means and fuzzy c-means which did not correctly identify the clusters (see Fig. 3). This is in part due to the fact that most of these indices use the centroids of clusters to compute dissimilarities, which is also at least in part the reason why Y-means and fuzzy c-means did not produce good results.

Nested Crescents Data Set. Figure 4 shows the result of clustering the nested crescents data set with our method. We can see that the two clusters are correctly identified.

Table 2. Nested circles data set validation results.

	XB	FS	Kwon	CWB	PBM	Silhouette
KMP (c=2)	0.45743	− 3367.37	274.708	0.422658	8.53994	0.340658
FCM (c=2)	0.318881	− 2173.72	191.578	0.427173	6.34011	0.338348
RAC	2600.64	− 0.908002	$1.56039e6$	25.7067	0.00151379	0.0477678
HDA	2600.64	− 0.908002	$1.56039e6$	25.7067	0.00151379	0.0477678

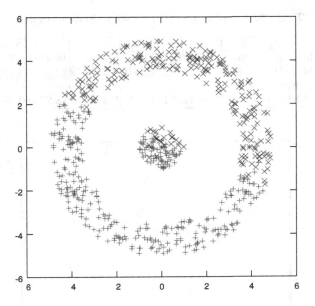

Fig. 3. Y-means result on nested circles data set.

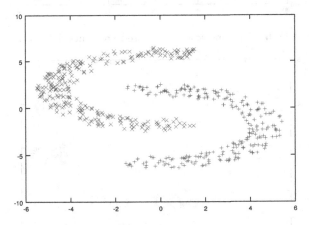

Fig. 4. Clustering result on nested crescents data set.

Table 3 shows the validation results of our method and the compared methods for the nested crescents data set. The RAC method has no values for this data set as it clustered the entire data set into a single cluster.

Again, Y-means and fuzzy c-means obtain better values with validity indices while producing inferior results (see Fig. 5).

5 Groups Data Set. Figure 6 shows the result of clustering the five groups data set with our method. We can observe seven clusters, two of which contain one

Table 3. Nested crescents data set validation results.

	XB	FS	Kwon	CWB	PBM	Silhouette
KMP (c=2)	0.318794	−5495.04	159.647	0.302265	19.4784	0.342838
FCM (c=2)	0.142008	−5212.3	71.2541	0.269979	22.4437	0.472577
RAC	−	−	−	−	−	−
HDA	0.304331	−5071.23	152.415	0.312159	18.2753	0.377258

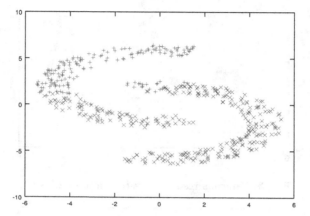

Fig. 5. Y-means result on nested crescents data set.

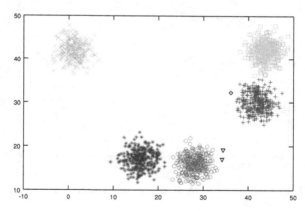

Fig. 6. Clustering result on five groups data set.

and two member points respectively. These two clusters are treated as outliers and the remaining five clusters then correspond to the expected result.

Table 4 shows the validation results of our method and the compared methods for the five groups data set. The results for the proposed method were calculated after removing all outliers. Similarly to the Iris data set, our method outperforms the other hard clustering methods in the XB and Kwon indices as well as in the

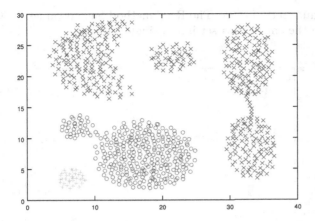

Fig. 7. Clustering result on Aggregation data set.

CWB index for this data set. Our method also produced the best values for the PBM and Silhouette indices.

Aggregation Data Set. Figure 7 shows the result of clustering the Aggregation data set with our method. We can observe that the three clusters produced are not ideal. The top 3 clusters have been grouped together yet should be separate. After adjusting the merging constant K from its default value of 2.0 to 0.8, we obtain the clustering seen in Fig. 8. This new clustering is better but still not perfect as the upper-left and upper-center clusters are still grouped together and some outliers are produced. Reducing K to 0.7 produced the clustering seen in Fig. 9. Reducing K further produced no improvement as the clusterings produced were under-merged and represented the data even more poorly.

Table 5 shows the validation results of our method and the compared methods for the Aggregation data set. The results for the proposed method were calculated after removing all outliers. With the exception of the FS index, our method performed best with a merging constant of 0.7. With these values, our method outperformed the Y-means method in all but the CWB index. For the other indices, our method performed similarly but slightly worse than fuzzy c-means with 5 clusters with the exception of the PBM index where our method

Table 4. Five groups data set validation results.

	XB	FS	Kwon	CWB	PBM	Silhouette
KMP (c=5)	7.64716	−540692	11494.1	0.608328	427.786	0.532795
FCM (c=5)	0.0506787	−523890	78.7023	0.155432	2101.97	0.730427
RAC	0.05887	−581968	91.0297	0.15704	3985.93	0.730427
HDA	0.0583695	−581594	90.1025	0.156968	4012.27	0.73118

performed significantly better. The RAC method has no values for this data set as it clustered the entire data set into a single cluster.

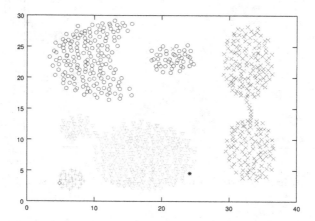

Fig. 8. Clustering result on Aggregation data set.

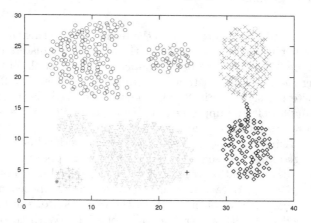

Fig. 9. Clustering result on Aggregation data set.

Table 5. Aggregation data set validation results.

	XB	FS	Kwon	CWB	PBM	Silhouette
KMP (c=5&7)	0.320621	−86011.4	253.039	0.100997	98.4674	0.272249
FCM (c=5)	0.185912	−78942.4	148.601	0.215287	117.716	0.500565
FCM (c=7)	0.26758	−73294.5	214.788	0.34196	66.6775	0.467089
RAC	—	—	—	—	—	—
HDA (K=2.0)	1.98674	−128690	1567.92	0.329051	142.117	0.241994
HDA (K=0.7)	0.489475	−121270	387.897	0.295478	302.46	0.468173
HDA (K=0.8)	0.723013	−138778	571.038	0.310896	284.03	0.455008

5 Conclusion

In this paper, an automatic clustering method based on a heuristic divisive approach has been proposed and implemented. The method is based on the DIANA algorithm interrupted by a heuristic stopping function. As this process alone generally produces too many clusters, its result is then passed on to a merging method. The advantage of this two phase approach being that with the splitting and merging using different criteria for determining if data belong in a same cluster, the merged clusters can take non elliptical shapes. This advantage sets our method apart from the majority of hard clustering methods in that it can handle data which is not linearly separable fairly well.

Five data sets have been used to evaluate the proposed clustering method. The proposed method was also compared against an automatic hard clustering method, a fuzzy clustering method (for which a known number of clusters was provided), and an automatic clustering method based on fuzzy c-means using multiple cluster validity indices. The proposed method was shown to be roughly equivalent in effectiveness as the others to which it was compared when clustering linearly separable data sets and equivalent or better when clustering non linearly separable data sets without ever needing to be provided a number of clusters.

There remains work to be done in finding more appropriate validation methods to evaluate the proposed method as the validity indices used fall victim to the same pitfalls as most hard clustering methods when the data set is not linearly separable. There also remains to further optimize the proposed method and it's heuristics, and to attempt modifying it for specific applications.

In conclusion, the proposed clustering method not only identifies a desired number of clusters, but produces valid clustering results.

Acknowledgements. We gratefully acknowledge the support from NBIF's (RAI 2012-047) New Brunswick Innovation Funding granted to Dr. Nabil Belacel.

References

1. Gan, G.: Data Clustering in C++: An Object-Oriented Approach. Chapman and Hall/CRC, Boca Raton (2011)
2. Jain, A.K., Murty, M.N., Flynn, P.J.: Data clustering: a review. ACM Comput. Surv. **31**, 264–323 (1999)
3. Guan, Y., Ghorbani, A., Belacel, N.: Y-means: a clustering method for intrusion detection. In: Canadian Conference on Electrical and Computer Engineering 2003, IEEE CCECE 2003, vol. 2, pp. 1083–1086, IEEE (2003)
4. Mok, P., Huang, H., Kwok, Y., Au, J.: A robust adaptive clustering analysis method for automatic identification of clusters. Pattern Recogn. **45**, 3017–3033 (2012)
5. Wu, K.L., Yang, M.S.: A cluster validity index for fuzzy clustering. Pattern Recogn. Lett. **26**, 1275–1291 (2005)
6. Xie, X., Beni, G.: A validity measure for fuzzy clustering. IEEE Trans. Pattern Anal. Mach. Intell. **13**, 841–847 (1991)
7. Pal, N., Bezdek, J.: On cluster validity for the fuzzy c-means model. IEEE Trans. Fuzzy Syst. **3**, 370–379 (1995)

8. Fukuyama, Y., Sugeno, M.: A new method of choosing the number of clusters for the fuzzy c-means method. In: Proceedings of Fifth Fuzzy Systems Symposium, pp. 247–250 (1989)

9. Pakhira, M.K., Bandyopadhyay, S., Maulik, U.: Validity index for crisp and fuzzy clusters. Pattern Recogn. **37**, 487–501 (2004)

10. Rezaee, M.R., Lelieveldt, B., Reiber, J.: A new cluster validity index for the fuzzy c-mean. Pattern Recogn. Lett. **19**, 237–246 (1998)

11. Rousseeuw, P.J.: Silhouettes: a graphical aid to the interpretation and validation of cluster analysis. J. Comput. Appl. Math. **20**, 53–65 (1987)

12. Kaufman, L.R., Rousseeuw, P.: Finding Groups in Data: An Introduction to Cluster Analysis. Wiley, New York (1990)

13. MacNaughton-Smith, P.: Dissimilarity analysis: a new technique of hierarchical sub-division. Nature **202**, 1034–1035 (1964)

14. Bezdek, J.C., Ehrlich, R., Full, W.: Fcm: the fuzzy c-means clustering algorithm. Comput. Geosci. **10**, 191–203 (1984)

15. Fisher, R.A.: The use of multiple measurements in taxonomic problems. Ann. Eugen. **7**, 179–188 (1936)

16. Gionis, A., Mannila, H., Tsaparas, P.: Clustering aggregation. ACM Trans. Knowl. Discov. Data **1**, Article 4 (2007)

Robust Signature Discovery for Affymetrix GeneChip® Cancer Classification

Hung-Ming Lai[1](✉), Andreas Albrecht[2], and Kathleen Steinhöfel[1]

[1] Algorithms and Bioinformatics Research Group, Department of Informatics,
King's College London, Strand, London WC2R 2LS, UK
{hung-ming.lai,kathleen.steinhofel}@kcl.ac.uk
[2] School of Science and Technology, Middlesex University,
Burroughs, London NW4 4BT, UK
a.albrecht@mdx.ac.uk

Abstract. Phenotype prediction is one of the central issues in genetics and medical sciences research. Due to the advent of high-throughput screening technologies, microarray-based cancer classification has become a standard procedure to identify cancer-related gene signatures. Since gene expression profiling in transcriptome is of high dimensionality, it is a challenging task to discover a biologically functional signature over different cell lines. In this article, we present an innovative framework for finding a small portion of discriminative genes for a specific disease phenotype classification by using information theory. The framework is a data-driven approach and considers feature relevance, redundancy, and interdependence in the context of feature pairs. Its effectiveness has been validated by using a brain cancer benchmark, where the gene expression profiling matrix is derived from Affymetrix Human Genome U95Av2 GeneChip®. Three multivariate filters based on information theory have also been used for comparison. To show the strengths of the framework, three performance measures, two sets of enrichment analysis, and a stability index have been used in our experiments. The results show that the framework is robust and able to discover a gene signature having a high level of classification performance and being more statistically significant enriched.

Keywords: Affymetrix · Cancer classification · Feature interdependence · Feature selection · Gene expression profiles · Gene signature discovery

1 Introduction

In recent biomedical research, transcriptome analysis using high-throughput screening (HTS) technologies, such as microarrays, has been a prevailing approach to obtain gene expression profiles of cells of interest in response to physiological and genetic changes in several tissues. Since HTS is capable of interrogating many thousands of oligonucleotide probes simultaneously, the analysis

© Springer International Publishing Switzerland 2015
B. Duval et al. (Eds.): ICAART 2014; LNAI 8946, pp. 329–345, 2015.
DOI: 10.1007/978-3-319-25210-0_20

of expression profiling data has shown enormous potential for the discovery of biological markers in carcinogenesis studies and in the diagnoses of diseases [1]. Different types of tumor cells can be marked by discriminating genes at expression level. Thus, biomarkers for distinct tumorigenesis stages and cancer classification under HTS experiments could be explored by selecting discriminating genes. The identification of subsets of these genes contributing to the predictive power is the process of finding so-called gene signatures and is subject to change [2]. Out of an abundance of transcripts in a tissue, a few genes are differentially expressed, while a tremendous amount of mRNAs would be regarded as noise. Also, biologists favor a small number of candidate genes to achieve greater efficiency for *in vitro* validation.

Identification of differentially expressed genes in bioinformatics can be referred to as feature selection, the domain of dimensionality reduction techniques, commonly termed in the context of data mining, machine learning and pattern recognition [3]. In particular, feature subset selection is a technique not only to reduce the feature dimension of data points without changing their initial representation, but also to select the minimal subset that maximizes the classification performance. In terms of knowledge discovery, this is actually based on the principle of parsimony [4], leading to a preferred model having as few as possible variables that sufficiently fit with the data this is very similar to the need of gene signature determination. Unfortunately, a typical microarray-based cancer experiment might only consist of tens to a hundred of clinical samples, but each sample has thousands to tens of thousands of genes to be questioned [5]. The presence of experimental noise is another widely criticised issue in the experimental design of microarrays. The noise is unavoidable and doomed to existence from the early stages of sample preparation, extraction and hybridization, largely due to the principles of microarray technology. Feature subset selection is known to be an NP-complete problem [6], and the curse of dimensionality and the common occurrence of experimental noise would make the procedure of discriminating gene selection and the process of finding a parsimony model even more challenging.

Over the past decade, one can categorize feature selection methods into three groups: filters, wrappers, and embedded techniques, depending on how they interact with a classification method [3]. A filter method measures features with respect to different phenotypes by considering the intrinsic properties of the data and does not make use of a classification algorithm within its selection scheme. There are two types of filters, univariate and multivariate methods. Univariate filters disregard feature interaction and evaluate features individually. Both parametric statistics (e.g. paired/unpaired student t-test &ANOVA) and nonparametric statistical tests like Wilcoxon rank sum are univariate. On the other hand, multivariate filters that consider feature-feature correlations to some extent are sometimes referred to as space search methods [7]. A wrapper employs a classification method to evaluate the prediction performance of a selected feature subset and an iterative selection process is wrapped around the classifier. The procedure is terminated with a stop criterion in order to obtain the best predictive model. Although the wrapper is able to manage feature-to-class relevance and feature-to-feature dependence, it seems prone to overfitting and is computationally time-consuming because of a small sample size and a large feature dimension. According to search strategies, the wrapper can be deterministic or randomized. Sequential forward selection and sequential backward elimination are two typical examples of deterministic wrappers, whereas simulated annealing or genetic algorithms serve as an illustration of randomized ones [8,9]. Similar to the wrapper, an embedded approach is also dependent on a classification method and takes feature correlations into account. However, the embedded is less computationally intensive than the wrapper as feature subset selection is embedded in a base classifier. As soon as a classifier is built, features are about to be ranked or weighted. SVM-RFE and its variants are one of the most representative examples of embedded feature selection [10–12]. The main idea is to rank features by the weight vector of a linear SVM hyperplane and to select features using a recursive feature elimination strategy.

In recent years, several feature selection methods based on information theory have been developed to deal with feature-to-feature dependence and the correlation between a feature and the selected feature subset in large scale gene expression data. Moreover, more recently a probabilistic interpretation has been established, derived from optimizing the conditional likelihood, for unifying information theoretic feature selection [13]. Three space search feature selection methods are now briefly described and then compared to a new gene selection filter proposed in the present paper. The three multivariate methods are all based on information theory and focus on the issue of feature-feature dependence and feature-phenotype correlation. Ding and Peng proposed the minimum-Redundancy and Maximum-Relevance framework (**mRMR**) to explore high order gene interactions [14]. This method uses mutual information to cope with a tradeoff between the reduction of feature redundancy within a feature subset and the strength of feature-to-class correlation. Their experimental results show that the defined criterion could lead to features with least redundancy. Using conditional mutual information as an evaluation criterion, Fleuret proposed a fast binary feature selection (**cmim**) to select features having the largest association with respect to sample classes conditioned on the selected feature subset [15]. As the **cmim** criterion would select features having more information about sample classes evaluated only by pairwise feature statistics, some informative features, in which biologists could be interested, would be removed, even though the author claims that the selected features are informative and weakly pairwise dependent. The third feature selector, **fcbf**, was designed by Yu and Liu to efficiently eliminate a considerable number of irrelevant and redundant features [16]. While **mRMR** and **cmim** define evaluation criteria, **fcbf** introduces an approximate Markov blanket as a search strategy for an efficient feature removal by using symmetrical uncertainty. This method therefore selects much fewer features than **mRMR** and **cmim** and is very prone to removing 'less informative' but important features that might be of interest to the domain expert. Although feature relevance and feature redundancy are well addressed by the three multivariate filters just discussed, feature interdependence is discarded in favour of reduced computational complexity. Feature inter-dependence may point to an important feature that is strongly discriminative together with other features in the selected feature subset, but is individually less informative relative to a class. The approach could be biologically meaningful within gene signatures in post-genomics.

In this article, we introduce a multivariate approach based on information theory for cancer related gene signature discovery under high-throughput gene expression profiles. The approach is a four-step framework considering several types of feature properties (feature relevance, feature redundancy, and feature interdependence) in the context of feature pairs. This complete framework has been refined by appropriately fine-tuning parameter settings, by establishing an aggregation scheme for gene signatures, and by proposing an RC plot to illustrate how coupled genes could bring more information about sample classes than individual genes can do. To show the effectiveness of the framework, brain cancer gene expression profiling data [17] generated by Affymetrix Human Genome U95Av2 GeneChip® has been used. The brain cancer expression matrix is composed of 12,625 interrogated genes and 50 samples from two cell-lines. It implies that the discovery of biologically discriminative genes in the brain experiment would be more demanding than that in some easy-classified benchmarks, such as the famous Leukemia dataset [18] . Through the brain cancer study, we will demonstrate the great strengths of this novel filter in terms of classification performance, robustness, and enrichment analysis.

2   Preliminaries

2.1   Domain Description

In this section, the domain of HTS gene selection for phenotype prediction is briefly described. Given a gene expression dataset $D = X \in \mathbb{R}^m, C \in \mathbb{R} = \{(x_i, c_i)\}_{i=1}^n$, where D includes n samples X labeled by a class vector C, and each sample is profiled over m gene expressions, i.e. $x_i = \{x_{i1}, \cdots, x_{im}\}_{i=1}^n$, $m \gg n$. The domain expert expects to find a small number of discriminating genes (from tens to a hundred) for clinical classification to be validated *in vitro* and to identify a gene signature for a specific disease. To address the issue of HTS-based gene signatures, we can refer to it as a feature selection problem. Let F be a full set of features (genes) $F = \{f_i\}_{i=1}^m$, then feature selection aims at choosing a feature subset $G \subset F$ that maximizes the prediction performance; moreover, if G is aimed at a minimum, a parsimonious subset is sought for.

2.2   Information Theory Basics

Entropy is the rationale behind information theory and is an intuitive measure to evaluate the uncertainty of a random variable. Given a variable, it is computed at the level of probability distributions [19]. Let X be a nominal random variable, Shannon *entropy* is defined as

$$H(X) = -\sum_{x \in X} p(x) \log p(x), \tag{1}$$

where x denotes the values of the random variable X over its alphabet X (the domain), and $p(x)$ is the marginal probability distribution of X. Without loss of generality, the domain X will be ignored in the rest of the paper. Unlike conventional statistics, an entropy-based measure does not make any *a priori* assumption. For instance, one is required to ask whether data is normally distributed before using the student's *t*-test. Additionally, other information quantities can also be defined through applying probability theory to entropy. The *conditional entropy* of X given Y is represented as

$$H(X|Y) = -\sum p(y) \sum p(x|y) \log p(x|y), \tag{2}$$

where $p(x|y)$ is the conditional probability of X given the observed values of Y. This quantity evaluates how much uncertainty of X is left given that the value of another random variable Y is known. Similarly, the *joint entropy* of two random variables X and Y is denoted as

$$H(X|Y) = -\sum \sum p(x, y) \log p(x, y), \tag{3}$$

where $p(x, y)$ is the joint probability distribution of X and Y. It quantifies the amount of information needed to describe the outcome of two jointly distributed random variables. Another important information theoretic measure, *mutual information*, quantifies the amount of information shared by two random variables X and Y, and can be obtained by the definition of entropy and conditional entropy

$$MI(X|Y) = H(X) - H(X|Y). \tag{4}$$

The mutual information is the reduction of entropy of one variable, if the other is known. This measure is symmetric and non-negative, and the value of zero implies that the two variables are statistically independent. The mutual information of X and Y can also be conditioned on Z, *conditional mutual information*, and defined by

$$CMI(X, Y|Z) = H(X|Z) - H(X|Y, Z). \tag{5}$$

The quantity measures the information amount shared between X and Y, if Z is known. Finally, we introduce *symmetrical uncertainty* that will be heavily utilized in our gene selection framework throughout the paper. The measure could be viewed as one type of normalized mutual information and defined as

$$SU_{X,Y} = 2 \left[\frac{H(X) - H(X|Y)}{H(X) + H(Y)} \right]. \tag{6}$$

If X is a joint random variable, the joint symmetrical uncertainty could be acquired by exactly the same idea as the joint entropy.

2.3 Feature Relevance

Given a full set of features F and a feature f_i, then let $F_i = F \backslash f_i$ denote that the feature f_i is removed from the set F. Kohavi and John (hereafter KJ) defined three feature types of relevance to sample classes via the probability distribution of the class C conditioned on the features of f_i and F_i, as in the following Definitions 1–3 [20].

Definition 1. KJ-Strong Relevance:
A feature f_i is strongly relevant to C iff

$$p(C|f_i, F_i) \neq p(C|F_i). \tag{7}$$

Definition 2. KJ-Weak Relevance:
A feature f_i is weakly relevant to C iff

$$p(C|f_i, F_i) = p(C|F_i) \text{ and } \exists F_i' \subset F_i \text{ such that } p(C|f_i, F_i') \neq p(C|F_i'). \tag{8}$$

Definition 3. KJ-Irrelevance:
A feature f_i is irrelevant to C iff

$$\forall F_i' \subseteq F_i, \ p(C|f_i, F_i') = p(C|F_i'). \tag{9}$$

The three definitions imply that an ideal feature subset should include all strongly relevant features and some weakly relevant features with least feature redundancy, and all irrelevant features should be removed. Given two jointly distributed random variables $f_i f_j$ (or f_{ij}), similar to KJ definitions, we can define a strongly relevant feature pair f_{ij} by the conditional probability distribution of the class C.

Definition 4. Strongly Relevant Feature Pair:
A feature pair f_{ij} is strongly relevant to C iff

$$p(C|f_{ij}, F_{ij}) \neq p(C|F_{ij}), \tag{10}$$

where F_{ij} denotes the feature set F with the features f_i and f_j both together eliminated from F. Therefore, a feature pair is referred to as a united-individual and must be selected or removed together during the process of selection. The strong relevance of a feature pair will be the basis for the framework presented in our paper for finding HTS gene signatures.

2.4 Feature vs Feature Pair

We propose a 'Ratio by Correlation' (RC) plot in order to demonstrate if feature pairs can reveal more information about the class C than single features could do and whether or not the coupled features can provide insight into feature interdependence, revealing potentially some genetic regulatory interactions between features. Out of the probe-sets (features) designed in the Affymetrix Human

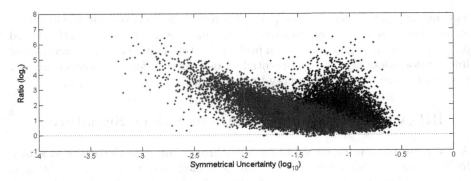

Fig. 1. Ratio by correlation plot.

Genome U95Av2 Array with a real gene expression data set [17], ten thousand feature pairs f_{ij} were randomly selected to generate the RC plot as shown in Fig. 1. Given the population of the selected pairs and a gene expression matrix with the corresponding sample class vector, we used symmetrical uncertainty to calculate two correlation measures between two features and C ($SU_{i,c}$ and $SU_{j,c}$, respectively), and additionally one correlation value between a feature pair and C ($SU_{ij,c}$). Then the mean of $SU_{i,c}$ and $SU_{j,c}$ was computed and represented by M, followed by three computations displayed below:

$$SU_{IF,C} = \log_{10} M; \tag{11}$$

$$SU_{FP,C} = \log_{10}(SU_{ij,c}); \tag{12}$$

$$R = \log_2(SU_{ij,c}) - \log_2 M. \tag{13}$$

The RC plot is constructed by plotting R against $SU_{IF,C}$ and against $SU_{FP,C}$, respectively. Here, $SU_{IF,C}$ represents the average correlation between individual feature and C, whereas $SU_{FP,C}$ denotes correlation between a feature pair and C. Thus, R is the ratio between the feature pair correlation and individual feature correlation. For the convenience of visualization, a logarithmic scale is used, with twofold changes for the ratios and tenfold increases for correlations.

While most coupled features have a significantly increased joint effect, there still exist many cases where two features coupled together do not provide more information about a class, and this happens especially for those features that might potentially be considered as strongly relevant to the class. Moreover, a few cases have been found where a feature pair has a decrementing joint effect if two strong features are joined together. It might imply a common phenomenon in gene regulation where one gene can be silenced or deactivated by another gene or its products. On the other hand, there are only a few coupled features from very weak features having exceptionally high strength of association between them and the class due to some kind of their underlying interdependence. Overall, we observe that single features (grey scatter plot) move rightwards towards feature pairs (black scatter plot) in Fig. 1. It means that there is a clear tendency for features to combine to stronger pairs. Therefore, we believe that feature pairs would play more important roles than individual features in gene selection based on high-throughput gene expression profiles. Feature pair could either bring more information about C or have a potential for dealing with feature interdependency that could take more gene synergy into account.

3   iRDA – A Framework for Finding Gene Signatures

A complete framework for selecting high-throughput gene signatures is shown in Algorithm 1. This novel gene selector is named **iRDA**, abbreviated by gene selection guided by interdependence with redundant-dependent analysis and a gene aggregation. The framework is based on information-theoretic measures, an appropriate search strategy, a suitable parameter estimation criterion, a mixture of forward-backward phases, and a simple aggregation scheme. The rationale for devising such a framework is to select a gene signature around gene synergy that could potentially discover genetic regulatory modules or disease-related factors. Interdependence between features is, therefore, a matter of concern.

The proposed gene selection method is a four-step framework with a vast body of feature pairs, including a set of analyses of feature relevance, feature interdependence, feature redundancy and dependence, and feature aggregation. Features relevant to C defined by KJ looks sensible in theory, but it would hardly work in practice, specifically for the analysis of large-scale gene expression profiles. In general, high throughput gene expression profiling has only a relatively small number of differentially expressed genes, and correlations between features and labels are exponentially distributed. In this paper, we estimate the degree of features relevant to a target class via an analysis of a partition method working on a selected correlation measure. Given a random value for feature f_i, symmetrical uncertainty $SU_{i,c}$ is used to quantify the strength of association between features and labels. After sorting all of the calculated correlations in descending order, k-mean clustering is proceed upon the sorted list of $SU_{i,c}$ to partition features into five groups. We label the five clusters as $R_1 \cdots R_5$ in descending order according to their centroids, to gradually make the way down the scale of KJ-relevance/-irrelevance to C. These feature types will be a prerequisite to conduct our main idea of feature interdependent analysis.

The consideration of multi-way gene interactions would have the potential for a road map of feature interdependence. However, because of the immense complexity of gene regulatory mechanisms, it would not be a good strategy to infer multi-way feature interdependence in a direct way. Unlike traditional feature selection filters working on a search space of individual features, feature pairs will be our main body throughout the framework and individual features with various scale of relevance to sample classes will only be an indicator in the subsequent analyses. It is impractical to perform exhaustive search for visiting all feature pairs if the number of features is very large. Therefore, feature relevance partitions from the previous step could be an indicator to produce potential feature pairs that are KJ relevant to C. In the second step, given a joint random

Algorithm 1. iRDA Gene Selector.

Given: $D = \{X \in \mathbb{R}^m, C \in \mathbb{R}\} = \{(x_i, c_i)\}_{i=1}^n$, and $F = \{f_i\}_{i=1}^m$
Parameter: ε
Find: gene signature G

RELEVANCE:
1    $\forall f_i$, calculate $SU_{i,c}$
2   Sort $SU_{i,c}$ into descending order
3   Perform k-mean clustering (k=5) on the sorted $SU_{i,c}$
4   Label 5 clusters R_1-R_5 whose centroids are in descending order
INTERDEPENDENCE:
5   **Forward Phase**
6   t=1
7   for i=1 to sup(R_1)
8      $G_s^t = \emptyset$
9   for j=i+1 to sup(R_4)
10   if $SU_{ij,c} > \varepsilon$, where ε is estimated by Eq. (14)
11   add feature pairs $f_i f_j$ to G_s^t, f_i followed by f_j
12   where f_i is a seed (f_s^t) and added once only
13   end
14   t=t+1
15   end
16   Let $G_{pre} = \{G_s^t | G_s^t \neq \emptyset, G_s^t$ led by $f_s^t\}$
REDUNDANCY AND DEPENDENCE:
17   **Backward Phase**
18   for each G_s^t do
19   for each $f_i \in G_s^t$ do first in last check
20   f_i is removed instantly if CMI($f_i, C | G_s^t$)=0
21      $G_s^t = \emptyset$ if CMI($f_s^t, C | G_s^t$) = 0
22   end
23   end
24   **Insertion Phase**
25    $\forall f_{s,j} \in G_{pre}$, add f_s to G_j^t if applicable
26    $G_{pre}^{'}$ is then established
27   Perform backward phase on $G_{pre}^{'}$
28   Let $G_{post} = \{G_s^t | G_s^t \neq \emptyset, \#G_s^t > 1, SU_{s,c}^{t-1} > SU_{s,c}^t\}$
AGGREGATION:
29   t=1, $G = \emptyset$
30   do
31      $G = G \cup G_s^t$
32   t=t+1
33   while $G = G_{post}$ or G is defined

---

variable of two features $f_i f_j$ (or f_{ij}), joint symmetrical uncertainty $SU_{ij,c}$ is used to measure the strength of correlation between coupled features and a class variable. The aim of this step is to search for those strongly relevant feature pairs whose joint symmetrical uncertainty values are greater than a threshold

ε. We assume that one feature in R_1 partition colliding with the other feature in the partitions of R_1, R_2, R_3 and R_4 might have a positive joint effect for producing potential coupled features. Based on this assumption, an estimation of the threshold ε will be a critical task for exploring feature pairs. We propose to estimate the critical value by the following equation:

$$\varepsilon = \overline{SU_{ij,c}} , \tag{14}$$

where $f_{ij} \in \Omega$, $\Omega = \{f_{ij}^{(T)} | T = 1, \cdots , T^*; SU_{ij,c} > SU_{i,c}; SU_{i,j} < SU_{j,c}\}$. Given the number of trials T, two features (f_i, f_j) in the sorted list of $SU_{i,c}$ are coupled in turn, where $SU_{i,c} > SU_{j,c}$, to test if the conditions of $SU_{ij,c} > SU_{i,c}$ and $SU_{i,j} < SU_{j,c}$ are satisfied. Then when T^* successful feature pairs that meet the conditions are executed, the mean of their $SU_{ij,c}$ is computed to be the estimation of the threshold ε. The conditions reveal that a feature pair has positive joint effect and less redundancy between the two coupled features even though a feature correlation sometimes does not necessarily mean redundancy. Once a feature pair succeeds in the examination of $SU_{ij,c} > \varepsilon$, the feature is then added to a subset of G_s^t led by a seed feature f_s. It means, every feature pair f_{sj} in G_s^t has the same feature f_s, and every feature belongs to the subset in the order of its relevance to C. Finally, there could be a collection of G_s^ts led by various seed features. Through the approximation of high-order feature interdependence led by seed features and their coupled features, feature interdependence could be extended from mutual dependence on feature pairs to high dimensional gene interactions.

If a subset is formed, it is necessary to ask if there are any redundancies among features within a selected feature subset. A minimal feature subset must include the most discriminative features, but avoid redundant features. Thus the third step is mainly to check and remove redundant features as many as possible to form a parsimonious set of features. Given a collection of subsets derived from interdependent analysis, G_{pre}, the conditional mutual information $\mathrm{CMI}(f_i, C | G_s^t)$ of a feature f_i and label C conditioned on a subset $G_s^t \in G_{pre}$ is used for this purpose by using an approximation of backward elimination with first in last check policy. For any G_s^t, we test if the value of $\mathrm{CMI}(f_i, C | G_s^t)$ is zero for every feature checked one by one and from the end of G_s^t to the beginning of G_s^t. A feature whose CMI value is zero will instantly be removed and the next less relevant feature will then be checked until the features in G_s^t have all been tested. If a seed feature is eliminated, the subset G_s^t led by this feature will be discarded; otherwise, features that remain in a retained subset are considered to be dependent on the seed feature. When redundancy analysis of G_{pre} is finished, for any coupled features with seed features (f_{sj}) in G_{pre}, a seed feature f_s might be added to the subset led by feature f_j if applicable. This procedure is in order to complement the greedy formulization of G_s^t. Therefore, we might have a new collection of subsets G'_{pre} so that a second round of redundant analysis would be required for G'_{pre}. **iRDA** actually includes a forward phase and a backward phase. Interdependent analysis carries out forward addition and more false positive features might be selected in this phase while redundant and dependent analysis performs backward elimination to identify and to remove false positive features. An insertion phase included in redundant and dependent analysis increases a chance that true positive features might enter some potential subsets. Through these phases, a final collection of parsimonious subsets G_{post} is able to safely accomplish.

As biomedical researchers are always more interested in candidate genes regarding a specific disease, in a word, a gene signature is a main issue to find potentially biomarkers, biological process, molecular function, cellular mechanism, and regulatory motifs. Conventional gene selectors usually allow people select genes as many as they can to define a gene signature so the final step of our method is to aggregate genes to establish a gene list where an appropriate gene signature might be found. Since each subset $G_s^t \in G_{post}$ is built by a seed feature f_s, the strength of relevance between f_s and C is able to use as an indicator for gene aggregation. We first select a subset having the most relevant seed feature, and aggregating genes by considering next subset whose seed feature is the next most relevant to C. This procedure proceeds until no genes can be accumulated ($G = G_{post}$) or a preferred gene signature G is defined.

4   Experimental Results

To show the proposed framework is potentially capable of selecting the most discriminative gene signature for phenotype prediction and of finding significant genetic regulation within the selected signature, a publicly available microarray-based brain cancer classification data was used [17]. The experiment was designed to investigate whether high-throughput gene expression profiling could classify high grade gliomas better than histological classification. The data set consists of 50 samples and 12,625 probe-sets using Affymetrix Human Genome U95Av2 Array. Out of 50 high grade gliomas, there are 28 glioblastomas (GBM) and 22 anaplastic oligodendrogliomas (AO). Upon this gene expression matrix, features were discretized to three bins as suggested by [14] and each bin was then designated by a discrete value such as 1, 3 and 5 for the better calculation of information theoretic measures. We evaluated the proposed framework with three model-free feature selection filters (**mRMR, cmim** and **fcbf**) to know the capacities of four gene selectors in terms of classification performance and enrichment analysis. While classification performance reveals how good a selected model could predict, enrichment analysis could display whether a gene signature actively involves gene synergy.

Because of the curse of dimensionality, the conventional training-test data partition given a ratio (say 60-40 %) is not very appropriate for the assessment of gene selection approaches in the domain of high-throughput gene expression data. Thus, the procedure of leave-one-out cross-validation (LOOCV) was used in our experiments. Three performance measures were chosen to assess the predictive power of selectors. They are the generalisation error rate (ERR), the area under a receiver operating characteristic curve (AUC) and the Matthews correlation coefficient (MCC). Besides, a reference classifier is required to induct filter-based feature selectors into a learning process. This is due to their independence of learning methods. We utilized the k-nearest-neighbour (k-NN) classifier (k=3) to establish classification models after gene selectors had been performed.

There were three feasible subsets generated by **iRDA** for the binary classification of the brain dataset and 8 unique features in total were involved in these subsets. Hence, three sets of features were established by **mRMR, cmim** and **fcbf** and each set had eight features to be compared with **iRDA**. From the viewpoint of parsimony, a minimal feature subset is selected to evaluate how well the chosen features could dedicate themselves to a class versus those features of the other three selectors, results shown in Table 1. The optimal subset was the one led by the first seed feature and its cardinality was just three. The misclassification rate was 0.4 that was also the lowest one and only **fcbf** could reach the same level using four genes. In addition, the four-gene set of **fcbf** had the highest of MCC performance, 92.26 %, very slightly better than the three-gene set of **iRDA** by 0.15 %. The best AUC performance went to the **cmim** gene signature at the level of 100 % whereas the minimal gene set of **iRDA** had nearly approached the same level by 0.32 %; however, **cmim** had employed seven genes more than twice the features produced by **iRDA**. In sum, a parsimony model of the three-gene set built by **iRDA** had very good predictive power from all aspects of performance measures.

Other than the selection of a parsimonious subset, it is also an essential matter to select one of the best gene signatures with a reasonable gene size that could have the strongest classification performance and a strong possibility for biological findings regarding a certain disease or cancer such as biomarkers or regulatory modules. To do so, three feature subsets of **iRDA** were aggregated in order of the seed feature relevance to C and eventually an eight-gene signature could be established. We then compared our gene sets with those of **mRMR**, **cmim** and **fcbf** to know the performance among them and to see what the best gene signature would be. Figure 2 shows that when **iRDA** aggregated the other features led by the other seeds into a parsimony model, classification performance was stronger and stronger, finally leading to no misclassification. Meanwhile, both AUC &MCC performance could also approach the highest level of 100 % even though the parsimony model initially could not greatly outperform all of the other methods as discussed in Table 1. Furthermore, it is observed that except the proposed method no the other selectors here were able to dominate all of the three performance measures. For instance, when we only compared **mRMR**,

Table 1. Prediction performance in terms of parsimony.

	iRDA		mRMR		cmim		fcbf	
	%	#Gene	%	#Gene	%	#Gene	%	#Gene
ERR	4.00	3	6.00	4	6.00	6	4.00	4
AUC	99.68	3	98.54	8	100	7	94.72	5
MCC	92.11	3	87.96	4	88.32	6	92.26	4

cmim and **fcbf, fcbf** could have the best level of error and the MCC performance but its AUC level was undoubtedly the lowest. Similarly, **cmim** was able to reach the AUC value at 100 % but was decidedly inferior to **fcbf** in either error or MCC. In a word, **iRDA** was far superior to the compared gene filters in terms of classification error and performance and an eight-gene signature (the full feature set of **iRDA**) was recognised as the best one.

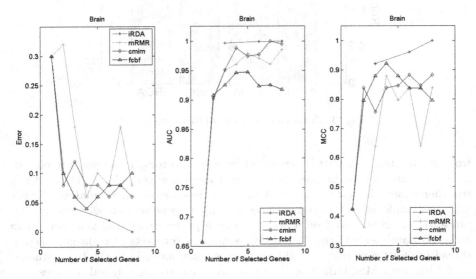

Fig. 2. Classification performance of four sets of gene signatures.

Robustness is another crucial issue in the context of high dimensional gene selection for cancer classification. One of likely scenarios relating to finding biologically potential biomarkers under large scale gene expression profiles is that discriminative genes selected by a feature selection method could be many and varied given minor instance variations. It is, therefore, important to know how robust a gene selector can be. The more robust the signature is, the more convinced the domain expert feels. We assess robustness of the four multivariate filters by using the pairwise similarity measure of the Jaccard Index [21] with leave-one-out sample perturbation based on the same brain dataset. The experimental result is shown in Fig. 3. The bar chart reveals that **iRDA** is by far the most robust method (reaching at the level of 0.8), followed by **fcbf** (inferior to **iRDA** by 21 %), and nearly twice better than **mRMR** and **cmim**. Both **mRMR** and **cmim** have impressive classification performance (especially in AUC), but surprisingly, they are less robust (<0.5) even in this very minor variation experiment. In a word, **iRDA** could provide great performance in terms of classification and stability analysis, and could be reliable for downstream biological validations.

Since the proposed method has paid attention to feature interdependence, it is an essential issue to know if there is any molecular information extracted

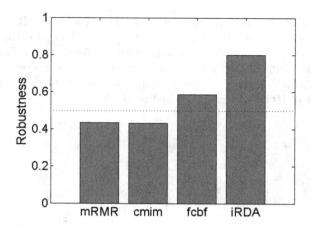

Fig. 3. Robustness of four multivariate filters.

from a gene signature that is generated by gene selectors. This relates to enrichment analysis that might provide an insight as to how genes inter-actively work together about biological process. A tool of gene set enrichment analysis, GSEA [22], was employed in this paper to see how many gene sets are statistically significantly enriched based on a collection of *a priori* annotated gene sets, here MSigDB database was considered. We generated four sets of eight-gene signatures from four gene selectors, the same as we mentioned in Fig. 2, to be studied on GSEA with MSigDB. After the process of collapsing original features into gene symbols, there were 7 genes in **iRDA** and **cmim** signatures and 8 genes in **mRMR** and **fcbf** signatures. Based on these collapsed features, Table 2 shows that given an **iRDA** gene signature, there were 6 and 4 gene sets recognised as statistically significant enrichment in two phenotypes of GBM and AO respectively (p-value<0.05) while the other three gene signatures had far fewer enriched gene sets. Moreover, out of up-regulated gene sets in two phenotypes, 6 gene sets were statistically significantly enriched in total (FDR<0.25) for the **iRDA** gene signature by far the most number of enrichment in this study.

In addition to GSEA, we have also adopted WebGestalt [23] to carry out a functional genomic enrichment analysis that biological themes of gene lists could be open to interpretation. The same sets of four gene signatures as used in GSEA were once again submitted to WebGestalt. After transferring probe-set id into gene symbol, the number of genes remained in their original gene sets was 7, 7, 6 and 8 for the signatures of **iRDA**, **mRMR**, **cmim** and **fcbf**, respectively. These remained genes would be the basis to see how many functional factors could be found and how many genes have been involved in those identified biological factors at the statistical significance level of 0.05. Gene synergy is initially one of our main ideas to develop a new gene selector; therefore it is important to understand if there are any relationships between gene regulatory modules and a gene set. We have found that 5 genes in **iRDA** seven-gene signature were

Table 2. Gene set enrichment analysis.

	iRDA	mRMR	cmim	fcbf
Native features	8	8	8	8
Collapsed features	7	8	7	8
Enrichment in GBM				
FDR < 25 %	3	0	0	0
p-value < 5 %	6	0	0	1
Enrichment in AO				
FDR < 25 %	3	1	1	0
p-value < 5 %	4	1	2	0
Enrichment in total				
FDR < 25 %	6	1	1	0
p-value < 5 %	10	1	2	1

Table 3. Functional genomic enrichment analysis.

	iRDA		mRMR		cmim		fcbf	
Selected probe-sets	8		8		8		8	
Mapped genes	7		7		6		8	
	#Gene	#Factor	#Gene	#Factor	#Gene	#Factor	#Gene	#Factor
Transcription factor	5	5	0	–	2	2	0	–
MicroRNA	5	8	2	1	2	1	2	1
Disease	2	1	0	–	0	–	0	–

p-value<0.05

connected to five transcription factors and eight microRNA targets, respectively; and there were only two interactions found between two transcription factors and two genes within the **cmim** six-gene signature while no interactive relationships with transcription factor were found in the gene signatures of **mRMR** and **fcbf** (see Table 3). Although one microRNA-mRNA interaction was found with two genes in the gene lists of **mRMR**, **cmim** and **fcbf**, the discovered interaction was actually the same one and included in **iRDA** microRNA-mRNA findings. To reveal cancer-related genes, disease association analysis was performed. Out of the **iRDA** selected genes, a report has statistically significantly related two genes to the disease of inflammation - one of key factors in tumour development [24].

5   Conclusions

A framework for high throughput gene signatures, named **iRDA**, is presented in this paper. Whereas individual features are searched in conventional gene selection in a either univariate or multivariate manner, the proposed filter is mainly focused on feature pairs. Single feature relevance to a class variable is just used as an indicator throughout the framework. By using a number of information theoretic measures and through a series of analysis of feature characteristics including relevance, interdependence, redundancy and dependence, the **iRDA** gene selector is devised around gene synergy based on feature pairs and seed features that lead to various possible parsimonious set of feature. With a simple aggregation scheme, a gene signature is eventually able to be defined for finding biological information related to a certain disease in different phenotypes.

To demonstrate the effectiveness of this newly developed gene selector in the domain of high-throughput gene signatures, a brain cancer gene expression profiling data was examined. This expression matrix was derived from Affymetrix Human Genome U95Av2 Array, and having 50 labelled samples and 12,625 interrogated genes. The curse of dimensionality implicates that the task of gene selection is an enormous computing challenge. Based on the brain cancer data set, we have compared **iRDA** with three filters (**mRMR**, **cmim** and **fcbf**) that are widely discussed in the research community. Also, these methods all use information theoretic measures. The experimental results show that an 8-gene signature was defined by **iRDA** and it outperformed the other three methods in terms of classification performance with three performance measures. By using the pairwise Jaccard Index, we have also demonstrated that iRDA is the most robust gene selector among the other three filters. In addition to classification and stability analysis, we finally performed two sets of enrichment analysis to see how effectively feature interdependence has been tackled in the framework. The results show that more statistically significant gene sets and genetic regulatory interactions could be found in our gene signature. Furthermore, within the **iRDA** 8-gene signature, there were two genes associated with a disease of inflammation at the statistical significance level. And no the other filters could find disease-related genes. The rationale behind these significant findings is that our method is able to find an important feature which is individually weakly relevant to a class but might have strong interdependence between features. This type of genes accompanied by other genes in a selected gene list would more contribute to the phenotype than they appear solely at the expression level. Except for **iRDA**, however, most recent filter-based feature selectors could not search for these features that may attract the interest of the domain user.

We think that our **iRDA** framework can have the capacity of finding robust and small size gene signatures with a potentially high predictive power that, in turn, could disclose biological information regarding gene synergy.

References

1. Nevins, J.R., Potti, A.: Mining gene expression profiles: expression signatures as cancer phenotypes. Nature Rev. Genet. **8**, 601–609 (2007)
2. Kim, S.-Y.: Effects of sample size on robustness and prediction accuracy of a prognostic gene signature. BMC Bioinform. **10**, 147 (2009)
3. Saeys, Y., Inza, I., Larrañaga, P.: A review of feature selection techniques in bioinformatics. Bioinformatics **23**, 2507–2517 (2007)
4. Bell, D.A., Wang, H.: A formalism for relevance and its application in feature subset selection. Mach. Learn. **41**, 175–195 (2000)
5. Ein-Dor, L., Zuk, O., Domany, E.: Thousands of samples are needed to generate a robust gene list for predicting outcome in cancer. Proc. Nat. Acad. Sci. **103**, 5923–5928 (2006)

6. Davies, S., Russell, S.: NP-completeness of searches for smallest possible feature sets. In: Proceedings of the 1994 AAAI Fall Symposium on Relevance, pp. 37–39 (1994)

7. Lazar, C., Taminau, J., Meganck, S., Steenhoff, D., Coletta, A., Molter, C., De Schaetzen, V., Duque, R., Bersini, H., Now, A.: A survey on filter techniques for feature selection in gene expression microarray analysis. IEEE/ACM Trans. Comput. Biol. Bioinf. (TCBB) **9**, 1106–1119 (2012)

8. Albrecht, A., Vinterbo, S.A., Ohno-Machado, L.: An Epicurean learning approach to gene-expression data classification. Artif. Intell. Med. **28**, 75–87 (2003)

9. Gheyas, I.A., Smith, L.S.: Feature subset selection in large dimensionality domains. Pattern Recogn. **43**, 5–13 (2010)

10. Guyon, I., Weston, J., Barnhill, S., Vapnik, V.: Gene selection for cancer classification using support vector machines. Machine Learning **46**, 389–422 (2002)

11. Zhou, X., Tuck, D.P.: MSVM-RFE: extensions of SVM-RFE for multiclass gene selection on DNA microarray data. Bioinformatics **23**, 1106–1114 (2007)

12. Mundra, P.A., Rajapakse, J.C.: SVM-RFE with MRMR filter for gene selection. IEEE Trans. NanoBiosci. **9**, 31–37 (2010)

13. Brown, G., Pocock, A., Zhao, M.-J., Luj, N.M.: Conditional likelihood maximisation: a unifying framework for information theoretic feature selection. The. J. Mach. Learn. Res. **13**, 27–66 (2012)

14. Ding, C., Peng, H.: Minimum redundancy feature selection from microarray gene expression data. J. Bioinf. Comput. Biol. **3**, 185–205 (2005)

15. Fleuret, F.: Fast binary feature selection with conditional mutual information. J. Mach. Learn. Res. **5**, 1531–1555 (2004)

16. Yu, L., Liu, H.: Efficient feature selection via analysis of relevance and redundancy. J. Mach. Learn. Res. **5**, 1205–1224 (2004)

17. Nutt, C.L., Mani, D., Betensky, R.A., Tamayo, P., Cairncross, J.G., Ladd, C., Pohl, U., Hartmann, C., Mclaughlin, M.E., Batchelor, T.T.: Gene expression-based classification of malignant gliomas correlates better with survival than histological classification. Cancer Res. **63**, 1602–1607 (2003)

18. Golub, T.R., Slonim, D.K., Tamayo, P., Huard, C., Gaasenbeek, M., Mesirov, J.P., Coller, H., Loh, M.L., Downing, J.R., Caligiuri, M.A.: Molecular classification of cancer: class discovery and class prediction by gene expression monitoring. Science **286**, 531–537 (1999)

19. Cover, T.M., Thomas, J.A.: Elements of Information Theory. Wiley, New York (2012)

20. Kohavi, R., John, G.H.: Wrappers for feature subset selection. Artif. Intell. **97**, 273–324 (1997)

21. Saeys, Y., Abeel, T., Van de Peer, Y.: Robust feature selection using Ensemble feature selection techniques. In: Daelemans, W., Goethals, B., Morik, K. (eds.) ECML PKDD 2008, Part II. LNCS (LNAI), vol. 5212, pp. 313–325. Springer, Heidelberg (2008)

22. Subramanian, A., Tamayo, P., Mootha, V.K., Mukherjee, S., Ebert, B.L., Gillette, M.A., Paulovich, A., Pomeroy, S.L., Golub, T.R., Lander, E.S.: Gene set enrichment analysis: a knowledge-based approach for interpreting genome-wide expression profiles. Proc. Nat. Acad. Sci. **102**, 15545–15550 (2005)

23. Wang, J., Duncan, D., Shi, Z., Zhang, B.: WEB-based GEne SeT AnaLysis Toolkit (WebGestalt): update 2013. Nucleic Acids Res. **41**, W77–W83 (2013)

24. Coussens, L.M., Zitvogel, L., Palucka, A.K.: Neutralizing tumor-promoting chronic inflammation: a magic bullet? Science **339**, 286–291 (2013)

Influence of the Interest Operators
in the Detection of Spontaneous
Reactions to the Sound

A. Fernández$^{(\boxtimes)}$, J. Marey, M. Ortega, and M.G. Penedo

Departamento de Computación, Universidade da Coruña, Coruña, Spain
{alba.fernandez,jorge.marey,mortega,mgpenedo}@udc.es

Abstract. Hearing plays a key role in our social participation and daily activities. In health, hearing loss in one of the most common conditions, so its diagnosis and monitoring is highly important. The standard test for the evaluation of hearing is the pure tone audiometry, which is a behavioral test that requires a proper interaction and communication between the patient and the audiologist. This need of understanding is which makes this test unworkable when dealing with patients with severe cognitive decline or other communication disorders. In these particular cases, the audiologist base the evaluation in the detection of spontaneous facial reaction that may indicate auditory perception. With the aim of supporting the audiologist, a screening method that analyzes video sequences and seeks for eye gestural reactions was proposed. In this paper, a comprehensive survey about one of the crucial steps of the methodology is presented. This survey determines the optimal configuration for all of them, and evaluates in detail their combination with different classification techniques. The obtained results provide a global vision of the suitability of the different interest operators.

Keywords: Hearing assesment · Gesture information · Eye movement analysis

1 Introduction

Hearing loss occurs when the sensitivity to the sounds normally heard is diminished. It can affect to all age ranges, however, there is a progressive loss of sensitivity to hear high frequencies with increasing age. Considering that population aging is nowadays a global phenomenon [1,2] and that the studies of Murlow [3] and A. Davis [4] point out that hearing loss is the disability more closely related to aging, the number of elder people with hearing impairment is increasingly higher.

Hearing plays a key role in the process of "active aging" [5]. Active aging is the attempt to maximize the physical, mental and social well-being of our elders. Hearing plays a key role in the process of active aging. This high impact of the hearing on the aging process makes necessary to conduct regular hearing checks if any symptom of decreased hearing is noticed.

© Springer International Publishing Switzerland 2015
B. Duval et al. (Eds.): ICAART 2014; LNAI 8946, pp. 346–361, 2015.
DOI: 10.1007/978-3-319-25210-0_21

Pure Tone Audiometry (PTA) is unequivocally described as the gold standard for audiological evaluation. Results from pure-tone audiometry are used for the diagnosis of normal or abnormal hearing sensitivity, namely, for the assessment of hearing loss. It is a behavioral test so it may involve some operational constraints, especially among population with special needs or disabilities.

In the standard protocol for a pure-tone audiometry the audiologist delivers auditory stimuli to the patient at different frequencies and intensities. The patient is wearing earphones and the auditory stimuli are delivered through an audiometer to these earphones. The patient must indicate somehow (typically by raising his hand) when he perceives the stimuli. In the case of patients with cognitive decline or other communication disorders, this protocol becomes unenforceable, since the interaction audiologist-patient is practically impossible. Taking in consideration that cognitive decline is highly related to age (and hearing loss is also related to age), the number of patients with communication difficulties to be assessed is potentially substantial.

Since a typical interaction question-answer is not possible, the audiologist focuses his attention on the patient's behavior, trying to detect spontaneous reactions that can be a signal of perception of the auditory stimuli. These reactions are shown by facial expression changes, mainly expressed in the eye region. Typically, changes on the gaze direction or excessive opening of the eyes could indicate perception of the auditory stimuli. The interpretation of these reactions requires broad experience from the audiologist. The reactions are totally dependent on the patient, each patient may react differently and even a same patient may show different reactions at different times, since these reactions are completely unconscious. Moreover, although the audiologist has experience enough, it is a entirely subjective procedure. This subjectivity greatly limits the reproducibility and robustness of the measurements performed in different sessions or by different experts leading to inaccuracies in the assessments.

The development of an automatic method capable of analyzing the patient facial reactions will be very helpful for assisting the audiologists in the evaluation of patients with cognitive decline and, this way, reducing the imprecision associated to the subjectivity of the problem. It is important to clarify at this point that other techniques aimed at the interpretation of facial expressions are not applicable in this domain. Most of these techniques (such as [6] or [7]) are focused on the classification of the facial expressions into one of the typical expressions (anger, surprise, happiness, disgust, etc.). The facial expressions of this particular group of patients do not directly correspond to any of those categories. They are specific to each patient, without following a fixed pattern, and, as commented before, they can even vary within the same patient.

Some initial researches have already been developed in [8] considering the particularities of this domain. In this work, one of the most important steps of this methodology is going to be addressed in detail: the selection of the interest points used for the optical flow (whose behavior affects every subsequent step of the methodology). Different interest point detectors are going to be studied in order to find the most appropriate for this specific problem.

The remainder of this paper is organized as follows: Section 2 presents the methodology used as base for this work and introduces the parts over which this study will be focused, Sect. 3 is devoted to the experimental results and their interpretation. Finally, in Sect. 4 some conclusions and future work ideas are presented.

2 Methodology

As depicted in the Introduction, the development of an automatic solution capable of detecting facial movements as a response to auditory stimuli could be very helpful for the audiologists in the evaluation of patients with cognitive decline. An initial approach was proposed in [8], which is going to be the base for this study. A general scheme of the original methodology is shown in Fig. 1.

Fig. 1. Schematic representation of the methodology.

This method focuses its attention on the eye region, which has been highlighted by the audiologists as the most representative for the facial reactions of these patients. This methodology is addressed in a global way since movements can occur in any area of the region. In order to address the problem from a global viewpoint but having a manageable descriptor, interest points are going to be used.

Therefore, the first steps of the methodology are aimed to the location of this particular area. The proposed approach previously locate the face region in order to reduce the search area, and then locates the eye region within the face region. Both regions are located applying the Viola-Jones object detector [9], the face by the application of a cascade provided by the OpenCV tool, and the eye region with a cascade specifically trained for this region. Once the eye area is located, the motion estimation begins. To that end, two separated frames are analyzed to determine the movement produced between them. The motion is estimated by applying the iterative Lucas-Kanade [10] optical flow method with pyramids [11]. Once the motion has been detected, it is characterized based on several descriptors, this characterization will allow to apply a classifier which will determine the type of movement occurred. All these stages are further detailed in [8].

Since the results of the optical flow depend on the interest points that the method receives as input, choosing these interest points is a crucial step, since the following steps will be highly affected by the results of this stage of the methodology.

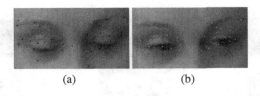

(a)   (b)

Fig. 2. Sample of the optical flow results: (a) Reference image with the interest points represented in blue. (b) Second image with the corresponding points obtained by the optical flow represented in red (Color figure online).

Firstly, it is important to describe the characteristics that define an interest point. Usually, these points are defined by qualities like: well-defined position on the image, mathematically well-founded, rich in terms of local information and stable to global perturbations. These properties are assigned regularly to corners or to locations where the color of the region suffers a big change.

Considering this, we want to choose those interest points than can be easily matched by the optical flow. To select them, an analysis between different interest operators was conducted. Each of these methods has different foundations, and consequently, a different way of working, so the results that one provides can be very different from those provided by the others. The analysis performed in this work is further explained in Sect. 3.

Once the interest points are detected over the reference image, the application of the optical flow method will provide the location of these points on the second image. This means that we obtain a correspondence between the two images. In Fig. 2 we can see a sample of this: Fig. 2(a) is the reference image where the detected interest points are represented in blue, Fig. 2(b) is the second image showing the correspondence of the interest points obtained by the optical flow method.

With this information, we can build vectors where the interest point detected on the reference image is the origin, and the corresponding point on the second image is the end of the vector. These vectors represent the direction and the amount of movement of each point of the reference image. Figure 3 shows the obtained vectors for the movement between Fig. 2(a) and (b). Globally, this information is going to be interpreted as the movement produced in the eye region between two images.

This representation can be modified into a more intuitive way with vectors depicted as arrows. The arrow for a particular point represents its movement

Fig. 3. Movement vectors between Fig. 2(a) and (b).

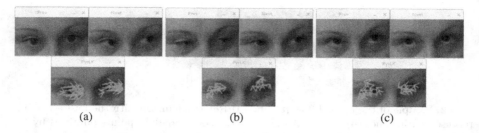

(a)   (b)   (c)

Fig. 4. Samples of the movement vectors for different changes on the eye region: (a) gaze shift to the left, (b) eye opening and (c) gaze shift to the right.

from the initial time considered to the final one. With this representation it is possible to visually analyze the results obtained by the optical flow method.

In Fig. 4 several samples of this representation can be observed. Vectors shown in this figure are the vectors with a length greater than a established threshold (those that represent significant movements), as they are a good example of how with this technique it is possible to detect the changes on the eye region. In Fig. 4(a) the gaze direction is moving towards the patient's left, this movement is detected by the optical flow and represented by the vectors that are pointing to the patient's left. In Fig. 4(c) the movement is the opposite, the gaze direction moves slightly to the patient's right and the optical flow is still capable of detecting it. In the case of Fig. 4(b) the eyes open slightly, so in this case, vector are pointing up following the movement occurred within the region.

The obtained vectors are characterized according to some features, so it is possible to obtain a descriptor that can be classified into one of our considered movements. After reaching consensus with the experts, four typical movements were identified as the most relevant: eye opening, eye closure, gaze shift to the left or gaze shift to the right. The features used for obtaining these descriptors are related with the strength, orientation and dispersion of the movement vectors, the specific way in which the descriptors are formed is detailed in [8].

As mentioned, the final aim is to classify the movements produced within this region into one of the four categories previously presented. To that end, different classifiers have been tested too. The four established classes serve as an initial test set that allow to draw conclusions about the most suitable interest operator for this domain. These conclusions will allow to establish a foundation for moving forward and then including new classes that may be deemed relevant by the audiologists. The analysis of the different alternatives for the classification will be addressed on the next section.

3   Experimental Results

As commented before, several interest points detectors were tested in order to find the most appropiate for this domain. The detectors tested are: Harris corner detector [12], Good Features to Track [13], SIFT [14], SURF [15], FAST [16,17]

Fig. 5. Sample of the particular setup of the video sequences.

and a particular version of Harris with a little modification. Also different classification techniques were tested, in order to find the best detector-classifier combination.

Video sequences show patients seated in front of the camera as in Fig. 5. As showed in the picture, the audiometer is also recorded so the audiologist can check when he was delivering the stimuli. Video sequences are Full HD resolution (1080 × 1920 pixels) and 25 FPS (frames per second). Despite the high resolution of the images, it is important to take into account that the resolution of the eye region will not be as optimal, and moreover, lighting conditions will affect considerably.

Test were conducted with 9 different video sequences, each one from a different patient. Each audiometric test takes between 4 and 8 min. Considering that video sequences have a frame rate of 25 FPS, an average video sequence of 6 min will have 9000 frames, implying a total number of 81000 frames for the entire video set. Taking into account that reactions only occur in a timely, we finally have 128 pairs of frames to be considered. Since each eye is considered separately, the test set will consist of 256 movements. These movements are labeled into four classes depending in the movement they represent (see Table 1).

Three different experiments were conducted in order to find the best detector for this domain. The three experiments are:

1. Find the best classifiers.
2. Find the best configuration parameters for each interest points detector.
3. Evaluate the detector-classifier results.

Table 1. Number of samples for each class of movement.

Eye opening	Eye closure	Gaze left	Gaze right
80	82	46	48

3.1   Classifier Selection

In this part of the research, different classifiers were tested with the aim of selecting the three best methods for applying them on the following tests. The considered classifiers are provided by the WEKA tool [18], and they are: Naive Bayes, Logistic, Multilayer Perceptron, Random Committee, Logistic Model Trees (LMT), Random Tree, Random Forest and SVM.

To obtain these results, 18 tests were conducted for each pair detector-classifier, where each one of these tests is the result of a 10-fold cross validation. Computing the average per method (without considering the detector used) we obtain the results shown in Fig. 6. As it can be observed on this graph, all the methods obtain an accuracy between 60 and 75 %. Worst results are observed for Naive Bayes, Logistic and Random Tree. Best results are obtained with SVM, followed by Random Committee and Random Forest, so these are going to be the three classifiers considered for the next survey.

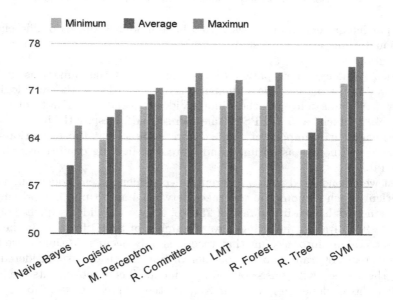

Fig. 6. Minimum, maximum and average success percentage by classifier.

3.2   Adjustment of Parameters

This methodology makes use of different parameters, that are going to be adjusted according to this experiment. The parameter adjustment is performed dependently on the method used. The parameters studied in this section are:

– Number of detected points: it indicates the number of points that the detector needs to select. Very few points may not be enough to create a correct motion descriptor and a number too high may introduce too much noise.

– Minimum percentage of equal points to remove the movement: sometimes, the detected motion may be due to global motion between the two frames and not to a motion within the region. This will imply a high number of vectors with the same direction and strength. With the aim of removing this offset component, the parameter λ is introduced. This parameter indicates the required minimum percentage of equal vectors to be considered a global motion, and consequently, discard them.

– Minimum Length: very short vectors will not be representative of motion. In order to select the representative vectors three classes were established depending on the length of the vector: u_1 for vectors smaller than 1.5 pixels, u_2 for vectors between 1.5 and 2.5 pixels and u_3 for vector between 2.5 and 13 pixels (vectors larger than 13 pixels will be considered erroneous). Vectors in u_1 are considered too small and are not taked into account for the descriptor, while vectors in u_3 are considered relevant and are always part of the descriptors. The inclusion or not of vectors in u_2 is going to be studied on this section.

Harris. Harris has a particular behavior, it detects few points concentrated in areas with high contrast. The obtained results are represented in Fig. 7. Each line represents a classifier (Random Committee, Random Forest and SVM), distinguishing between using only u_3 vectors (green lines) and u_2 and u_3 vectors (blue lines).

It can be observed that, the higher λ is, the better the results are. Moreover, the inclusion of the vectors in u_2 shows worse results. It can be noticed that in Fig. 7(c) there is a value nearly the 100 % of accuracy. This value is an outlier that may not be repeatable, since it breaks the tendency of the other values. However, it confirms the tendency that with higher λ values the accuracy increases.

Good Features to Track. This detector was specifically designed for the calculation of the optical flow. Figure 8 shows the obtained results for this classifier. As it can be observed, results are quite consistent regardless of the values of the

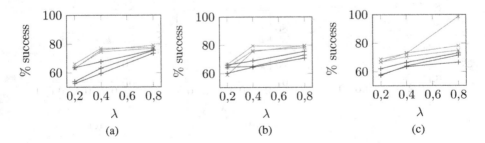

Fig. 7. Classification results for Harris. Green lines for u_3 vectors and blue lines for u_3 and u_2 vectors. Each of the three lines for each color corresponds to a different classifier. (a) For 40 points of interest. (b) 80 points. (c) 160 points (Color figure online).

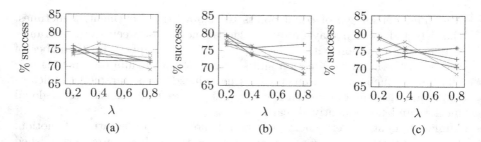

Fig. 8. Classification results for Good Features to Track. Green lines for u_3 vectors and blue lines for u_3 and u_2 vectors. Each of the three lines for each color corresponds to a different classifier. (a) For 40 points of interest. (b) 80 points. (c) 160 points (Color figure online).

parameters. The behavior is better for low values of λ, and also considering 80 points of interest. Although the results are very similar, including vectors in u_2 slightly increases the success rate.

SIFT. The SIFT detections are quite similar to the detections of Good Features to Track. Its results are also broadly similar (see Fig. 9). Unlike the previous method, in this case the results for 80 points of interest are slightly worse than for 40 or 160. The λ parameter does not affect the results too much. Inclusion of the intermediate vectors (u_2) offers also better results.

SURF. SURF detector is a very particular method, since it is very selective about the detected points. With these images, it is not possible to select more than 35–40 points, even with very permissive thresholds. Due to this particularity, the only results obtained are the ones shown in Fig. 10. Better results are obtained when including vectors in u_2, for which the most appropriate value of λ is 0.8.

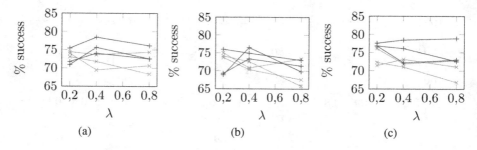

Fig. 9. Classification results for SIFT. Green lines for u_3 vectors and blue lines for u_3 and u_2 vectors. Each of the three lines for each color corresponds to a different classifier. (a) For 40 points of interest. (b) 80 points. (c) 160 points (Color figure online).

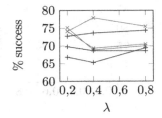

Fig. 10. Classification results for SURF. Green lines for u_3 vectors and blue lines for u_3 and u_2 vectors. Each of the three lines for each color corresponds to a different classifier (Color figure online).

FAST. The interest points detected by FAST are quite significant for this domain. Charts with the results can be observed in Fig. 11. Regarding the length of the vectors, results vary according to the number of points considered. For 40 and 80 points, best results are obtained only considering the strong vectors (u_3), while for 160 points best results are obtained when considering vectors in u_3 and u_2. For 40 points of interest the most appropriate is a high value for λ, for 80 points the results are quite stable regardless of the value of λ, and for 160 points low values for λ offer better results.

Harris Modified. The original Harris detector detects few points in areas with high contrast. To achieve a greater separation between the points, and therefore more representative points, a location of the local maximums is conducted. Also a thresholding is applied over the Harris image, and finally, the *and* operation is computed with these two images, obtaining this way more distributed interest points.

Results for this alternative version of Harris are charted in Fig. 12. These results are similar to the ones obtained with FAST. In the general case, better results are obtained considering only vectors in u_3. For 80 and 160 interest points,

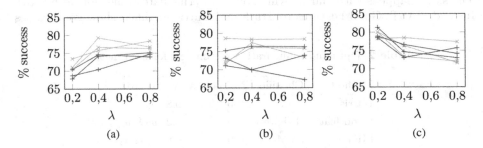

(a)   (b)   (c)

Fig. 11. Classification results for FAST. Green lines for u_3 vectors and blue lines for u_3 and u_2 vectors. Each of the three lines for each color corresponds to a different classifier. (a) For 40 points of interest. (b) 80 points. (c) 160 points (Color figure online).

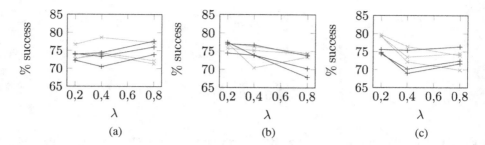

Fig. 12. Classification results for Harris modified. Green lines for u_3 vectors and blue lines for u_3 and u_2 vectors. Each of the three lines for each color corresponds to a different classifier. (a) For 40 points of interest. (b) 80 points. (c) 160 points (Color figure online).

the best behavior occurs for the lower value of λ (0.2). In the case of considering 40 points, best results occur for λ equal to 0.4.

3.3   Final Evaluation of the Results

Once the behavior of the different methods in relation to the configuration of their parameters has been analyzed, we are going to compare here the results of the different methods with their best configuration. The optimum configuration parameters and classifiers for each method are detailed in Table 2.

Results are shown graphically for better understanding. In order to assess the capacity of each one of the interest operators in the detection of the relevant movements the obtained descriptors are compared with the ground truth of movements previously labeled by the experts.

The graph below (Fig. 13) shows the true positive and false positive rate ($T_{tp}(d)$ and $T_{fp}(d)$ respectively). It can be note that SURF has a good value for the false positive rate, but a poor value for the true positive rate. SIFT is the opposite case, it has a good value for the $T_{tp}(d)$ but poor for the $T_{fp}(d)$. The same happens with Harris, which offers intermediate values for both rates. Instead, FAST, Good Features to Track and Harris modified show good values

Table 2. Optimum configuration parameters for each method.

Method	Classifier	No. points	λ	Vectors
Harris	SVM	160	0.8	u_3
Good feat	R. Forest	80	0.2	u_2 & u_3
SIFT	SVM	160	0.8	u_2 & u_3
SURF	SVM	40	0.4	u_3
FAST	R. Forest	160	0.2	u_2 & u_3
Harris mod	SVM	160	0.2	u_3

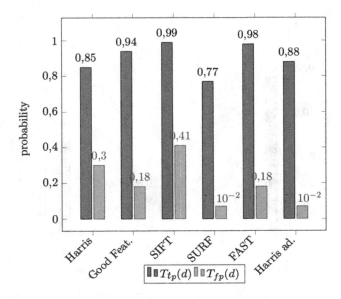

Fig. 13. True positive rate $(T_{tp}(d))$ and false negative rate $(T_{fp}(d))$ for the different methods.

for both rates. Good Features and FAST offer almost equivalent results, while Harris modified has a worst $T_{tp}(d)$ but it is compensated with a optimum $T_{fp}(d)$ rate.

Given the previous results, only FAST, Good Features to Track and Harris modified are considered for the last evaluation. Figure 14 shows the true positive rate in detection $(T_{tp}(d))$, the specificity $(1 - T_{fp}(d))$ and the true positive rate in classification $(T_{tp}(c))$. All the methods have a similar value for the true positive rate in classification $(T_{tp}(c))$. FAST offers better results than Good Feature for the three evaluated measures; so between these two methods, FAST would be chosen. Comparing between FAST and Harris modified, it can be observed that the $T_{tp}(c)$ is quite similar, while the $T_{tp}(d)$ and the specificity are slightly opposite. FAST offers better results for the $T_{tp}(d)$ while with Harris better results are obtained for the specificity. The decision of choosing one or another depends on the suitable results for this domain. If we want to reduce the number of false positives Harris is the best solution, while if the true positive detections are more important, FAST is the method that should be chosen.

3.4   Classification by Classes

A more detailed survey about the results was conducted in order to analyze the classification results by classes. Tables 3, 4 and 5 represent the confusion matrices for the three best interest operators selected from the previous results combined with the best classifier for each one of them, i.e., *FAST - Random*

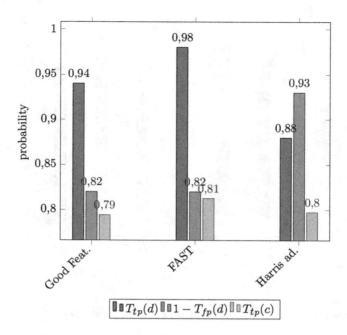

Fig. 14. True positive rate ($T_{tp}(d)$) and specificity ($1 - T_{tp}(d)$) for detection and true positive rate for classification ($T_{tp}(c)$).

Forest (Table 3), *Good Features to Track - Random Forest* (Table 4) and *Harris ad. - SVM* (Table 5). Where *Open*, *Close*, *Left* and *Right* correspond with the considered movement categories: eye opening, eye closure, gaze shift to the left and gaze shift to the right.

As it can be inferred from these confusion matrices, the classification accuracy for the eye closure movement is always over the 90 %. Instead, the worst results are obtained for the movements of gaze shift to the right and gaze shift to the left. None of them is able to achieve the 80 % of accuracy; which worsens in the case of *Harris ad.* and *Good Features to Tack* where the accuracy falls to 60 %. This negative effect for *Harris ad.* and *Good Features to Tack* is offset by a better behavior for the eye opening movements, where they outperform *FAST*.

Table 3. Confusion matrix for *FAST - Random Forest*.

	Open	Close	Left	Right
Open	0.772	0.050	0.076	0.101
Close	0.012	0.927	0.060	0
Left	0.068	0.205	0.727	0
Right	0.174	0	0.065	0.761

Table 4. Confusion matrix for *Good Features to Track - Random Forest.*

	Open	Close	Left	Right
Open	0.818	0.039	0.026	0.117
Close	0.049	0.89	0.061	0
Left	0.075	0.3	0.625	0
Right	0.25	0	0.023	0.727

Table 5. Confusion matrix for *Harris ad. - SVM.*

	Open	Close	Left	Right
Open	0.811	0.04	0.027	0.122
Close	0.012	0.901	0.086	0
Left	0.033	0.2	0.767	0
Right	0.357	0	0.048	0.595

It can be observed that the main problem in classification occurs when distinguishing between the classes eye opening and gaze shift to the right, in this case, the misclassification occurs in both directions, sometimes samples of the class eye opening are classified as gaze shift to the right and sometimes the opposite happens. It can be also observed that there is a considerable number of misclassifications between classes eye closure and gaze shift to the left, but in this case, it only occurs when classifying samples of the class gaze shift to the left, and not in the opposite direction. Between the remaining classes there are also some errors in the classification, but they are almost negligible.

For a more general understanding, in Table 6 we show the true positive rate in classification $T_{tp}(c)$ for each class (the diagonal of the confusion matrices previously presented) and the false positive rate in classification $T_{fp}(c)$ computed from the misclassifications. Likewise, we show the ROC area value, which provides a measure of how good the classification is.

Table 6. Results by classes for *FAST*, *Good Features to Track* and *Harris ad.*

		Open	Close	Left	Right	Average
FAST	$T_{tp}(c)$	0.772	0.927	0.727	0.761	0.813
	$T_{fp}(c)$	0.07	0.077	0.068	0.039	0.066
	ROC area	0.896	0.969	0.88	0.934	0.924
Good Features	$T_{tp}(c)$	0.818	0.89	0.625	0.727	0.794
	$T_{fp}(c)$	0.108	0.093	0.039	0.045	0.08
	ROC area	0.928	0.966	0.929	0.941	0.943
Harris ad.	$T_{tp}(c)$	0.811	0.901	0.767	0.595	0.797
	$T_{fp}(c)$	0.111	0.062	0.056	0.049	0.0075
	ROC area	0.923	0.974	0.915	0.907	0.94

As it can be inferred from Table 6 all the three methods show successfully results for the ROC area, near to the ideal, where the best of them is *Good Features to Track* with a 0.943 value. From the evaluation of the false positive rate $T_{fp}(c)$ the results confirm the same idea that was deducted from the evaluation of the confusion matrices, which is that the class eye opening is the worst in terms of classification, since it has the highest $T_{fp}(c)$ value. In general terms, the obtained results are promising for both the detection and the classification for any of the interest operators presented here. This indicates that any of these three interest operators could provide optimal results when integrated with in the final methodology.

4   Conclusions

A methodology for supporting the audiologists in the detection of gestural reactions within the eye region was developed in previous research, but interest operator analysis for motion detection was not studied in detail. This paper analyzes different methods for the selection of the interest points, determines the best configuration parameters for each one of them, and it also analyzes its behavior according to different classifiers. We have also analyzed the behavior of the best combinations interest operator - classifier by classes, in order to evaluate the performance of each one of these combinations for each one of the movement categories established by the audiologists. The results obtained with new interest operators surpass the previous approach in terms of accuracy.

In clinical terms, the choice of a suitable interest points detector for this domain may improve the accuracy in the detection and interpretation of the gestural reactions.

Future works will involve an extension of the training dataset so a robust classifier can be trained with the configurations established by this work. This classifier may then by applied over the video sequences in order to detect the relevant movements and, thus, serve to assist the audiologists.

Acknowledgements. This research has been partially funded by Ministerio de Ciencia e Innovación of the Spanish Government through the research projects TIN2011-25476.

References

1. Davis, A.: The prevalence of hearing impairment and reported hearing disability among adults in Great Britain. Int. J. Epidemiol. **18**, 911–917 (1989)
2. IMSERSO: Las personas mayores en España. In: Instituto de Mayores y Servicios Sociales (2008)
3. Murlow, C., Aguilar, C., Endicott, J., Velez, R., Tuley, M., Charlip, W., Hill, J.: Association between hearing impairment and the quality of life of elderly individuals. J. Am. Geriatr. Soc. **38**, 45–50 (1990)
4. Davis, A.: Prevalence of hearing impairment. In: Davis, A. (ed.) Hearing in Adults, Chapter 3, pp. 45–46. Whurr Ltd., London (1995)

5. Espmark, A., Rosenhall, U., Erlandsson, S., Steen, B.: The two faces of presbiacusia: hearing impairment and psychosocial consequences. Int. J. Audiol. **42**, 125–135 (2002)
6. Happy, S., George, A., Routray, A.: A real time facial expression classification system using local binary patterns. In: 2012 4th International Conference on Intelligent Human Computer Interaction (IHCI), pp. 1–5 (2012)
7. Chew, S.W., Rana, R., Lucey, P., Lucey, S., Sridharan, S.: Sparse temporal representations for facial expression recognition. In: Ho, Y.-S. (ed.) PSIVT 2011, Part II. LNCS, vol. 7088, pp. 311–322. Springer, Heidelberg (2011)
8. Fernandez, A., Ortega, M., Penedo, M.G., Cancela, B., Gigirey, L.M.: Automatic eye gesture recognition in audiometries for patients with cognitive decline. In: Kamel, M., Campilho, A. (eds.) ICIAR 2013. LNCS, vol. 7950, pp. 27–34. Springer, Heidelberg (2013)
9. Viola, P., Jones, M.: Robust real-time object detection. Int. J. Comput. Vis. **4**, 85–107 (2001)
10. Lucas, B.D., Kanade, T.: An iterative image registration technique with an application to stereo vision. In: Proceedings of the 7th International Joint Conference on Artificial Intelligence, IJCAI 1981 - vol. 2, pp. 674–679. Morgan Kaufmann Publishers Inc., San Francisco, CA, USA (1981)
11. Bouguet, J.Y.: Pyramidal implementation of the Lucas-Kanade feature tracker: description of the algorithm. Intel Corporation, Microprocessor Research Labs (2000)
12. Harris, C., Stephens, M.: A combined corner and edge detector. In: Proceedings of the 4th Alvey Vision Conference, pp. 147–151 (1988)
13. Shi, J., Tomasi, C.: Good features to track. In: 1994 IEEE Computer Society Conference on Computer Vision and Pattern Recognition 1994, Proceedings CVPR 1994, pp. 593–600 (1994)
14. Lowe, D.G.: Distinctive image features from scale-invariant keypoints. Int. J. Comput. Vis. **60**(2), 91–110 (2004)
15. Bay, H., Ess, A., Tuytelaars, T., Van Gool, L.: Speeded-up robust features (surf). Comput. Vis. Image Underst. **110**(3), 346–359 (2008)
16. Rosten, E., Drummond, T.: Fusing points and lines for high performance tracking. In: Tenth IEEE International Conference on Computer Vision 2005, ICCV 2005, vol. 2, pp. 1508–1515 (2005)
17. Rosten, E., Drummond, T.W.: Machine learning for high-speed corner detection. In: Leonardis, A., Bischof, H., Pinz, A. (eds.) ECCV 2006, Part I. LNCS, vol. 3951, pp. 430–443. Springer, Heidelberg (2006)
18. Hall, M., Frank, E., Holmes, G., Pfahringer, B., Reutemann, P., Witten, I.H.: The WEKA data mining software: an update. SIGKDD Explor. Newsl. **11**(1), 10–18 (2009)

Confidence Measure for Experimental Automatic Face Recognition System

Pavel Král[1,2(⊠)] and Ladislav Lenc[1,2]

[1] Department of Computer Science and Engineering, Faculty of Applied Sciences,
University of West Bohemia, Plzeň, Czech Republic
[2] New Technologies for the Information Society, Faculty of Applied Sciences,
University of West Bohemia, Plzeň, Czech Republic
{pkral,llenc}@kiv.zcu.cz
http://nlp.kiv.zcu.cz

Abstract. This paper deals with automatic face recognition in order
to propose and implement an experimental face recognition system. It
will be used to automatically annotate photographs taken in completely
uncontrolled environment. Recognition accuracy of such a system can be
improved by identification of incorrectly classified samples in the post-
processing step. However, this step is usually missing in current systems.
In this work, we would like to solve this issue by proposing and integrating a confidence measure module to identify incorrectly classified
examples. We propose a novel confidence measure approach which combines four partial measures by a multi-layer perceptron. Two individual
measures are based on the *posterior* probability and two other ones use
the *predictor* features. The experimental results show that the proposed
system is very efficient, because almost all erroneous examples are successfully detected.

Keywords: Face recognition · Czech News Agency · Confidence measure · Multi-layer perceptron · Scale Invariant Feature Transform (SIFT)

1 Introduction

Automatic face recognition consists in the use of a computer for identification of a person from a digital photograph. This area has been focused on by
many researchers and many algorithms have been proposed during the past two
decades. Nowadays, face recognition can be seen as one of the most progressive
biometric authentication methods and represents a key task in several commercial or law enforcement applications as for example surveillance of wanted
persons, access control to restricted areas, automatic annotation of the photos
used in the recently very popular photo sharing applications or in the social
networks, etc.

The majority of the current applications achieves high recognition accuracy
only in the particular conditions (sufficiently aligned faces, similar face pose and
lighting conditions, etc.). Unfortunately, their recognition results are significantly

© Springer International Publishing Switzerland 2015
B. Duval et al. (Eds.): ICAART 2014; LNAI 8946, pp. 362–378, 2015.
DOI: 10.1007/978-3-319-25210-0_22

worse when these constraints are not fulfilled. Many approaches to resolve this issue have been proposed, however only few of them perform well in a fully uncontrolled environment.

In our previous work, we proposed the SIFT based Kepenekci face recognition method [19] and showed that it significantly outperforms other approaches particularly on uncontrolled face images. However, its recognition accuracy is still not perfect. Therefore, we proposed in [17] two Confidence Measure (CM) approaches in order to detect and handle incorrectly recognized examples. These approaches are based on the *posterior* class probability. We experimentally showed that these approaches are very promising in our task. However, the further improvement of the results is beneficial.

The main goal of this paper consists in proposing a novel confidence measure approach which will be integrated into our experimental automatic face recognition system. This method combines two previously proposed measures with two novel ones in a supervised way using a multi-layer perceptron classifier. The novel measures are based on the *predictor* features which characterize our face model.

The proposed system will be used by the Czech News Agency (ČTK)[1] to annotate people in photographs during insertion into the ČTK database[2]. Its main strength is to successfully process photos of a great number of different persons taken in a totally uncontrolled environment. The system (with the source code) is publicly available for research purposes for free.

The following section gives a brief overview of important face recognition and confidence measure approaches. Some existing face recognition systems are also mentioned at the end of this section. Section 3 describes our face recognition and confidence measure methods. Section 4 details the architecture of the proposed system. Section 5 evaluates and compares the performance of our system on the ČTK corpus. In the last section we discuss the achieved results and give some further research directions.

2   Related Work

This section is composed of three parts. The successful face recognition approaches are described in the first part, while the second part is focused on the confidence measure task itself. This section further summarizes some existing face recognition systems.

2.1   Face Recognition

One of the first successful approaches is Eigenfaces [34]. This approach is based on the Principal Component Analysis (PCA). Unfortunately, it is sensitive on variations in lighting conditions, pose and scale. However, the PCA based

---

[1] http://www.ctk.eu.
[2] http://multimedia.ctk.cz/en/foto/.

approaches are still popular, as shown in [24]. Another method, the Fisher-faces [6], is derived from Fisher's Linear Discriminant (FLD). According to the authors, this approach should be less sensitive to changing lighting conditions than Eigenfaces.

Independent Component Analysis (ICA) can be also successfully used in the automatic face recognition field [3]. Contrary to Eigenfaces, ICA uses higher order statistics. It thus provides more powerful data representation. The authors showed that ICA performs slightly better than PCA method on the FERET corpus.

Another efficient face recognition approach is the Elastic Bunch Graph Matching (EBGM) [7]. This approach uses features constructed by the Gabor wavelet transform. Several other successful approaches based on Gabor wavelets have been introduced [30]. Some approaches [29] combine the pre-processing with Gabor wavelets with well-established methods such as Eigenfaces, Fish-erfaces, etc. Kepenekci proposes in [15] an algorithm that addresses the main issue of Elastic Bunch Graph Matching, manual labelling of the landmarks. The proposed method outperforms the classical EBGM.

Other successful approaches [1,33] use so-called, Local Binary Patterns (LBP) for facial feature extraction. A modification of the original LBP app-roach called Dynamic Threshold Local Binary Pattern (DTLBP) is proposed in [21]. It takes into consideration the mean value of the neighbouring pixels and also the maximum contrast between the neighbouring points. It is stated there that this variation is less sensitive to the noise than the original LBP method. Another extension of the original method is Local Ternary Patterns (LTP) pro-posed in [32]. It uses three states to capture the differences between the center pixel and the neighbouring ones. Similarly to the DTLBP the LTP is less sen-sitive to the noise. The so called Local Derivative Patterns (LDP) are proposed in [37]. The difference from the original LBP is that it uses features of higher order. It thus can represent more information than the original LBP.

Speeded-Up Robust Features (SURF) [4] is another recent method used for automatic face recognition. This method is invariant to face rotation. To ensure rotation invariance, one orientation is assigned to each key-point. The computa-tion is based on the circular neighbourhood of the key-points.

Recently, the Scale Invariant Feature Transform (SIFT) is successfully used for face recognition [2]. The main advantage of this approach is the ability to detect and describe local features in images. The features are invariant to image scaling, translation and rotation. Moreover, they are also partly invariant to changes in illumination. Therefore, this approach is beneficial for face recognition in real conditions where the images differ significantly. Another approach based on the SIFT, called Fixed-key-point-SIFT (FSIFT), is presented in [16].

For further information about the face recognition, please refer to the survey [5].

2.2   Confidence Measure

Confidence measure is used as a post-processing of the recognition to determine whether the result is correct or not. The incorrectly recognized samples should be removed from the recognition set or another processing (e.g. manual correction) can be further realized.

This technique is mainly used in the automatic speech processing field [27, 36] and is mostly based on the *posterior* class probability. However, two other groups of approaches exist [14]. The first one uses a classifier in order to decide whether the classification is correct or not. This classifier uses a set of the so-called *predictor* features which should have a maximal discriminability between the correct and incorrect classes. The second group uses a likelihood ratio between the *null* (a correct recognition) and the *alternative* (an incorrect recognition) hypotheses.

The confidence measure can be successfully used in other research areas as shown in [28] for genome maps construction, in [12] for stereo vision or in [23] for handwriting sentence recognition.

Another approach related to the confidence measure is proposed by Proedrou et al. in the pattern recognition task [26]. The authors use a classifier based on the nearest neighbours algorithm. Their confidence measure is based on the algorithmic theory of randomness and on transductive learning.

Unfortunately, only few works about the confidence measure in the face recognition domain exist. Li and Wechsler propose a face recognition system which integrates a confidence measure [20] in order to reject unknown individuals or to detect incorrectly recognized faces. Their confidence measure is, as in the previous case, based on the theory of randomness. The proposed approaches are validated on the FERET database.

Eickeler et al. propose and evaluate in [11] five other CMs also in the face recognition task. They use a pseudo 2-D Hidden Markov Model classifier with features created by the Discrete Cosine Transform (DCT). Three proposed confidence measures are based on the *posterior* probabilities and two others on ranking of results. Authors experimentally show that the *posterior* class probability gives better results for the recognition error detection task.

Note that the most of the proposed approaches are unsupervised. However, the supervised [31] and semi-supervised [10] methods have been also proposed.

2.3 Face Recognition Systems

As already shown above, numerous papers presented in the face recognition domain concentrate only on the recognition task itself. Unfortunately, to the best of our knowledge, relatively few works about whole face recognition systems exist.

One example of such a system is proposed in [22]. The system compensates the face position and also solves partial occlusion and different facial expressions. Only one training example per person is used.

Another face recognition system is described in [35]. The training images are well aligned (acquired in controlled conditions) whereas the recognized images are real-world photos. The system is based on the Sparse Representation and Classification (SRC) [13] algorithm. It achieves very good results on the FERET database.

Campadelli et al. present in [9] another face recognition system. The authors localize the face in the images and then compute the facial features. Their face recognition algorithm is based on the EBGM, but the fiducial points are detected completely automatically. The system is evaluated on the FERET corpus. The authors show that their system has recognition scores comparable to the elastic bunch graph matching.

For additional information about the face recognition and the face recognition systems, please refer to the survey [38]. The authors also mention some commercial face recognition systems. Unfortunately, neither the system architecture nor the approaches used are usually reported.

3   Confidence Measure for Face Recognition System

3.1   Face Recognition

For the face recognition task, we use our previously proposed SIFT based Kepenekci method [19] which uses the efficient SIFT algorithm for parametrization and adapted Kepenekci matching [18] for recognition. This method was chosen, because as proven previously, it significantly outperforms other approaches particularly on lower quality real data.

SIFT Parametrization. This algorithm creates an image pyramid with re-sampling between each level to determine potential key-point positions. Each pixel is compared with its neighbours. Neighbours in its level as well as in the two neighbouring levels are analysed. If the pixel is maximum or minimum of all neighbouring pixels, it is considered to be a potential key-point.

For the resulting set of key-points their stability is determined. The locations with low contrast and unstable locations along edges are deleted.

The orientation of each key-point is computed next. The computation is based on gradient orientations in the neighbourhood of the pixel. The values are weighted by the magnitudes of the gradient.

The last step consists in the descriptor creation. The computation involves the 16×16 neighbourhood of the pixel. Gradient magnitudes and orientations are computed in each point of the neighbourhood. Their values are weighted by a Gaussian. For each sub-region of size 4×4 (16 regions), the orientation histograms are created. Finally, a vector containing 128 (16×8) values is created.

Adapted Kepenekci Matching. This approach combines two methods of matching and uses the weighted sum of the two results.

Let T be a test image and G a gallery image. For each feature vector t of face T we determine a set of relevant vectors g of face G. Vector g is relevant iff:

$$\sqrt{(x_t - x_g)^2 + (y_t - y_g)^2} < distanceThreshold \tag{1}$$

where x and y are the coordinates of the feature vector points.

If no relevant vector to vector t is identified, vector t is excluded from the comparison procedure. The overall similarity of two faces OS is computed as the average of similarities between each pair of corresponding vectors as:

$$OS_{T,G} = mean\{S(t,g), t \in T, g \in G\} \tag{2}$$

Then, the face with the most similar vector to each of the test face vectors is determined. The C_i value denotes how many times gallery face G_i was the closest one to some of the vectors of test face T. The similarity is computed as C_i/N_i where N_i is the total number of feature vectors in G_i. The weighted sum of these two similarities is used for similarity measure:

$$FS_{T,G} = \alpha OS_{T,G} + \beta \frac{C_G}{N_G} \tag{3}$$

The face is recognized by the following equation:

$$F\hat{S}_{T,G} = \arg \max_G (FS_{T,G}) \tag{4}$$

The cosine similarity is used for vector comparison.

3.2   Confidence Measure

Posterior Class Probability Approaches. Let $P(F|C)$ be the output of the classifier, where C is the recognized face class and F represents the face features. The values $P(F|C)$ are normalized to compute the *posterior* class probabilities as follows:

$$P(C|F) = \frac{P(F|C).P(C)}{\sum_{I \in \mathcal{FIM}} P(F|I).P(I)} \tag{5}$$

$\mathcal{FIM}$ represents the set of all individuals and $P(C)$ denotes the *prior* probability of the individual's (face) class C.

We propose two different approaches. In the first approach, called **absolute confidence value**, only faces $\hat{C}$ complying with

$$\hat{C} = \arg \max_C (P(C|F)) \tag{6}$$

$$P(\hat{C}|F) > T \tag{7}$$

are considered as being recognized correctly.

The second approach, called **relative confidence value**, computes the difference between the *best* score and the *second best* one by the following equation:

$$P\Delta = P(\hat{C}|F) - \max_{C \neq \hat{C}} (P(C|F)) \tag{8}$$

Only the faces with $P\Delta > T$ are accepted. This approach aims to identify the "dominant" faces among all the other candidates. T is the acceptance threshold and its optimal value is adjusted experimentally.

Note that these two measures working separately were already presented in [17]. However, their description is important in the context of the whole composed approach.

Predictor Feature Approaches. As already stated, this type of approaches uses the features with a maximal discriminability between the correct and incorrect classes to classify the recognition results. Two measures are proposed next.

The first one is based on the number of vectors in the model with the highest output value during the recognition task (i.e. the recognized face model). The number of vectors is given by the results of the SIFT algorithm. A face model with a high number of vectors is more general and it can be more likely identified as a good one. Conversely, a few vector face model is more accurate. Therefore, when this model is chosen as a good one (the highest output value) we assume that it is very probable that the recognition is correct.

Let V be the number of vectors in the face model and let T be the acceptance threshold. Only the faces where $V < T$ are accepted. The optimal value of the threshold T will be set experimentally. This measure is hereafter called the *vector number* approach.

The second measure uses a standard deviation of the similarities among images in the recognized face model. Let the recognized model M be composed of the images $I_1, I_2, ..., I_N$. The S measure is defined as follows:

$$S = \sqrt{\frac{1}{N}\sum_{i=1}^{N}(FS_{I_i,M\setminus I_i} - \mu)^2} \tag{9}$$

where $FS_{I_i,M\setminus I_i}$ is the similarity (see Eq. 3) of the image I_i and a model $M \setminus I_i$ created from the remaining images from model M and μ is computed by the following equation:

$$\mu = \frac{1}{N}\sum_{i=1}^{N} FS_{I_i,M\setminus I_i} \tag{10}$$

Similarly as in the case of the *vector number* measure we suppose that higher standard deviation characterizes a more general face model and vice versa. Therefore, only the recognition results where $S < T$ are accepted. The optimal value of the acceptance threshold T will be set experimentally. This measure is hereafter called the *standard deviation* approach.

Composed Supervised Approach. Let R_k be the score obtained by a partial unsupervised measure k described above and let variable H determines whether the face image is classified correctly or not. A Multi-layer Perceptron (MLP) which models the *posterior* probability $P(H|R_1, .., R_N)$ is used to combine all partial measures in a supervised way. Note that the variable N represents the number of measures to combine

In order to identify the best performing topology, several combinations and MLP configurations are built and evaluated. The MLP topologies will be described in detail in the experimental section.

4   System Description

The proposed system has (as shown in Fig. 1) a modular architecture. It is composed of five modules (see the rectangles) connected by dependencies (see the oriented edges). The input image and the recognition results are represented by parallelograms. The storage of the face representation is shown by the *Face Gallery* sign.

The first module $M1$ deals with face extraction. This module converts a color image into its grey-scale representation, then it performs face detection. The detected face is further extracted from the image in the next step. This module also detects the eyes in the detected face region and transforms and resizes the face.

The second module $M2$ is used to create the face representation. It detects the SIFT key-points and creates a set of SIFT descriptors for a representation of the face image.

The next module $M3$ is used to create a face model M. It uses the SIFT representations of the face images (output of module $M2$) and saves them to the face gallery.

The fourth module $M4$ deals with face recognition. A recognized face is compared to the face models stored in the *Face Gallery* and the most similar model is chosen as the recognized one.

The last *confidence measure* module $M5$ is dedicated to identifying whether the recognition result is correct or not. This unique step is particularly important, because when the user knows that the recognition is probably not correct, he can manually correct the recognition result.

Modules $M1$ and $M2$ are used in both face representation (or modelling) and face recognition tasks. However, module $M3$ is used only for face representation and modules $M4$ and $M5$ are used only for recognition. The last remark is that every module should be used separately in order to create another face processing system.

5   Experiments

5.1   Czech News Agency Corpus

This corpus is composed of images of individuals in an uncontrolled environment that were randomly selected from the large ČTK database. All images were taken over a long time period (20 years or more). The corpus contains grey-scale images of 638 individuals of size 128×128 pixels. It contains about 10 images for each person. The orientation, lighting conditions and image backgrounds differ significantly.

Figure 2 shows four examples of one face from this corpus. This corpus is available for free for research purposes at http://home.zcu.cz/~pkral/sw/ or upon request to the authors.

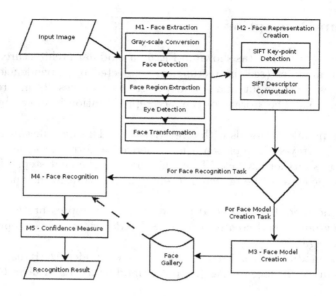

Fig. 1. System Architecture.

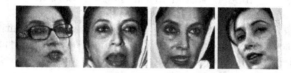

Fig. 2. Examples of one face from the ČTK face corpus.

5.2   Recognition Results with Confidence Measure

Three experiments are described next. The first experiment analyses the discriminability of the proposed partial measures by histograms. This experiment is realized in order to show the suitability of the proposed measures. The second experiment reports the results of the measures also used separately. In the last experiment, we show the classification results of the whole composed approach.

Discriminability of the Proposed Measures. In the first experiment, we would like to analyse the discriminability of the proposed partial measures. We created two histograms for every measure in order to analyse the distribution of the correctly and incorrectly classified faces. The reported output densities of the measures are based on the 638 values (the number of individuals in the corpus). Note that all output values are normalized to the interval [0..1].

Figure 3 shows the output densities of the correctly and incorrectly classified faces when the *absolute confidence value* measure is used. These histograms show that the majority of the correctly recognized face examples has higher output values than the incorrectly recognized ones. This fact confirms our assumption

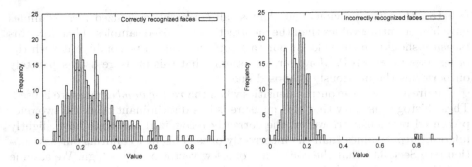

Fig. 3. Histograms of the correctly (left) and incorrectly (right) classified faces using the *absolute confidence value* measure.

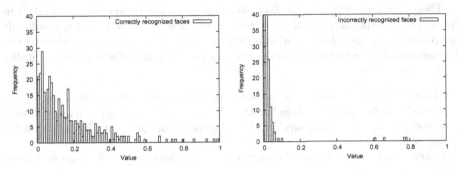

Fig. 4. Histograms of the correctly (left) and incorrectly (right) classified faces using the *relative confidence value* measure.

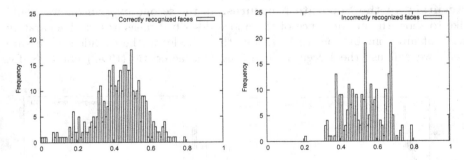

Fig. 5. Histograms of the correctly (left) and incorrectly (right) classified faces using the *vector number* measure.

that the first measure is suitable for our task and should be useful to be integrated to the whole composed method.

Figure 4 plots the output densities when the *relative confidence value* measure is used. These histograms show clearly that the discriminability of this measure is better than the previous one. Almost all correctly recognized face examples have higher output values than the incorrectly recognized samples. Therefore this measure should be suitable for our task and we decided to combine it with the other ones by an MLP. Moreover, we assume that this measure used separately outperforms the previously proposed one.

Figure 5 depicts the output densities when the *vector number* measure is used. These histograms show that this measure is less discriminant than the two ones presented previously. However, the correctly recognized examples have slightly inferior output values than the incorrectly ones. This fact confirms our assumption (see Sect. 3.2) that the confidence of a few vector model is high. We assume that this measure will bring poor results if used separately. However, it can add some further information when it will be combined with the other approaches. Therefore, we decided to integrate it into the whole composed approach.

The output densities of the last *standard deviation* measure are reported in Fig. 6. The discriminability of these two histograms are limited and it is difficult to propose some conclusions about this measure. However, we decided to use this measure in the further experiments and verify its usefulness experimentally.

To summarize:

- *relative confidence value (rel)* measure is the best proposed one;
- *absolute confidence value (abs)* method has also very good separation abilities;
- *vector number(vect)* measure can bring some complementary information for our task;
- contribution of the *standard deviation (sd)* measure is questionable and must be confirmed experimentally.

Accuracy of the Separate Measures. In the second experiment we would like to show the performance of the above described measures used separately without any combination. As in many other articles in the confidence measure field, we will use the Receiver Operating Characteristic (ROC) curve [8] for

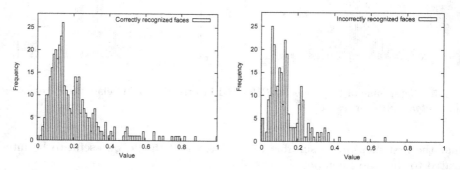

Fig. 6. Histograms of the correctly (left) and incorrectly (right) classified faces using the *standard deviation* measure.

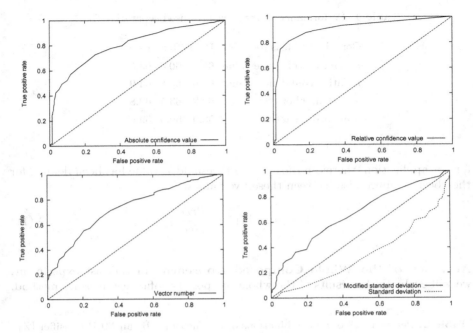

Fig. 7. ROC curves of the four proposed measures used separately. The corrected *standard deviation* measure is reported with the *modified standard deviation* label.

evaluation of this experiment. This curve clearly shows the relationship between the true positive and false positive rates for the different *acceptance* threshold.

Figure 7 shows the results of the proposed *absolute confidence value*, *relative confidence value*, *vector number* and *standard deviation* measures used separately. This experiment shows that the *relative confidence value* method significantly outperforms the all other approaches.

We can further deduce that our assumption in the fourth proposed measure was not correct. Based on this experiment we can consider that the dependence between the value of the standard deviation and the correctly recognized faces is reversed. We modify the definition of such measure as follows: only the faces where $S > T$ are accepted.

After this modification we can conclude that all proposed measures are suitable for our task in order to identify incorrectly recognized faces. Note that the corrected version of the ROC curve of the fourth *standard deviation* measure is reported with the *modified standard deviation* caption.

We will further compare the results of the separate measures with the whole composed approach. Therefore, we created Table 1 to show the scores of the separate measures with optimal threshold configurations. The F-measure (F-mes) [25] is used as an evaluation metric, the Precision (Prec) and Recall (Rec) are also reported in this table. Precision and recall have the similar importance

Table 1. Performance of the measures used separately [%].

Confidence Measure	Prec	Rec	F-mes
Absolute confidence value	65.7	60.6	63.0
Relative confidence value	69.6	60.8	64.9
Vector number	62.2	63.5	62.8
Standard deviation	58.9	60.3	59.6

for our application, therefore the optimal threshold $\hat{T}$ value has been defined for the "best" compromise between these two values:

$$\hat{T} = \arg\min_{T} \left| 1 - \frac{Prec}{Rec} \right| \tag{11}$$

Accuracy of the Whole Composed Approach. In the last experiment, we will evaluate the results of the whole composed confidence measure method.

Table 2. Performance of all combinations of the measures by an MLP classifier [%].

Confidence Measure	Prec	Rec	F-mes
1. Separate measures			
abs. confidence value (abs)	92.5	64.8	76.2
rel. confidence value (rel)	96.2	80.4	87.6
vector number (vect)	55.4	84.9	67.0
standard deviation (sd)	54.0	65.3	59.1
2. Combinations of two measures			
abs, rel	97.2	83.5	89.8
abs, sd	70.4	55.8	62.2
abs, vect	95.8	75.8	84.6
rel, sd	95.8	84.3	89.7
rel, vect	97.7	85.6	91.2
sd, vect	67.6	90.6	77.4
3. Combinations of three measures			
abs, rel, sd	96.7	90.0	93.2
abs, rel, vect	97.2	93.7	95.4
abs, sd, vect	93.4	90.5	91.9
rel, sd, vect	94.8	94.8	94.8
4. Combination of all measures (the whole approach)			
abs, rel, sd, vect	100	99.5	99.8

First, we will show the impact of the use of an MLP classifier with the separate measures. Then, we compare and evaluate all possible combinations of the proposed measures in order to show the complementarities among them.

Several MLP configurations are tested. The best MLP topology uses three layers. The number of the input neurons corresponds to the number of measures to combine, 10 neurons are in the hidden layer and two outputs are used to identify the *correctly* and *incorrectly* recognized faces. This MLP topology was set empirically on a small development corpus which contains 120 examples (i.e. 120 confidence values).

The results of this experiment are reported in Table 2. These results show that the separate measures used with an MLP have better F-measure values (except *sd* approach) than used in the unsupervised way. A successive addition of the measures improves progressively the F-measure value. When all measures are combined, the resulting F-measure is close to 100 %. This figure also shows that all measures bring complementary relevant information and are thus useful to be integrated to the whole composed approach (i.e. the whole combined approach gives the best recognition score).

6   Conclusions and Perspectives

We proposed and evaluated a novel confidence measure approach and integrated it in the experimental automatic face recognition system as a new module. The proposed approach combines two measures based on the *posterior* probability and two ones based on the *predictor* features in a supervised way with an MLP. We experimentally showed that the proposed approach is very efficient, because it detects almost all erroneous examples. We further showed that it is possible to use all four proposed measures separately. However, every measure brings complementary information and it is thus beneficial to combine all measures in the composed approach.

To summarize, the main scientific contribution of this paper consists in:

1. proposing two novel measures based on the *predictor* features;
2. proposing a combined supervised confidence measure approach which combines the measures from two groups of methods; two ones based on the *posterior* class probability and the other two ones on the *predictor* features;
3. evaluation of the proposed method in the face recognition task on the real ČTK data;
4. integration of the proposed confidence measure approach into the experimental automatic face recognition system.

The first perspective consists in proposing of semi-supervised confidence measures. In this approach, the CM model will be progressively adapted according to the recognized data. We will further integrate other more suitable features into our model. Another perspective consists in the use of our confidence measure approach in the task of automatic creation of the face corpora.

Acknowledgements. This work has been partly supported by the UWB grant SGS-2013-029 Advanced Computer and Information Systems and by the European Regional Development Fund (ERDF), project "NTIS - New Technologies for Information Society", European Centre of Excellence, CZ.1.05/1.1.00/02.0090. We also would like to thank Czech New Agency (ČTK) for support and for providing the photographic data.

References

1. Ahonen, T., Hadid, A., Pietikäinen, M.: Face recognition with local binary patterns. In: Pajdla, T., Matas, J.G. (eds.) ECCV 2004. LNCS, vol. 3021, pp. 469–481. Springer, Heidelberg (2004). http://dx.doi.org/10.1007/978-3-540-24670-1_36
2. Aly, M.: Face recognition using sift features (2006)
3. Bartlett, M.S., Movellan, J.R., Sejnowski, T.J.: Face recognition by independent component analysis. IEEE Trans. Neural Netw. **13**, 1450–1464 (2002)
4. Bay, H., Ess, A., Tuytelaars, T., Van Gool, L.: Speeded-up robust features (surf). Comput. Vis. Image Underst. **110**(3), 346–359 (2008). doi:10.1016/j.cviu.2007.09.014
5. Beham, M.P., Roomi, S.M.M.: A review of face recognition methods. Int. J. Pattern Recogn. Artif. Intell. **27**(4), 1–35 (2013)
6. Belhumeur, P.N., Hespanha, J.A.P., Kriegman, D.J.: Eigenfaces vs. fisherfaces: recognition using class specific linear projection. IEEE Trans. Pattern Anal. Mach. Intell. **19**(7), 711–720 (1997)
7. Bolme, D.S.: Elastic bunch graph matching. Ph.D. thesis, Colorado State University (2003)
8. Brown, C.D., Davis, H.T.: Receiver operating characteristics curves and related decision measures: a tutorial. Chemometr. Intell. Lab. Syst. **80**(1), 24–38 (2006)
9. Campadelli, P., Lanzarotti, R.: A face recognition system based on local feature characterization. In: Tistarelli, M., Bigun, J., Grosso, E. (eds.) Advanced Studies in Biometrics. LNCS, vol. 3161, pp. 147–152. Springer, Heidelberg (2005)
10. Deng, J., Schuller, B.: Confidence measures in speech emotion recognition based on semi-supervised learning. In: INTERSPEECH (2012)
11. Eickeler, S., Jabs, M., Rigoll, G.: Comparison of confidence measures for face recognition. In: FG, pp. 257–263. IEEE Computer Society (2000). http://dblp.uni-trier.de/db/conf/fgr/fg2000.html#EickelerJR00
12. Hu, X., Mordohai, P.: A quantitative evaluation of confidence measures for stereo vision. IEEE Trans. Pattern Anal. Mach. Intell. **34**(11), 2121–2133 (2012)
13. Huang, K., Aviyente, S.: Sparse representation for signal classification. Adv. Neural Inf. Process. Syst. **19**, 609 (2007)
14. Jiang, H.: Confidence measures for speech recognition: a survey. Speech Commun. **45**(4), 455–470 (2005)
15. Kepenekci, B.: Face recognition using Gabor wavelet transform. Ph.D. thesis, The Middle East Technical University (2001)
16. Križaj, J., Štruc, V., Pavešić, N.: Adaptation of SIFT features for robust face recognition. In: Campilho, A., Kamel, M. (eds.) ICIAR 2010. LNCS, vol. 6111, pp. 394–404. Springer, Heidelberg (2010)
17. Lenc, L., Král, P.: Confidence measure for automatic face recognition. In: International Conference on Knowledge Discovery and Information Retrieval. Paris, France, 26–29 October 2011

18. Lenc, L., Král, P.: Novel matching methods for automatic face recognition using SIFT. In: Iliadis, L., Maglogiannis, I., Papadopoulos, H. (eds.) Artificial Intelligence Applications and Innovations. IFIP AICT, vol. 381, pp. 254–263. Springer, Heidelberg (2012)

19. Lenc, L., Král, P.: Face recognition under real-world conditions. In: International Conference on Agents and Artificial Intelligence. Barcelona, Spain, 14–18 February 2013

20. Li, F., Wechsler, H.: Open world face recognition with credibility and confidence measures. In: Kittler, J., Nixon, M.S. (eds.) AVBPA 2003. LNCS, vol. 2688, pp. 462–469. Springer, Heidelberg (2003)

21. Li, W., Fu, P., Zhou, L.: Face recognition method based on dynamic threshold local binary pattern. In: Proceedings of the 4th International Conference on Internet Multimedia Computing and Service, pp. 20–24. ACM (2012)

22. Martínez, A.M.: Recognizing imprecisely localized, partially occluded, and expression variant faces from a single sample per class. IEEE Trans. Pattern Anal. Mach. Intell. 24(6), 748–763 (2002)

23. Marukatat, S., Artières, T., Gallinari, P., Dorizzi, B.: Rejection measures for handwriting sentence recognition. In: Proceedings of the Eighth International Workshop on Frontiers in Handwriting Recognition, 2002, pp. 24–29. IEEE (2002)

24. Poon, B., Amin, M.A., Yan, H.: Performance evaluation and comparison of pca based human face recognition methods for distorted images. Int. J. Mach. Learn. Cybernet. 2(4), 245–259 (2011)

25. Powers, D.: Evaluation: from precision, recall and F-measure to ROC., informedness, markedness & correlation. J. Mach. Learn. Technol. 2(1), 37–63 (2011)

26. Proedrou, K., Nouretdinov, I., Vovk, V., Gammerman, A.J.: Transductive confidence machines for pattern recognition. In: Elomaa, T., Mannila, H., Toivonen, H. (eds.) ECML 2002. LNCS (LNAI), vol. 2430, pp. 381–390. Springer, Heidelberg (2002)

27. Senay, G., Linares, G., Lecouteux, B.: A segment-level confidence measure for spoken document retrieval. In: 2011 IEEE International Conference on Acoustics, Speech and Signal Processing (ICASSP), pp. 5548–5551. IEEE (2011)

28. Servin, B., de Givry, S., Faraut, T.: Statistical confidence measures for genome maps: application to the validation of genome assemblies. Bioinformatics 26(24), 3035–3042 (2010)

29. Shen, L.: Recognizing faces - an approach based on Gabor wavelets. Ph.D. thesis, University of Nottingham (2005)

30. Shen, L., Bai, L.: A review on gabor wavelets for face recognition. Pattern Anal. Appl. 9, 273–292 (2006)

31. Sukkar, R.A.: Rejection for connected digit recognition based on gpd segmental discrimination. In: 1994 IEEE International Conference on Acoustics, Speech, and Signal Processing, 1994, ICASSP 1994, vol. 1, pp. I-393. IEEE (1994)

32. Tan, X., Triggs, B.: Enhanced local texture feature sets for face recognition under difficult lighting conditions. IEEE Trans. Image Process. 19(6), 1635–1650 (2010)

33. Timo, A., Hadid, A., Pietikinen, M.: Face description with local binary patterns: application to face recognition. IEEE Trans. Pattern Anal. Mach. Intell. 28, 2037–2041 (2006)

34. Turk, M.A., Pentland, A.P.: Face recognition using eigenfaces. In: IEEE Computer Society Conference on Computer Vision and Pattern Recognition (1991)

35. Wagner, A., Wright, J., Ganesh, A., Zhou, Z., Mobahi, H., Ma, Y.: Toward a practical face recognition system: Robust alignment and illumination by sparse representation. IEEE Trans. Pattern Anal. Mach. Intell. 34(2), 372–386 (2012)

36. Wessel, F., Schluter, R., Macherey, K., Ney, H.: Confidence measures for large vocabulary continuous speech recognition. IEEE Trans. Speech Audio Process. **9**(3), 288–298 (2001)
37. Zhang, B., Gao, Y., Zhao, S., Liu, J.: Local derivative pattern versus local binary pattern: face recognition with high-order local pattern descriptor. IEEE Trans. Image Process. **19**(2), 533–544 (2010)
38. Zhao, W., Chellappa, R., Phillips, P.J., Rosenfeld, A.: Face recognition: a literature survey. ACM Comput. Surv. (CSUR) **35**(4), 399–458 (2003)

Author Index

Printed in the United States
By Bookmasters